Bayesian Inference

Bayesian Inference: Theory, Methods, Computations provides a comprehensive coverage of the fundamentals of Bayesian inference from all important perspectives, namely theory, methods and computations.

All theoretical results are presented as formal theorems, corollaries, lemmas etc., furnished with detailed proofs. The theoretical ideas are explained in simple and easily comprehensible forms, supplemented with several examples. A clear reasoning on the validity, usefulness, and pragmatic approach of the Bayesian methods is provided. A large number of examples and exercises, and solutions to all exercises, are provided to help students understand the concepts through ample practice.

The book is primarily aimed at first or second semester master students, where parts of the book can also be used at Ph.D. level or by research community at large. The emphasis is on exact cases. However, to gain further insight into the core concepts, an entire chapter is dedicated to computer intensive techniques. Selected chapters and sections of the book can be used for a one-semester course on Bayesian statistics.

Key Features:

- Explains basic ideas of Bayesian statistical inference in an easily comprehensible form
- Illustrates main ideas through sketches and plots
- Contains large number of examples and exercises
- Provides solutions to all exercises
- Includes R codes

Silvelyn Zwanzig is a Professor for Mathematical Statistics at Uppsala University. She studied Mathematics at the Humboldt University of Berlin. Before coming to Sweden, she was Assistant Professor at the University of Hamburg in Germany. She received her Ph.D. in Mathematics at the Academy of Sciences of the GDR. She has taught Statistics to undergraduate and graduate students since 1991. Her research interests include theoretical statistics and computer-intensive methods.

Rauf Ahmad is Associate Professor at the Department of Statistics, Uppsala University. He did his Ph.D. at the University of Göttingen, Germany. Before joining Uppsala University, he worked at the Division of Mathematical Statistics, Department of Mathematics, Linköping University, and at Biometry Division, Swedish University of Agricultural Sciences, Uppsala. He has taught Statistics to undergraduate and graduate students since 1995. His research interests include high-dimensional inference, mathematical statistics, and U-statistics.

Bayesian Inference
Theory, Methods, Computations

Silvelyn Zwanzig and Rauf Ahmad

CRC Press
Taylor & Francis Group
Boca Raton London New York

CRC Press is an imprint of the
Taylor & Francis Group, an **informa** business

A CHAPMAN & HALL BOOK

Designed cover image: © Silvelyn Zwanzig and Rauf Ahmad

First edition published 2024
by CRC Press
2385 NW Executive Center Drive, Suite 320, Boca Raton FL 33431

and by CRC Press
4 Park Square, Milton Park, Abingdon, Oxon, OX14 4RN

CRC Press is an imprint of Taylor & Francis Group, LLC

ISBN: 978-1-0321-09497 (hbk)
ISBN: 978-1-0321-18093 (pbk)
ISBN: 978-1-0032-21623 (ebk)

DOI: 10.1201/9781003221623

Typeset in CMR10
by KnowledgeWorks Global Ltd.

Publisher's note: This book has been prepared from camera-ready copy provided by the authors.

Contents

Preface

Our main objective of writing this book is to present Bayesian statistics at introductory level, in pragmatic form, specifically focusing on the exact cases. All three aspects of Bayesian inference, namely theory, methods, and computations, are covered.

The book is typically written as a textbook, but it also contains ample material for researchers. As textbook, it contains more material than needed for a one-semester course, whereas the choice of chapters also depends on the background and interest of the students, and specific aims of the designed course. A number of examples and exercises are embedded in all chapters, and solutions to all exercises are provided, often with reasonable details of solution steps.

Another salient feature of the book is cartoon-based depictions, aimed at a leisurely comprehension of otherwise intricate mathematical concepts. For this contribution, the authors are indebted to the first author's daughter, Annina Heinrich, who so meticulously and elegantly transformed arcane ideas into amusing illustrations.

The book also adds a new piece of joint output in the authors' repository of long-term team work, mostly leading to research articles or book chapters.

Most material of the book stems from the Master and Ph.D. courses on Bayesian Statistics over the last several years. The courses were offered through the Department of Mathematics and the Center for Interdisciplinary Mathematics (CIM), Uppsala University. Special thanks are due to the colleagues and students who have read the preliminary drafts of the book and provided valuable feedback. Any remnants of mistakes or typos are of course the authors' responsibility, and any information in this regard will be highly appreciated.

We wish the readers a relishing amble through the pleasures and challenges of Bayesian statistics!

Silvelyn Zwanzig and Rauf Ahmad
December 2023, Uppsala, Sweden

Chapter 1

Introduction

Bayesian theory has seen an increasingly popular trend over the recent few decades. It can be ascribed to a combination of two factors, namely, the fast growth of computational power, and the applicability of the theory to a wide variety of real life problems.

Reverend Thomas Bayes (1702–1761), the English nonconformist Presbyterian minister, laid the foundation stone of the theory in terms of a theorem which duly carries his name in probability theory. For any two events, A and B, defined in a sample space Ω, with marginal probabilities $\mathsf{P}(A)$, $\mathsf{P}(B)$ and conditional probability $\mathsf{P}(B|A)$, the theorem gives the conditional probability

$$\mathsf{P}(A|B) = \frac{\mathsf{P}(B|A)\mathsf{P}(A)}{\mathsf{P}(B)}.$$

The theorem, thus, updates unconditional (prior) probability of A, $\mathsf{P}(A)$, into conditional (posterior) probabilities of A, $\mathsf{P}(A|B)$, using knowledge of B.

The theorem, however, remained unknown to the wider world until one of Bayes' closest friends, Thomas Price (1723–1791), posthumously dug it up in 1764, meticulously edited it, and sent it to the Royal Society which both Thomases were members of. Thomas Price also attached a detailed note with the article, arguing for the value and worth of the theory Bayes had put forth in it. For a detailed historical profile of the subject, and of Thomas Bayes, see e.g., Bellhouse (2004) and Stigler (1982).

In this context, it is indeed interesting to note that the axiomatic approach to probability theory, as developed by Kolmogorov, and used as a standard probability framework today, emerged about two centuries after Bayes introduced his theorem.

What was introduced by Bayes and popularized by Price, was later formalized by Laplace, modernized by Jeffreys, and mathematicised by de Finnetti. Harold Jeffreys rightly accoladed that *the Bayes theorem is to the theory of*

DOI: 10.1201/9781003221623-1

probability what Pythagoras theorem is to geometry. Over time, Bayesian theory has occupied a prominent place in the comity of researchers.

We may recall ourselves that Statistics is more a research tool than a discipline like others, and as such, Statistics helps develop and apply techniques to measure and assess uncertainty in inductive inference. This implies applicability of Statistics in any research context involving random mechanism where the researcher intends to inductively draw conclusions.

Frequentist way of inference, based on likelihood theory, is one way to achieve this end. In likelihood theory, the uncertainty, captured through the data x, is formulated in terms of densities or mass functions, $f(x|\theta)$, $\theta \in \Theta$, where Θ is the parameter space, so that the underlying model can be stated as

$$\mathcal{P} = \{f(x|\theta); \theta \in \Theta\}.$$

The target of inference is the unknown parameter, θ, which generates the data and is considered to be fixed. For Bayesian theory, on the other hand, the uncertainty comes as a whole package, both in data - via likelihood – but also in the unknown parameter. The former is formulated in terms of densities or mass functions, $f(x|\theta)$, and the later in terms of setting prior distributions on the parameters, denoted $\pi(\theta)$, so that a Bayes model includes both the likelihood and the prior, i.e.,

$$\{\mathcal{P}, \pi\}.$$

Thus, the modus operandi of the frequentist inference consists of the science of collecting empirical evidence and the art of gleaning information from this evidence in order to support or refute the conjectures made about the unknown parameters. The modus operandi of the Bayesian inference consists of the art of setting a prior and the science of updating the prior into the posterior, by exploiting the information in data.

The addition of prior as a measure of uncertainty shifts the focus in Bayesian theory, by considering the prior as the main source of information about the parameter, where data, or likelihood, helps improve the prior into a posterior. Applying Bayes theorem gives Posterior $\propto$ Likelihood $\times$ Prior; Formally,

$$\pi(\theta|X) \propto f(x|\theta)\pi(\theta).$$

The prior can be based on subjective beliefs. It mainly pertains to the fact, one that cannot be overemphasized, that the scientific inquiry often incorporates subjectively formulated conjectures as essential components in exploring the nature. This can be evidenced from the use of subjectivism by world renowned researchers excelled in a variety of scientific and philosophical domains. Press and Tamur (2001) list up many of them, in the context of their subjectivism, including giants such as Johannes Kepler, Gregor Mendel, Galileo Galilei,

Isaac Newton, Alexander von Humboldt, Charles Darwin, Sigmund Freud, and Albert Einstein.

The posterior follows therefore from an amalgamation of subjectively decided prior and data-based likelihood, a combination of belief and evidence. By this, the posterior logically tilts toward the component that carries the heavier weight. One might in fact find oneself tempted to imagine that the prior as belief in Bayesian inference possibly reflects the religious aptitude of the founder of the idea.

Certain specific advantages of Bayesian theory makes it specifically attractive for practical applications. In particular, although the prior can incorporate a subjective component into inference, it also provides a flexible mode of inference by picking the most suitable model while trying out a variety of feasible options. Further, a prior does not only reflect the researcher's knowledge, rather also ignorance, concerning the parameter.

The recent renaissance of Bayesian theory can indeed be considered from another, historical and philosophical, perspective. In the eighties of the last century, all possible and interesting models were considered, with explicit formulas derived, new distribution families introduced and studied. At that time, the splitting of the scientific community into Bayesian and fiducial statisticians seemed to be complete and definite.

During the last decades, however, the more pragmatic Bayesian approach has become popular, especially because of the iterative structure of the formulas which allow adaptive applications. An objective Bayesian analysis is developed, where the prior is determined by information–theoretic criteria, avoiding a subjective influence on the result of the study on one hand and exploring the advantages of Bayes analysis on the other.

A major reason for this renaissance lies in the efficient computational support. The availability of computer-intensive methods such as MCMC (Markov Chain Monte Carlo) and ABC (Approximative Bayesian Method) in the modern era have led to a second coming of Bayesian theory. This has obviously widened the spectrum of applications of Bayesian theory into the modern areas of statistical inference involving big data and its associated problems. The less reliance on mathematics, and more on computational algorithms, has rightfully enhanced the scope of interest in Bayesian way of inference for researchers and situations where mathematical intricacies can be avoided.

This book is primarily designed as a textbook, to present the fundamentals of Bayesian statistical inference, based on the Bayes model $\{\mathcal{P}, \pi\}$. It covers basic theory, supplemented with methodological and computational aspects. It may be emphasized that it is not our objective to weigh in for or against Bayesian or frequentist inference. We do not even compare the two in any precise context.

Our motivation stems from the fact that Bayesian inference provides a pragmatic statistical approach to discern information from real life data. This aspect indeed sets the book apart from many, if not all, others already in the literature on the same subject, in that we focus on the cases, both in univariate and multivariate inference, where exact posterior can be derived. Further, the solutions of all exercises are furnished, to make the book self-contained.

The book can be used for a one-semester Bayesian inference course at the undergraduate or master level, depending on the students' background, for which we expect a reasonable orientation to basic statistical inference, at the level of e.g., Liero and Zwanzig (2011).

After this brief introduction, Chapter 2 provides general Bayesian modelling framework, where the main principles are presented, and the difference between a usual statistical model $\mathcal{P}$ and a Bayesian model $\{\mathcal{P}, \pi\}$ is explained. Chapter 3 specifically addresses the issue of setting a prior, its variants, and their effects on the posterior. Methods for incorporating subjective information and the ideas and principles of objective priors are discussed.

Chapter 4 focuses on statistical decision theory. It explains the connection between Bayesian and frequentist forms of inference. It gives optimality properties of Bayes method for a statistical model $\mathcal{P}$, and explains, vice versa, how the frequentist methods can be expressed as Bayesian and how their optimality properties can be derived. Chapter 5 summarizes the main ingredients of Bayesian asymptotic theory. Chapters 4 and 5 are rather technical and can be skipped, if the reader so wishes, without loosing smooth transition to the rest of the book.

Chapter 6 deals with Bayesian theory of linear models under normal distribution. It includes univariate linear models, linear mixed models and multivariate models. The explicit formulas for the posteriors are derived. These results are also important for evaluating simulation methods.

Chapter 7 is on general estimation theory of Bayesian inference. Chapter 8 treats the Bayesian approach for hypothesis testing. Chapter 9 adds essential computational aspect of Bayesian theory, where the main ideas and principles of MC, MCMC, Importance sampling, Gibbs sampling and ABC are presented, including R codes for the examples. Solutions to all exercises in the book are given in Chapter 10. Apart from the exercises, all chapters are equipped with detailed examples to explain the methods.

## Chapter 2

Bayesian Modelling

The standard setting in statistics is that we consider the data $\mathbf{x}$ as a realization (observation) of a random variable $\mathbf{X}$. The data $\mathbf{x}$ may consist of numbers or vectors, tables, categorical quantities or functions. The set of all possible observations $\mathcal{X}$ is named **sample space** and the distribution $P^{\mathbf{X}}$ of the random variable $\mathbf{X}$ is defined for measurable subsets of the sample space:

> The data $\mathbf{x} \in \mathcal{X}$ is a realization of $\mathbf{X} \sim P^{\mathbf{X}}$.

The distribution $P^{\mathbf{X}}$ is unknown. It is called the true distribution or the underlying distribution or the data generating distribution. The goal of mathematical statistics is to extract information on the underlying distribution $P^{\mathbf{X}}$ from the data $\mathbf{x}$. This problem can only be solved when we can reduce the set of all possible distributions over $\mathcal{X}$. We have to assume that we have knowledge about the underlying distribution. Formulating this type of assumptions in a mathematical way is the purpose of modelling.

In this chapter we present the main principles of Bayesian modelling. First, we explain the difference between a statistical and a Bayes model. Then we introduce principles of determining a Bayes model.

2.1 Statistical Model

We start with the notion of a statistical model; see for example in Liero and Zwanzig (2011).

> **Definition 2.1 A parametric statistical model** is a set of probability measures over $\mathcal{X}$
> $$\mathcal{P} = \{\mathsf{P}_\theta : \theta \in \Theta\}.$$
> The set Θ has finite dimension and it is called the **parameter space**.

Furthermore, we pose a strong condition that the postulated model is the right one, i.e.,

DOI: 10.1201/9781003221623-2

> $\mathsf{P}^{\mathbf{X}} \in \mathcal{P}$. There exists θ_0 such that $\mathsf{P}^{\mathbf{X}} = \mathsf{P}_{\theta_0}$.

The parameter θ_0 is the true parameter or the underlying parameter.

Example 2.1 (Flowers)

Snake's head (Kungsängsliljan, *Fritillaria meleagris*) is one of the preserved sights in Uppsala. In May, whole fields along the river are covered by the flowers. They have three different colors: violet, white and pink. Suppose the color of n randomly chosen flowers are determined as n_1 white, n_2 violet and n_3 pink. The (n_1, n_2, n_3) triplets are realizations of the r.v. (N_1, N_2, N_3), with $\sum_{j=1}^{3} N_j = n$, which has a multinomial distribution $\mathsf{Mult}(n, p_1, p_2, p_3)$:

$$P(N_1 = n_1, N_2 = n_2, N_3 = n_3) = \frac{n!}{n_1! \, n_2! \, n_3!} p_1^{n_1} p_2^{n_2} p_3^{n_3}, \qquad (2.1)$$

with $n_3 = n - n_1 - n_2$, $p_3 = 1 - p_1 - p_2$ and $0 < p_j < 1$ for $j = 1, 2, 3$. The unknown parameter $\theta = (p_1, p_2, p_3)$ consists of the probabilities of each color. With only three colors, we have $p_1 + p_2 + p_3 = 1$. The parameter space is

$$\Theta = \{\theta = (p_1, p_2, p_3) : 0 < p_j < 1, \ j = 1, 2, 3, \ p_1 + p_2 + p_3 = 1\}.$$

Thus

$$\mathcal{P} = \{\mathsf{Mult}(n, p_1, p_2, p_3) : \theta = (p_1, p_2, p_3) \in \Theta\}.$$

$\square$

Example 2.2 (Measurements)

Suppose we carry out a physical experiment. Under identical conditions we take repeated measurements. The data consist of n real numbers x_i, $i = 1, \ldots, n$ and $\mathcal{X} = \mathbb{R}^n$. The sources of randomness are the measurement errors, which can be assumed as normally distributed. Thus we have a sample of i.i.d. r.v.'s and the model is given by

$$\mathcal{P} = \{\mathsf{N}(\mu, \sigma^2)^{\otimes n} : (\mu, \sigma^2) \in \mathbb{R} \times \mathbb{R}_+\}.$$

$\square$

Let us consider a different model for the experiment above.

Example 2.3 (Measurements with no distribution specification)

We consider the same experiment as above. But the sources of randomness are unclear. We are only interested in an unknown constant, namely the expected value. Further we assume that the variance is finite and the density f of P^X is unknown. In this case we can formulate the model for the distribution of sample of i.i.d. r.v.'s $(X_1, \ldots, X_n)$ by

$$\mathcal{P} = \{\mathsf{P}_\theta^{\otimes n} : \mathsf{E}X = \mu, \mathsf{Var}X = \sigma^2, (\mu, \sigma^2) \in \mathbb{R} \times \mathbb{R}_+, f \in \mathcal{F}\}.$$

The unknown parameter θ consists of μ, σ^2, f and the parameter space Θ is determined as $\mathbb{R} \times \mathbb{R}_+ \times \mathcal{F}$. In this case, $\mathcal{F}$ is some functional space of infinite dimension. This model does not fulfill the conditions of a statistical model in Definition 2.1. Otherwise when we suppose that $\mathcal{F}$ is a known parametric distribution family with density $f(x) = f_\theta(x)$, $\theta = (\mu, \sigma^2, \vartheta)$, $\vartheta \in \mathbb{R}^q$, then the parameter space has a dimension less or equal $q + 2$ and the model fulfills the conditions of Definition 2.1. $\qquad\qquad\qquad\qquad\qquad\qquad\qquad\qquad\qquad\qquad\qquad\quad$ $\square$

Let us consider one more example which is probably not realistic but still a good classroom example for understanding purposes; see also Liero and Zwanzig (2011).

Example 2.4 (Lion's appetite)
Suppose that the appetite of a lion has three different stages:

$$\theta \in \Theta = \{\text{hungry}, \text{moderate}, \text{lethargic}\} = \{\theta_1, \theta_2, \theta_3\}.$$

At night the lion eats x people with a probability $P_\theta(x)$ given by the following table:

x	0	1	2	3	4
θ_1	0	0.05	0.05	0.8	0.1
θ_2	0.05	0.05	0.8	0.1	0
θ_3	0.9	0.05	0.05	0	0

Thus the model consists of three different distributions over $\mathcal{X} = \{0, 1, 2, 3, 4\}$. $\square$

The Corona crisis, started in early 2020, led to an increased interest in statistics. We consider an example related to Corona mortality statistics.

Example 2.5 (Corona) The data consist of the number of deaths related to an infection with COVID-19 in the ith county in Sweden $i = 1, \ldots, 17$, during the first wave in week 17 (end of April) in 2020. We assume a logistic model for the probability of death p by COVID-19, depending on the covariates x_1 (age), x_2 (gender), x_3 (health condition) and x_4 (accommodation). Thus,

$$p = \frac{\exp(\alpha_1 x_1 + \alpha_2 x_2 + \alpha_3 x_3 + \alpha_4 x_4)}{1 + \exp(\alpha_1 x_1 + \alpha_2 x_2 + \alpha_3 x_3 + \alpha_4 x_4)}.$$

$\qquad\qquad\qquad\qquad\qquad\qquad\qquad\qquad\qquad\qquad\qquad\qquad\qquad\qquad\quad$ $\square$

Given a statistical model $\mathcal{P} = \{P_\theta : \theta \in \Theta\}$, the main tool for exploring the information in the data is the likelihood function.
For convenience, we introduce the **probability function** to handle both discrete and continuous distributions because measure theory is not a prerequisite of our book. Let $A \subseteq \mathcal{X}$. For a continuous r.v. $\mathbf{X}$ with density $f(\cdot | \theta)$

$$P_\theta(A) = \int_A f(\mathbf{x} | \theta) d\mathbf{x},$$

8

and for a discrete r.v.

$$P_\theta(A) = \sum_{x \in A} P_\theta(\{x\}),$$

the probability function is defined by

$$p(x|\theta) = \begin{cases} f(x|\theta) & \text{if } P_\theta \text{ is continuous} \\ P_\theta(\{x\}) & \text{if } P_\theta \text{ is discrete} \end{cases}.$$

Definition 2.2 (Likelihood function) For an observation x of a r.v. X with $p(\cdot|\theta)$, the likelihood function $\ell(\cdot|x) : \Theta \to \mathbb{R}_+$ is defined by

$$\ell(\theta|x) = p(x|\theta).$$

If $X = (X_1, \ldots, X_n)$ is a sample of independent r.v.'s, then

$$\ell(\theta|x) = \prod_{i=1}^n P_{i,\theta}(x_i) \quad \text{in the discrete case} \tag{2.2}$$

and

$$\ell(\theta|x) = \prod_{i=1}^n f_i(x_i|\theta) \quad \text{in the continuous case,} \tag{2.3}$$

where X_i is distributed according to $P_{i,\theta}$ and $f_i(\cdot|\theta)$, respectively. The **likelihood principle** says that all information is contained in the likelihood function and a statistical procedure should be based on maximizing the likelihood function. Let us quote two variants of the likelihood principle. In Robert (2001, p.16), the likelihood principle is formulated as following:

"The information brought by an observation x is entirely contained in the likelihood function $\ell(\theta|x)$.
Moreover, if x_1 and x_2 are two observations depending on the same parameter θ, such that there exists a constant c satisfying

$$\ell_1(\theta|x_1) = c\,\ell_2(\theta|x_2), \text{ for every } \theta,$$

then they bring the same information about θ and must lead to identical inferences."

Example 2.6 (Bernoulli trials)
Consider two different sampling strategies for the sequences of Bernoulli trials

$$(X_1, \ldots, X_n, \ldots), \text{ i.i.d. from } X \sim \text{Ber}(p).$$

First the number of trials are fixed by n and the number of successes x_1 is counted. Then

$$\mathbf{X}_1 = \sum_{i=1}^{n} X_i \sim \text{Bin}(n, p),$$

with likelihood function at $\theta = p$

$$\ell_1(\theta|x_1) = \binom{n}{x_1} \theta^{x_1} (1-\theta)^{n-x_1} \propto \theta^{x_1} (1-\theta)^{n-x_1}.$$

The other strategy is to fix the number k of successes and observe the sequence until k successes are attained. The number of failures x_2 is counted, where x_2 is a realization of $\mathbf{X}_2$ which follows a negative binomial distribution:

$$\mathbf{X}_2 \sim \text{NB}(k, p),$$

with likelihood function at $\theta = p$

$$\ell_2(\theta|x_2) = \binom{x_2 + k - 1}{k - 1} \theta^k (1-\theta)^{x_2} \propto \theta^k (1-\theta)^{x_2}.$$

Suppose 20 Bernoulli trials give 5 successes. Then $x_1 = 5$ and $x_2 = 15$, and we get

$$\ell_1(x_1|\theta) \propto \theta^5 (1-\theta)^{15} \propto \ell_2(x_2|\theta).$$

The likelihood principle is fulfilled, as both sampling strategies yield the same result.   □

The **maximum likelihood principle** is alternatively formulated in Lindgren (1962, p. 225) as follows:

> "A statistical inference or procedure should be consistent with the assumption that the best explanation of a set of data $\mathbf{x}$ is provided by $\widehat{\theta}_{\text{MLE}}$ a value of θ that maximizes $\ell(\theta|\mathbf{x})$."

Example 2.7 (Lion's appetite)
For $x = 3$, the likelihood function in lion's example is:

	θ_1	θ_2	θ_3
$\ell(\theta|x = 3)$	0.8	0.1	0

This leads to the conclusion that a lion, having eaten 3 persons, was hungry.□

Let us consider the likelihood function of a normal sample.

Example 2.8 (Normal sample)
Let $X_1, \ldots, X_n$ be i.i.d. r.v.'s according to $\mathsf{N}(\mu, \sigma^2)$. Then for $\theta = (\mu, \sigma^2)$

$$\ell(\theta|\mathbf{x}) = \prod_{i=1}^{n} \frac{1}{\sqrt{2\pi}\sigma} \exp\left(-\frac{1}{2\sigma^2}(x_i - \mu)^2\right)$$

$$\propto (\sigma^2)^{-\frac{n}{2}} \exp\left(-\frac{1}{2\sigma^2}\sum_{i=1}^{n}(x_i - \mu)^2\right).$$

Since $\sum_{i=1}^{n}(x_i - \overline{x})(\overline{x} - \mu) = 0$, we obtain

$$\sum_{i=1}^{n}(x_i - \mu)^2 = \sum_{i=1}^{n}(x_i - \overline{x})^2 + \sum_{i=1}^{n}(\overline{x} - \mu)^2.$$

Using the sample variance

$$s^2 = \frac{1}{n-1}\sum_{i=1}^{n}(x_i - \overline{x})^2,$$

it follows that the likelihood function depends only on the sufficient statistic $(\overline{x}, s^2)$ since

$$\ell(\theta|\mathbf{x}) \propto (\sigma^2)^{-\frac{n}{2}} \exp\left(-\frac{1}{2\sigma^2}((n-1)s^2 + n(\overline{x} - \mu)^2)\right).$$

In particular, it holds that

$$\ell(\theta|\mathbf{x}) \propto \ell(\mu|\sigma^2, \mathbf{x})\ell(\sigma^2|\mathbf{x}),$$

with

$$\ell(\mu|\sigma^2, \mathbf{x}) \propto \exp\left(-\frac{n}{2\sigma^2}(\overline{x} - \mu)^2\right)$$

and

$$\ell(\sigma^2|\mathbf{x}) \propto (\sigma^2)^{-\frac{n}{2}} \exp\left(-\frac{(n-1)s^2}{2\sigma^2}\right).$$

The likelihood function of μ given σ^2 has the form of the Gaussian bell curve with center $\overline{x}$ and inflection points at $\overline{x} - \frac{1}{\sqrt{n}}\sigma$, $\overline{x} + \frac{1}{\sqrt{n}}\sigma$. The likelihood functions of μ and σ^2 are plotted in Figure 2.1.   □

2.2   Bayes Model

The Bayes model sharpens the statistical model with an essential additional assumption. It consists of two parts:

- The underlying unknown parameter $\theta \in \Theta$ is supposed to be a realization of a random variable with the distribution π over Θ.

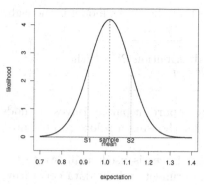 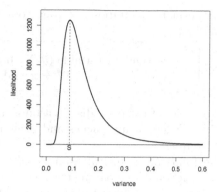

Figure 2.1: Likelihood functions in Example 2.12. Left: Likelihood function $\ell(\mu|\sigma^2, \mathbf{x})$ with maximum at the sample mean $\bar{x}$ and inflection points $S1 = \bar{x} - \frac{1}{\sqrt{n}}\sigma$ and $S2 = \bar{x} + \frac{1}{\sqrt{n}}\sigma$. Right: Likelihood function $\ell(\sigma^2|\mathbf{x})$ with maximum at $S = \frac{n-1}{n}s^2$, where s^2 is the sample variance.

- The distribution π is known,

$$\theta \sim \pi. \tag{2.4}$$

In literature we can find discussions on the meaning of (2.4). Some papers give philosophical interpretations up to religious foundations.

Here we follow a pragmatic point of view: We do not speculate who has carried out the parameter generating experiment. We just state that assumption (2.4) is not easy to interpret; may be it is much stronger than we can imagine, but it makes life much easier.

In Bayesian inference we have two random variables: θ and $\mathbf{X}$, and both play different roles. The random variable θ is not observed, but the **parameter generating distribution** π is known. The random variable $\mathbf{X}$ is observed. The **data generating distribution** is the conditional distribution of $\mathbf{X}$ given θ and it is known up to θ.

Summarizing we have the following definition:

Definition 2.3 (Bayes model) A Bayes model $\{\mathcal{P}, \pi\}$ consists of a set of conditional probability distributions over $\mathcal{X}$, $\mathcal{P} = \{\mathsf{P}_\theta : \theta \in \Theta\}$, where $\mathsf{P}^{\mathbf{X}|\theta} = \mathsf{P}_\theta$ and one distribution π over Θ.

The set Θ has finite dimension and is called the **parameter space**. The distribution π over Θ is called the **prior** distribution.

Furthermore, we impose a strong condition that the Bayes model is the right one, i.e.,

> The random variable $(\theta, \mathbf{X})$ has a distribution $\mathsf{P}^{(\theta, \mathbf{X})}$ where $\mathsf{P}^\theta = \pi$ and $\mathsf{P}^{\mathbf{X}|\theta} \in \mathcal{P}$.

The choice of the prior distribution is a very important part of Bayesian modelling. Different approaches and principles for choosing a prior are presented in Chapter 3.

The following historical example of Thomas Bayes is used to illustrate the construction of the parameter generating experiment and the data generating experiment; see the details in Stigler (1982).

Example 2.9 (Bayes' billiard table)

Consider a flat square table with length 1. It has no pockets. Bayes never specified it as a billiard table, but under this name the example is now well known in literature. The first ball W is rolled across the table and stopped at an arbitrary place, uniformly distributed over the unit square. The table is divided vertically through the position of the ball W. The position of W is not saved. A second ball O is rolled across the table in the same manner n times. It is counted how often the second ball O comes to rest to the left of W. Translating this description we have that the horizontal position of W is the underlying parameter θ with prior distribution $\mathsf{U}(0,1)$. The data $\mathbf{x}$ is the number of successes, where the probability of success equals θ. Thus

$$\theta \sim \mathsf{U}(0,1), \quad \mathbf{X}|\theta \sim \mathsf{Bin}(n,\theta).$$

$\square$

Note, the parameter is generated only one time, whereas the data can be generated repeatedly many times. The notion of an i.i.d. experiment $\mathbf{X} = (X_1, \ldots, X_n)$ in a Bayes set up is related to the conditional distribution of $\mathbf{X}$ given θ. The r.v. $(\theta, X_1, \ldots, X_n)$ is not an *i.i.d. sample*. Especially the r.vs. X_{i_1}, X_{i_2} are not independent.

We demonstrate it by the following example.

Example 2.10 (Normal i.i.d. sample)

Consider an i.i.d. sample from $\mathsf{N}(\theta, \sigma^2)$ with σ^2 known. The parameter of interest is θ with normal prior

$$\theta \sim \mathsf{N}(\mu_0, \sigma_0^2),$$

where μ_0 and σ_0^2 are known. Further, set the sample size $n = 2$. Then the random vector (θ, X_1, X_2) is three dimensional normally distributed with

$$\mathsf{E}X_i = \mathsf{E}_\theta(\mathsf{E}(X_i|\theta)) = \mathsf{E}\theta = \mu_0,$$

Figure 2.2: The billiard table for Example 2.9.

$$\mathrm{Var}X_i = \mathsf{E}_\theta(\mathrm{Var}(X_i|\theta)) + \mathrm{Var}_\theta(\mathsf{E}(X_i|\theta)) = \mathrm{Var}X_i + \mathrm{Var}\,\theta = \sigma^2 + \sigma_0^2.$$

Using the relation

$$\mathrm{Cov}(U,Z) = \mathsf{E}_Y(\mathrm{Cov}((U,Z)|Y)) + \mathrm{Cov}_Y(\mathsf{E}(U|Y), \mathsf{E}(Z|Y)), \qquad (2.5)$$

we have

$$\mathrm{Cov}(X_1, X_2) = \mathsf{E}_\theta(\mathrm{Cov}((X_1, X_2)|\theta)) + \mathrm{Cov}_\theta(\mathsf{E}(X_1|\theta), \mathsf{E}(X_2|\theta)).$$

As the data are conditionally i.i.d., it implies that $\mathrm{Cov}((X_1, X_2)|\theta) = 0$, so that

$$\mathrm{Cov}(X_1, X_2) = \mathrm{Cov}_\theta(\mathsf{E}(X_1|\theta), \mathsf{E}(X_2|\theta)) = \mathrm{Var}\,\theta = \sigma_0^2.$$

Further,

$$\mathrm{Cov}(X_1, \theta) = \mathsf{E}(X_1\theta) - \mu_0^2 = \mathsf{E}_\theta\left(\mathsf{E}(X_1\theta|\theta)\right) - \mu_0^2 = \mathrm{Var}\,\theta = \sigma_0^2.$$

Summarizing, we obtain

$$\begin{pmatrix} \theta \\ X_1 \\ X_2 \end{pmatrix} \sim \mathsf{N}\left(\begin{pmatrix} \mu_0 \\ \mu_0 \\ \mu_0 \end{pmatrix}, \begin{pmatrix} \sigma_0^2 & \sigma_0^2 & \sigma_0^2 \\ \sigma_0^2 & \sigma_0^2 + \sigma^2 & \sigma_0^2 \\ \sigma_0^2 & \sigma_0^2 & \sigma_0^2 + \sigma^2 \end{pmatrix} \right),$$

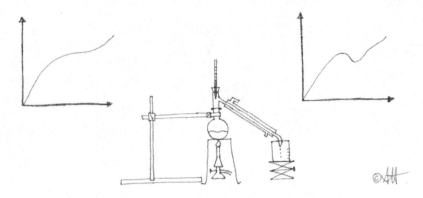

Figure 2.3: From prior to posterior via experiment.

which is not the distribution of an i.i.d. sample. But we still have

$$\begin{pmatrix} X_1 \\ X_2 \end{pmatrix} |\theta \sim N\left(\begin{pmatrix} \theta \\ \theta \end{pmatrix}, \begin{pmatrix} \sigma^2 & 0 \\ 0 & \sigma^2 \end{pmatrix} \right).$$

$\square$

In fact, we are not interested in the joint distribution of $(\theta, \mathbf{X})$; our interest is to learn about the underlying data generating distribution $\mathsf{P}^{\mathbf{X}|\theta}$, the main tool for which is the conditional distribution of θ given $\mathbf{X} = \mathbf{x}$. The information on the parameter based on prior and on the experiment is included in the conditional distribution of θ given $\mathbf{x}$, denoted by $\pi(.|x)$ and called **posterior distribution**. Roughly speaking the posterior distribution takes over the role of the likelihood. We formulate the **Bayesian inference principle** as follows.

> "The information on the underlying parameter θ is entirely
> contained in the posterior distribution.
> All statistical conclusions are based on $\pi(\theta|x)$."

In other words, when we know $\pi(.|x)$, we can do all inference, including estimation and testing. The derivation of the posterior distribution is the first step in a Bayes study, and the main tool to arrive at this is the **Bayes Theorem**. It holds that

$$\mathsf{P}^{(\theta, \mathbf{X})} = \mathsf{P}^{\theta} \mathsf{P}^{\mathbf{X}|\theta} = \mathsf{P}^{\mathbf{X}} \mathsf{P}^{\theta|\mathbf{X}}.$$

Thus

$$P^{\theta|X} = \frac{P^\theta P^{X|\theta}}{P^X}.$$

Assuming that the data and the parameter have continuous distributions, we rewrite this relation using the respective densities:

$$\pi(\theta|\mathbf{x}) = \frac{\pi(\theta)f(\mathbf{x}|\theta)}{f(\mathbf{x})}$$

where $f(\mathbf{x}|\theta)$ is the likelihood function $\ell(\theta|\mathbf{x})$. The joint density of $(\theta, \mathbf{X})$ can be written as

$$f(\theta, \mathbf{x}) = f(\mathbf{x}|\theta)\pi(\theta),$$

where $f(\mathbf{x})$ is the density of the marginal distribution, i.e.,

$$f(\mathbf{x}) = \int_\Theta f(\mathbf{x}|\theta)\pi(\theta)d\theta. \tag{2.6}$$

Hence for $\ell(\theta|\mathbf{x}) = f(\mathbf{x}|\theta)$, we have

$$\pi(\theta|\mathbf{x}) = \frac{\pi(\theta)\ell(\theta|\mathbf{x})}{\int_\Theta \ell(\theta|\mathbf{x})\pi(\theta)d\theta} \propto \pi(\theta)\ell(\theta|\mathbf{x}).$$

In case of a discrete distribution π over Θ, we have

$$f(\mathbf{x}) = \sum_{\theta \in \Theta} f(\mathbf{x}|\theta)\pi(\theta).$$

and

$$\pi(\theta|\mathbf{x}) = \frac{\pi(\theta)\ell(\theta|\mathbf{x})}{\sum_{\theta \in \Theta} \ell(\theta|\mathbf{x})\pi(\theta)} \propto \pi(\theta)\ell(\theta|\mathbf{x}),$$

where we use $\ell(\theta|\mathbf{x}) = P_\theta(\{\mathbf{x}\})$. In each case the most important relation for determining the posterior distribution is:

$$\boxed{\pi(\theta|\mathbf{x}) \propto \pi(\theta)\ell(\theta|\mathbf{x})}$$

The product $\pi(\theta)\ell(\theta|\mathbf{x})$ is the kernel function of the posterior. It includes the prior information (π) and the information from the data ($\ell(.|\mathbf{x})$). The posterior distribution is then determined up to a constant. To determine the complete posterior, we can apply different methods.

- Compare the kernel function with known distribution families,
- Calculate the normalizing constant analytically,
- Calculate the normalizing constant with Monte Carlo methods,
- Generate a sequence $\theta_{(1)}, \ldots, \theta_{(N)}$ which is distributed as $\pi(\theta|\mathbf{x})$, and
- Generate a sequence $\theta_{(1)}, \ldots, \theta_{(N)}$ which is distributed approximately as $\pi(\theta|\mathbf{x})$.

The computer intensive methods on Bayesian computations are included in Chapter 9. Hence we illustrate the first three items.

Example 2.11 (Binomial distribution and beta prior)

The Bayes model related to the historical billiard table Example 2.9 is given by $\theta \sim U(0,1)$, $\mathbf{X}|\theta \sim \text{Bin}(n,\theta)$ with the likelihood function

$$\ell(\theta|x) = \binom{n}{x}\theta^x(1-\theta)^{n-x} \propto \theta^x(1-\theta)^{n-x}.$$

The density of $U(0,1)$ is constant and equals one. Hence the posterior distribution is determined by

$$\pi(\theta|\mathbf{x}) \propto \theta^x(1-\theta)^{n-x}.$$

This is the kernel of a beta distribution $\text{Beta}(1+x, 1+n-x)$. The *beta distribution* is continuous, defined over the interval $[0,1]$, and depends on positive parameters α and β. Its density is given by

$$f(x|\alpha,\beta) = B(\alpha,\beta)^{-1}x^{\alpha-1}(1-x)^{\beta-1}, \tag{2.7}$$

where the normalizing constant $B(\alpha,\beta)$ is the beta function. Note that, $U(0,1)$ is the beta distribution with $\alpha = 1$, $\beta = 1$. In a more general case where the prior is a beta distribution $\text{Beta}(\alpha_0,\beta_0)$, we have

$$\pi(\theta|\mathbf{x}) \propto \theta^{\alpha_0-1}(1-\theta)^{\beta_0-1}\theta^x(1-\theta)^{n-x}$$
$$\propto \theta^{\alpha_0+x-1}(1-\theta)^{\beta_0+n-x-1}$$

and

$$\theta|x \sim \text{Beta}(\alpha_0+x, \beta_0+n-x).$$

$\square$

Example 2.12 (Normal i.i.d. sample and normal prior)

Consider an i.i.d. sample $\mathbf{X} = (X_1,\ldots,X_n)$ from $N(\mu,\sigma^2)$ with known variance σ^2. The unknown parameter is $\theta = \mu$. We assume that μ is a realization of normal distribution $N(\mu_0,\sigma_0^2)$. We have

$$\ell(\mu|\mathbf{x}) \propto \exp\left(-\frac{1}{2\sigma^2}\sum_{i=1}^{n}(x_i-\mu)^2\right)$$

and

$$\pi(\mu) \propto \exp\left(-\frac{1}{2\sigma_0^2}(\mu-\mu_0)^2\right).$$

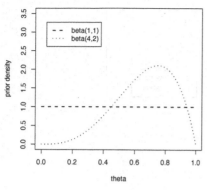
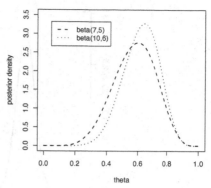

Figure 2.4: Example 2.11. Left: Prior distributions. Right: Posterior distributions, after observing $x = 6$ from $\text{Bin}(10, \theta)$.

Thus

$$\pi(\mu|\mathbf{x}) \propto \exp\left(-\frac{1}{2\sigma^2}\sum_{i=1}^{n}(x_i - \mu)^2 - \frac{1}{2\sigma_0^2}(\mu - \mu_0)^2\right).$$

Using the identity $\sum_{i=1}^{n}(x_i - \mu)^2 = (n-1)s^2 + n(\mu - \bar{x})^2$, where s^2 is the sample variance

$$s^2 = \frac{1}{n-1}\sum_{i=1}^{n}(x_i - \bar{x})^2$$

we obtain

$$\pi(\mu|\mathbf{x}) \propto \exp\left(-\frac{n}{2\sigma^2}(\mu - \bar{x})^2 - \frac{1}{2\sigma_0^2}(\mu - \mu_0)^2\right).$$

By completing the squares, we get

$$\pi(\mu|\mathbf{x}) \propto \exp\left(-\frac{n\sigma_0^2 + \sigma^2}{2\sigma_0^2\sigma^2}\left(\mu - \frac{\bar{x}n\sigma_0^2 + \mu_0\sigma^2}{n\sigma_0^2 + \sigma^2}\right)^2\right).$$

This is the kernel function of a normal distribution, so that the posterior distribution is:

$$N(\mu_1, \sigma_1^2), \text{ with } \mu_1 = \frac{\bar{x}n\sigma_0^2 + \mu_0\sigma^2}{n\sigma_0^2 + \sigma^2} \text{ and } \sigma_1^2 = \frac{\sigma_0^2\sigma^2}{n\sigma_0^2 + \sigma^2}. \qquad (2.8)$$

We see that the expectation is a weighted average of the prior mean and the sample mean. The variance of the posterior distribution σ_1^2 vanishes as n approaches infinity. The prior and posterior distributions are given in Figure 2.5, where the posterior distribution is a compromise between the information from the experiment and the prior. $\qquad\square$

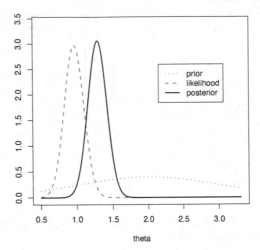

Figure 2.5: Example 2.12. Illustration of the posterior distribution as compromise between prior information and experimental information.

Let us consider a similar situation as in Example 2.12, but now focusing on an inference about the precision parameter $\tau = \frac{1}{\sigma^2}$.

Example 2.13 (Normal i.i.d. sample and gamma prior)
Consider an i.i.d. sample $\mathbf{X} = (X_1, \ldots, X_n)$ from $\mathsf{N}(0, \sigma^2)$ with unknown variance σ^2. The parameter of interest is the precision parameter $\theta = \tau = \sigma^{-2}$. As prior distribution of τ, we take a gamma distribution denoted by $\mathsf{Gamma}(\alpha, \beta)$, with the density

$$f(\tau | \alpha, \beta) = \frac{\beta^\alpha}{\Gamma(\alpha)} \tau^{\alpha - 1} \exp(-\beta \tau) \tag{2.9}$$

where $\alpha > 0$ is the shape parameter and $\beta > 0$ is the rate parameter. In this case, we have

$$\ell(\tau | \mathbf{x}) \propto \tau^{\frac{n}{2}} \exp\left(-\frac{\tau}{2} \sum_{i=1}^{n} x_i^2\right)$$

and

$$\pi(\tau | \alpha, \beta) \propto \tau^{\alpha - 1} \exp(-\beta \tau).$$

Then the posterior distribution has the kernel

$$\pi(\tau | \mathbf{x}, \alpha, \beta) \propto \tau^{\alpha - 1 + \frac{n}{2}} \exp\left(-\tau \left(\frac{1}{2} \sum_{i=1}^{n} x_i^2 + \beta\right)\right).$$

Hence the posterior distribution is also gamma distribution with shape parameter $\alpha + \frac{n}{2}$ and rate parameter $\frac{1}{2}\sum_{i=1}^{n} x_i^2 + \beta$:

$$\text{Gamma}\left(\alpha + \frac{n}{2}, \frac{1}{2}\sum_{i=1}^{n} x_i^2 + \beta\right). \tag{2.10}$$

$\square$

Let us consider the same set up once more, where now the parameter of interest is the variance.

Example 2.14 (Normal i.i.d. sample and inverse-gamma prior)
We consider the same sample as in Example 2.13. The unknown parameter is the variance $\theta = \sigma^2$. Let the prior distribution of θ be an *inverse-gamma distribution* InvGamma(α, β) with shape parameter α and scale parameter β, which has the density

$$f(\theta|\alpha, \beta) = \frac{\beta^\alpha}{\Gamma(\alpha)} \left(\frac{1}{\theta}\right)^{\alpha+1} \exp\left(-\frac{\beta}{\theta}\right). \tag{2.11}$$

Note that, if $X \sim \text{Gamma}(\alpha, \beta)$, then $X^{-1} \sim \text{InvGamma}(\alpha, \beta)$. We have

$$\ell(\theta|\mathbf{x}) \propto \left(\frac{1}{\theta}\right)^{\frac{n}{2}} \exp\left(-\frac{1}{2\theta}\sum_{i=1}^{n} x_i^2\right)$$

and

$$\pi(\theta|\alpha, \beta) \propto \left(\frac{1}{\theta}\right)^{\alpha+1} \exp\left(-\frac{\beta}{\theta}\right).$$

Then the posterior distribution has the kernel

$$\pi(\theta|\mathbf{x}, \alpha, \beta) \propto \left(\frac{1}{\theta}\right)^{\alpha+1+\frac{n}{2}} \exp\left(-\frac{\beta + \frac{1}{2}\sum_{i=1}^{n} x_i^2}{\theta}\right).$$

Hence the posterior distribution is the inverse-gamma distribution with shape parameter $\alpha + \frac{n}{2}$ and scale parameter $\frac{1}{2}\sum_{i=1}^{n} x_i^2 + \beta$ (see Figure 2.6), i.e.,

$$\sigma^2|\mathbf{x} \sim \text{InvGamma}\left(\alpha + \frac{n}{2}, \frac{1}{2}\sum_{i=1}^{n} x_i^2 + \beta\right). \tag{2.12}$$

$\square$

We revisit the lion example to illustrate a case where the constant can be easily calculated.

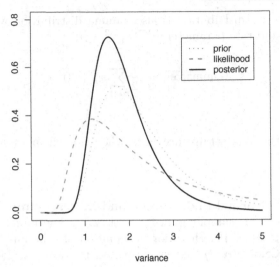

Figure 2.6: Example 2.14. The posterior distribution is more concentrated and slightly shifted towards the true parameter.

Example 2.15 (Lion's appetite)
For $x = 3$ the likelihood function is

$$
\begin{array}{c|ccc}
 & \theta_1 & \theta_2 & \theta_3 \\
\hline
\ell(\theta|x=3) & 0.8 & 0.1 & 0
\end{array}.
$$

The experience says that an adult male lion is lethargic with probability 0.8 where the probability that he is hungry is 0.1.

$$
\begin{array}{c|ccc}
 & \theta_1 & \theta_2 & \theta_3 \\
\hline
\pi(\theta) & 0.1 & 0.1 & 0.8
\end{array}.
$$

Thus

$$\pi(\theta_1|x=3) \propto (0.8)(0.1), \quad \pi(\theta_2|x=3) \propto (0.1)(0.1) \quad \pi(\theta_1|x=3) \propto (0.8)(0).$$

The normalizing constant is 0.09, and we obtain for the posterior distribution

$$
\begin{array}{c|ccc}
 & \theta_1 & \theta_2 & \theta_3 \\
\hline
\pi(\theta|x=3) & 0.889 & 0.111 & 0
\end{array}.
$$

After knowing that the lion has eaten 3 persons, the probability that he was hungry is really high; no surprise!   □

The following example is chosen to illustrate the need of computer intensive methods to calculate the posterior; see Figure 2.7.

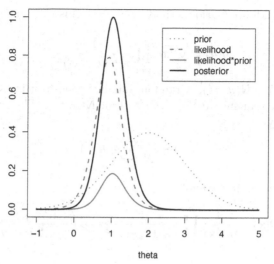

Figure 2.7: Example 2.16. The posterior distribution is normalized by an approximation of $m(\mathbf{x})$.

Example 2.16 (Cauchy i.i.d. sample and normal prior)

We consider an i.i.d. sample $\mathbf{X} = (X_1, \ldots, X_n)$ from a Cauchy distribution $\mathsf{C}(\theta, \gamma)$ with location parameter θ and scale parameter $\gamma > 0$. The density of $\mathsf{C}(\theta, \gamma)$ is

$$f(x|\theta, \gamma) = \frac{1}{\pi\gamma\left(1 + (\frac{x-\theta}{\gamma})^2\right)}. \qquad (2.13)$$

The parameter of interest is θ. We set $\gamma = 1$. Thus the likelihood is given by

$$\ell(\theta|\mathbf{x}) \propto \prod_{i=1}^{n} \frac{1}{1 + (x_i - \theta)^2}.$$

As prior distribution of θ we take the normal distribution $\mathsf{N}(\mu, \sigma^2)$, where μ and σ^2 are given. Using $\pi(\theta|\mathbf{x}) \propto \ell(\theta|\mathbf{x})\pi(\theta)$, we obtain

$$\pi(\theta|\mathbf{x}) \propto \prod_{i=1}^{n} \frac{1}{1 + (x_i - \theta)^2} \exp\left(-\frac{(\theta - \mu)^2}{2\sigma^2}\right). \qquad (2.14)$$

The normalizing constant

$$m(\mathbf{x}) = \int_{-\infty}^{\infty} \prod_{i=1}^{n} \frac{1}{1 + (x_i - \theta)^2} \exp\left(-\frac{(\theta - \mu)^2}{2\sigma^2}\right) d\theta$$

cannot be integrated analytically. Here in the introductory example we apply the independent Monte Carlo method, where $m(\mathbf{x})$ is taken as expected value of

$$g(\theta) = \sqrt{2\pi}\sigma \prod_{i=1}^{n} \frac{1}{1 + (x_i - \theta)^2}$$

with respect to $\theta \sim N(\mu, \sigma^2)$. A random number generator is used to generate independently a sequence $\theta^{(1)}, \ldots, \theta^{(N)}$ from $N(\mu, \sigma^2)$. Then the integral is approximated by

$$\widehat{m}(\mathbf{x}) = \frac{1}{N} \sum_{j=1}^{N} g(\theta^{(j)})$$

and the posterior is approximated by

$$\widetilde{\pi}(\theta|\mathbf{x}) = \frac{1}{\widehat{m}(\mathbf{x})} \prod_{i=1}^{n} \frac{1}{1 + (x_i - \theta)^2} \exp\left(-\frac{(\theta - \mu)^2}{2\sigma^2}\right).$$

$\square$

2.3   Advantages

In this section we present three settings, where the Bayesian approach gives easy computations in many useful models.

2.3.1   Sequential Analysis

Suppose we have successive data sets $x_{t_1}, \ldots, x_{t_n}$ related to the same parameter:

$$x_{t_j}, \text{ realization of } X_{t_j} \sim P^{X_{t_j}} \in \{P_{j,\theta}; \theta \in \Theta\}.$$

Further, we assume a common prior π:

$$\theta \sim \pi.$$

The posterior distribution based on the first data set is calculated by

$$\pi(\theta|x_{t_1}) \propto \pi(\theta)\ell_1(\theta|x_{t_1}),$$

where $\ell_1(\theta|x_{t_1})$ is the likelihood function corresponding to $P_{1,\theta}$. Knowing the results of the first experiment x_{t_1} we take the posterior distribution $\pi(\theta|x_{t_1})$ as new prior and calculate

$$\pi(\theta|x_{t_1}, x_{t_2}) \propto \pi(\theta|x_{t_1})\ell_2(\theta|x_{t_2}),$$

where $\ell_2(\theta|x_{t_2})$ is the likelihood function corresponding to $P_{2,\theta}$. Continuing similarly for each new independent data set, we finally obtain the posterior distribution based on all data sets

$$\pi(\theta|x_{t_1}, \ldots, x_{t_n}) \propto \pi(\theta)\ell(\theta|x_{t_1}, \ldots, x_{t_n}),$$

where $\ell(\theta|x_{t_1}, \dots, x_{t_n})$ is the likelihood corresponding to $P_{1,\theta} \times \dots \times P_{n,\theta}$. This procedure can also be generalized for cases with dependent data sets, using the likelihood functions of the conditional distributions $X_{t_j}|(X_{t_1}, \dots, X_{t_{(j-1)}})$.

Example 2.17 (Sequential data)

Consider i.i.d. data sets (x_1, x_2, x_3), (y_1, y_2), (z_1, z_2, z_3, z_4), independent of each other, where each single observation is a realization of $N(\theta, \sigma^2)$, σ^2 known. Suppose the prior $\theta \sim N(\mu_0, \sigma_0^2)$. Then the posterior distribution after the first data set is $N(\mu_1, \sigma_1^2)$ with

$$\mu_1 = \frac{\sum_{i=1}^{3} x_i \sigma_0^2 + \mu_0 \sigma^2}{3\sigma_0^2 + \sigma^2} \quad \text{and} \quad \sigma_1^2 = \frac{\sigma_0^2 \sigma^2}{3\sigma_0^2 + \sigma^2}.$$

The posterior distribution after the first two data sets is $N(\mu_2, \sigma_2^2)$ with

$$\mu_2 = \frac{(y_1 + y_2)\sigma_1^2 + \mu_1 \sigma^2}{2\sigma_1^2 + \sigma^2} \quad \text{and} \quad \sigma_2^2 = \frac{\sigma_1^2 \sigma^2}{2\sigma_1^2 + \sigma^2}.$$

Finally, the posterior distribution after all three data sets is $N(\mu_3, \sigma_3^2)$ with

$$\mu_3 = \frac{\sum_{i=1}^{4} z_i \sigma_2^2 + \mu_2 \sigma^2}{4\sigma_2^2 + \sigma^2} \quad \text{and} \quad \sigma_3^2 = \frac{\sigma_2^2 \sigma^2}{4\sigma_2^2 + \sigma^2}.$$

See Figure 2.8. □

2.3.2 Big Data

Another set up, where the Bayesian approach is really helpful, is the case where a huge amount of data is observed. The underlying data generating procedure is complicated, where the dimension p of the parameter space Θ exceeds the dimension n of the sample space. Using additional prior information makes the situation manageable. We demonstrate it by the following example.

Example 2.18 (Big Data)

Consider the linear model,

$$Y = X\beta + \epsilon, \tag{2.15}$$

where the $n \times 1$ vector Y is observed, the $n \times p$ matrix X is known. The unknown regression parameter is the $p \times 1$ vector β. The $n \times 1$ error vector ϵ is unobservable. We assume

$$\epsilon \sim N_n(0, \sigma^2 I_n).$$

Further we suppose that the variance σ^2 is known. The parameter of interest is $\theta = \beta \in \mathbb{R}^p$. Under the big data set up $n < p$, the $n \times p$ matrix X does

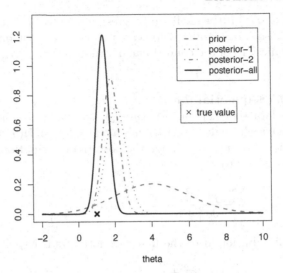

Figure 2.8: Example 2.17. The iteratively calculated posterior distributions are more and more concentrated around the unknown underlying parameter.

not have full rank so that the inverse of $\mathbf{X}^T\mathbf{X}$ does not exist. The likelihood function

$$\ell(\theta|\mathbf{Y}) \propto \left(\frac{1}{\sigma^2}\right)^{\frac{n}{2}} \exp\left(-\frac{1}{2\sigma^2}(\mathbf{Y} - \mathbf{X}\beta)^T(\mathbf{Y} - \mathbf{X}\beta)\right), \qquad (2.16)$$

has no unique maximum. The maximum likelihood estimators $\widehat{\beta}$ are the solutions of

$$\mathbf{X}^T\mathbf{X}\beta = \mathbf{X}^T\mathbf{Y}.$$

Using the Bayesian approach we assume a normal prior

$$\beta \sim \mathsf{N}_p(\beta_0, \sigma^2\Sigma_0),$$

where Σ_0 is positive definite, and obtain for the posterior

$$\pi(\beta|\mathbf{Y}) \propto \exp-\left\{\frac{1}{2\sigma^2}\left((\mathbf{Y} - \mathbf{X}\beta)^T(\mathbf{Y} - \mathbf{X}\beta) + (\beta - \mu_0)^T\Sigma_0^{-1}(\beta - \mu_0)\right)\right\}.$$

Completing squares we get

$$(\mathbf{Y} - \mathbf{X}\beta)^T(\mathbf{Y} - \mathbf{X}\beta) + (\beta - \beta_0)^T\Sigma_0^{-1}(\beta - \beta_0)$$
$$= (\beta - \beta_1)^T\Sigma_1^{-1}(\beta - \beta_1) + (\mathbf{Y} - \mathbf{X}\beta_1)^T(\mathbf{Y} - \mathbf{X}\beta_1) + (\beta_0 - \beta_1)^T\Sigma_0^{-1}(\beta_0 - \beta_1)$$

with

$$\beta_1 = \left(\mathbf{X}^T\mathbf{X} + \Sigma_0^{-1}\right)^{-1}\left(\mathbf{X}^T\mathbf{Y} + \Sigma_0^{-1}\beta_0\right)$$

and

$$\Sigma_1^{-1} = \mathbf{X}^T\mathbf{X} + \Sigma_0^{-1}.$$

Hence the posterior distribution is

$$\pi(\beta|\mathbf{Y}) \propto \exp\left(-\frac{1}{2\sigma^2}(\beta - \beta_1)^T\Sigma_1^{-1}(\beta - \beta_1)\right),$$

i.e.,

$$\beta|\mathbf{Y} \sim N_p(\beta_1, \sigma^2\Sigma_1).$$

Exploring the posterior instead of the likelihood function we avoid the complications due to $n < p$, since in this case the inverse of $\mathbf{X}^T\mathbf{X}$ does not exist, but the inverse of Σ_1 does. □

2.3.3 Hierarchical Models

Hierarchical models are based on stepwise conditioning. Consider non–identically distributed data sets $x_1, \ldots, x_m$, where each data set $x_j, j = 1, \ldots, m$ is the realization of a variable X_j with $X_j \sim P_{\theta_j}$. We assume

$$\theta_1, \ldots, \theta_m \text{ are i.i.d. } p(.|\alpha) \text{ distributed.}$$

In this case the distribution $p(.|\alpha)$ is a member of a parametric family, with parameter α. The dimension of α is less than the dimension of $\theta = (\theta_1, \ldots, \theta_m)$. It is a question of taste to consider $p(.|\alpha)$ as a prior distribution or as part of a hierarchical modelling. Furthermore the *hyperparameter* α can also be modelled as a realization of a prior

$$\alpha \sim \pi_{\mu_0},$$

where μ_0 is known. Setting $\mathbf{x} = (x_1, \ldots, x_m)$, the posterior distribution can be calculated as marginal distribution of

$$\pi(\theta, \alpha|\mathbf{x}) \propto p(\theta|\alpha)\pi_{\mu_0}(\alpha)\ell(\theta|\mathbf{x}),$$

which gives

$$\pi(\theta|x) \propto \pi(\theta)\ell(\theta|x)$$

where

$$\pi(\theta) = \int \pi(\theta|\alpha)\pi_{\mu_0}(\alpha)d\alpha.$$

The next example, taken from Dupuis (1995), illustrates a hierarchical model.

Example 2.19 (Capture–recapture)

In Dupuis (1995), a multiple capture–recapture experiment is analyzed by using the Arnason-Schwarz model. In Cévennes, France, on the Mont Lozère,

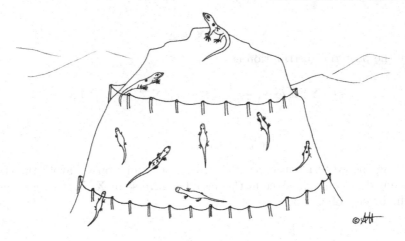

Figure 2.9: Capture–recapture for Example 2.19.

the migration behaviour of lizards, *Lacerta vivipara*, is studied. The lizard i, $i = 1, \ldots, n$ is captured (recaptured) and individually marked and returned in the stratum $r = 1, \ldots, k$ at times $j = 1, \ldots, 6$ (twice per year 1989, 1990, 1991). The number of marked juveniles is 96. The data for each animal i consists of $y_i = (x_i, z_i)$ with $x_i = (x_{(i,1)}, \ldots, x_{(i,6)})$ and $z_i = (z_{(i,1)}, \ldots, z_{(i,6)})$, where

$$x_{(i,j)} = \begin{cases} 1 & \text{if} \quad \text{lizard } i \text{ is captured at time } j \\ 0 & \text{otherwise} \end{cases}$$

$$z_{(i,j)} = \begin{cases} r & \text{if} \quad \text{lizard } i \text{ is captured in stratum } r \text{ at time } j \\ 0 & \text{otherwise} \end{cases}$$

Further it is assumed that $y_i = (x_i, z_i), i = 1, \ldots, n$, are i.i.d. The parameters are

$p_j(r) = $ probability that a lizard will be captured in stratum r at time j.

$q_j(r, s) = $ transition probability for moving from r to s at time j.

It is assumed that

$$q_j(r, s) = \phi_j(r) \psi_j(r, s)$$

where $\phi_j(r)$ is the survival probability and $\psi_j(r, s)$ is the probability of moving from r to s. The parameter of interest is given by

$$\theta = (p, \phi, \psi),$$

where $p = (p_1(1), \ldots, p_6(k))$, $\phi = (\phi_1(1), \ldots, \phi_6(k))$ and $\psi = (\psi_1(1,1), \ldots, \psi_6(k,k))$. Following assumptions are set:

$$p_j(r) \sim \text{Beta}(a, b), \text{ i.i.d.}$$

$$\phi_j(r) \sim \text{Beta}(\alpha, \beta), \text{ i.i.d.}$$

Set $\psi_j(r) = (\psi_j(r,1), \ldots, \psi_j(r,k))$ for the probabilities of moving from the location r at time j. They are assumed to follow Dirichlet distribution,

$$\psi_j(r) \sim \text{Dir}_k(e_1, \ldots, e_k), \text{ i.i.d.}$$

The *Dirichlet distribution* is a generalization of the beta distribution. In general, the probability function of $\mathbf{X} \sim \text{Dir}_k(p_1, \ldots, p_k)$ is given for $\mathbf{x} = (x_1, \ldots, x_k)$ with $\sum_{i=1}^{k} x_i = 1$. It depends on k positive parameters $p_1, \ldots, p_k$. Setting $\sum_{i=1}^{k} p_i = p_0$,

$$p(\mathbf{x}|p_1, \ldots, p_k) = \frac{\Gamma(p_0)}{\Gamma(p_1) \ldots \Gamma(p_k)} x_1^{p_1-1} \ldots x_k^{p_k-1}.$$

The components of $\psi_j(r)$ are dependent. The other parameters are assumed to be independent. The hyperparameters $e_1, \ldots, e_k$ are independent of the time j and the stratum s. In the study the experimenters set all hyperparameters as known, using the knowledge on the behavior of the lizards. The Bayesian inference is done by using Gibbs sampling, see Subsection 9.4.2.   □

2.4   List of Problems

1. Consider the following statistical model to analyze student's results in a written exam. Each student's result can be A, B, C, D, where A is the best grade and D means fail. The parameter θ is the level of preparation $0, 1, 2$, where 0 means nothing was done for the exam, 1 means that the student invested some time and the highest level 2 means the student was very well prepared. The probability $P_\theta(x)$ is given in the following table:

x	D	C	B	A
$\theta = 0$	0.8	0.1	0.1	0
$\theta = 1$	0.2	0.5	0.2	0.1
$\theta = 2$	0	0.1	0.5	0.4

The probability that a student has done nothing is 0.1 and that the student is very well prepared is 0.3. Calculate the posterior distribution for each x.

2. Assume $X|\theta \sim P_\theta$ and $\theta \sim \pi$. Show that:
The statistic $T = T(X)$ is sufficient iff the posteriors of T and X coincide: $\pi(\theta|X) = \pi(\theta|T(X))$.

3. Assume $X|n \sim \text{Bin}(n, 0.5)$. Find a prior on n such that $n|X \sim \text{NB}(x, 0.5)$.

4. Consider a sample $X = (X_1, \ldots, X_n)$ from $\mathsf{N}(0, \sigma^2)$, $n = 4$. The unknown parameter is the precision parameter $\theta = \tau = \sigma^{-2}$. As prior distribution of τ, we set $\mathsf{Gamma}(2, 2)$. Derive the posterior distribution.

5. In summer 2020, in a small town, 1000 inhabitants were randomly chosen and their blood samples are tested for Corona antibodies. 15 persons got a positive test result. From another study in Spring it was noted that the proportion of the population with antibodies is around 2%. Let X be the number of inhabitants tested positive in the study. Let θ be the probability of a positive test. Two different Bayes models are proposed.

   - $\mathcal{M}_0$: $X \sim \mathsf{Bin}(1000, p)$, $p \sim \mathsf{Beta}(1, 20)$
   - $\mathcal{M}_1$: $X \sim \mathsf{Poi}(\lambda)$, $\lambda \sim \mathsf{Gamma}(20, 1)$

   (a) Derive the posterior distributions for both models.
   (b) Discuss the differences between the two models.
   (c) Give a recommendation.

6. Consider a multiple regression model

$$y_i = \beta_0 + x_i \beta_1 + z_i \beta_2 + \varepsilon_i, \quad i = 1, \ldots, n$$

where ε_i are i.i.d. normally distributed with expectation zero and variance $\sigma^2 = 0.25$. Further $\sum_{i=1}^n x_i = 0$, $\sum_{i=1}^n z_i = 0$, $\sum_{i=1}^n x_i z_i = 0$, $\sum_{i=1}^n x_i^2 = n$, and $\sum_{i=1}^n z_i^2 = n$. The unknown three dimensional parameter $\beta = (\beta_0, \beta_1, \beta_2)^T$ is normally distributed with mean $\mu = (1, 1, 1)^T$ and covariance matrix $\Sigma = \mathbf{I}_3$. Determine the posterior distribution of β.

7. Hierarchical model. The observations belong to independent random variables $\mathbf{X} = (X_{ij})$, where

$$X_{ij} \sim \mathsf{N}(\theta_i, 1), \quad i = 1, \ldots, n, \quad j = 1, \ldots, k.$$

The parameters $\theta_1, \ldots, \theta_n$ are independent and normally distributed with

$$\theta_i | \mu \sim \mathsf{N}(\mu, 1), \quad \mu \sim \mathsf{N}(0, 1).$$

   (a) Determine the prior for $\theta = (\theta_1, \ldots, \theta_n)^T$.
   (b) Calculate the posterior distribution of θ given $\mathbf{x}$.
   (c) Calculate the posterior distribution of $\bar{\theta} = \frac{1}{n} \sum_{i=1}^n \theta_i$ given $\mathbf{x}$.

Chapter 3

Choice of Prior

The main difference between a statistical model and a Bayes model lies in the assumption that the parameter of the data generating distribution is the realization of a random variable from a known distribution. This distribution is called prior distribution. The parameter is not observed but its prior distribution is assumed to be known. The choice of the prior distribution is essential. In this chapter we present different principles for determining the prior.

Principles of Modelling

We begin by summarizing a general set of *principles of statistical modelling*. These principles are valid for a statistical model $\mathcal{P} = \{P_\theta : \theta \in \Theta\}$ following Definition 2.1 where we aim to determine a family of distributions. Concerning Bayes model $\{\mathcal{P}, \pi\}$, given in Definition 2.3, we need to model both a family of distributions and a single prior distribution. Thus the principles become even more important.

- First, all statistical models need assumptions for their valid applications. It is, therefore, worth keeping in mind, while developing the mathematical properties of models, that these properties will only hold as long as the assumptions do.

- Second, the development or selection of a statistical model should be objective-oriented. A model which is good for prediction, for example, may not be appropriate to explore certain relationships between variables.

- Third, since models are generally only approximative, it is highly recommended to be pragmatic rather than perfectionist. One should thus take the risk of setting necessary assumptions and, more so, keep track of them while analyzing the data under the postulated model.

- Fourth, it makes much sense to fit different models to the same problem and compare them.

- Finally, even with all the aforementioned requisites fully taken care of, it is always befitting to recall G.E.P. Box's well-known adage: Essentially all models are wrong, but some are useful. The practical question is how wrong they have to be to not be useful. In a nutshell, as in everyday life, one should

DOI: 10.1201/9781003221623-3

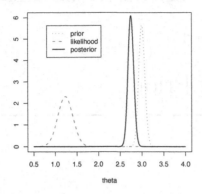 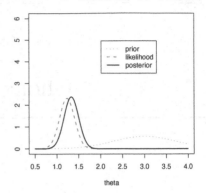

Figure 3.1: Example 3.1. Left: The prior has a small variance and a bad guess, and it dominates the likelihood. Right: The prior involves the same bad guess but also a high uncertainty. It does not dominate.

tread the path of pragmatism, trying to reduce the number of wrongs. It often suffices for good statistical modelling practice.

The new computer tools make it possible to handle almost all combinations of likelihood functions and prior distributions for deriving a posterior distribution by using

$$\pi(\theta|\mathbf{x}) \propto \pi(\theta)\ell(\theta|\mathbf{x}). \tag{3.1}$$

From this point of view we have big freedom in our choice, but of course it has to be reasonable. That the prior influences the posterior can be acceptable, but a prior which completely dominates the likelihood, makes the statistical study useless. The most extreme example is that the prior is the one point distribution in θ_0. Then we know the data generating distribution and no experiment is needed. Let us discuss an example.

Example 3.1 (Dominating prior)

Let $\mathbf{X} = (X_1, \ldots, X_n)$ be an i.i.d. sample in Example 2.12, where the true underlying data generating distribution is $\mathsf{N}(\mu, \sigma^2)$. First we set as prior $\mathsf{N}(\mu_0, \sigma_a^2)$, where μ_0 is far away from μ and the variance σ_a^2 is small. Roughly speaking, we are sure that the parameter is close to μ_0. Alternatively we use a normal prior $\mathsf{N}(\mu_0, \sigma_b^2)$ with the same expectation but with much larger variance σ_b^2, in order to express our uncertainty. In Figure 3.1 we see that the first prior dominates the likelihood. The more carefully chosen second prior looks reasonable. □

There are two main approaches for the choice of a prior.

Figure 3.2: Example 3.2. Young lions are more active.

(i) The prior is chosen in a pragmatic way to incorporate additional knowledge. Sometimes it is called the *subjective prior*.

(ii) The prior is derived from a theoretical background. It should fulfill desired properties, such as invariance, or it belongs to a preferred distribution family. Here some times the name *objective Bayesianism* is used.

In the following we explain the main methods of both approaches.

3.1 Subjective Priors

First we consider the following toy example.

Example 3.2 (Lion's appetite)
Recall Example 2.4. The appetite of a lion has three levels: hungry (θ_1), moderate (θ_2), lethargic (θ_3). An adult lion is only some times hungry, then he eats so much that he is lethargic for the next time. The moderate stage is unusual. Asking a zoologist, she proposed without doubt the following prior distribution for an adult animal:

	θ_1	θ_2	θ_3
$\pi(\theta)$	0.1	0.1	0.8

.

Young lions are more active. In this case her prior is:

	θ_1	θ_2	θ_3
$\pi(\theta)$	0.3	0.1	0.6

.

□

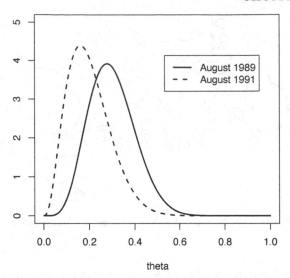

Figure 3.3: Example 3.3. The subjective priors are determined by experiences.

Let us continue with a real study, where biological knowledge is explored. It is a practical approach to explore the variability of a parametric distribution family. The parameters of the prior are called *hyperparameters*. The idea is to determine hyperparameters in such a way that the prior curve reflects the subjective knowledge. We will demonstrate this method by the following examples.

Example 3.3 (Capture probability)
Let us again consider the study in Example 2.19 published in Dupuis (1995). The parameter θ_j is the probability that a marked lizard is caught at time j. Let $j = 2$, which is August in the same year, and $j = 6$, the August two years later. The probability to catch a lizard after two years again is smaller than in the attempt. The expected values are 0.3 and 0.2 respectively, the variances are around 0.01. In Dupuis (1995) beta distributions are applied and the hyperparameters are chosen by taking into account biological rhythms of this species; see Figure 3.3.   □

Example 3.4 (Determining hyperparameters)
Consider an i.i.d. sample from a normal distribution $N(0, \sigma^2)$. The parameter of interest θ is the variance σ^2. We want to incorporate the additional information that the variance is most probably around 2 and with relatively high

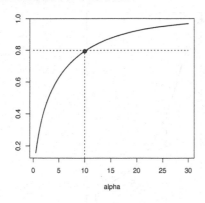 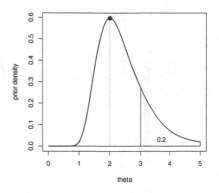

Figure 3.4: Example 3.4. Left: Determining α by $P(\theta > 3) \approx 0.8$. The distribution function of $\mathsf{InvGamma}(\alpha, 2(\alpha + 1))$ at 3 and the line at level 0.8 are plotted. Right: The prior density with $\alpha = 10$ and $\beta - 22$ reflects the subjective knowledge.

probability less that 3. This can be expressed as

$$\mathsf{mode} = 2 \text{ and } \mathsf{P}(\sigma^2 < 3) > 0.8.$$

As prior we assume an inverse-gamma distribution, $\mathsf{InvGamma}(\alpha, \beta)$ with density given in (7.26). The hyperparameters are α and β. The mode of $\mathsf{InvGamma}(\alpha, \beta)$ is

$$\mathsf{mode} = \frac{\beta}{\alpha + 1}$$

Thus $\beta = 2(\alpha + 1)$. The inverse-gamma distribution is included in R. By a simple graphical method we determine $\alpha = 10$ such that the second condition is fulfilled; see following R Code and Figure 3.4. The proposed subjective prior is $\mathsf{InvGamma}(10, 22)$.   □

R Code 3.1.1. Determining hyperparameters, Example 3.4.

```
library(invgamma) # package for inverse-gamma distribution
a<-seq(0.5,30,0.01) # hyperparameter alpha
b<-2*(a+1) # hyperparameter beta; the mode is defined as 2
plot(a,pinvgamma(3,a,b),"l",xlab="alpha",ylab="")
# distribution function at 3 as function of alpha
lines(0:30,rep(0.8,31),lty=2) # the desired level
points(10,pinvgamma(3,10,22)) # approximative crossing point
segments(10,0,10,pinvgamma(3,10,22),lty=2) # determining alpha
```

In the following example we are interested in the expected value of a normal distribution. We have only vague information and choose a heavy tailed distribution as prior.

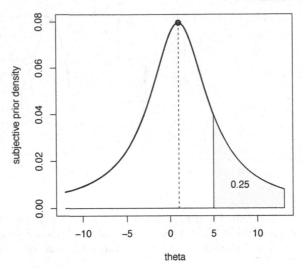

Figure 3.5: Example 3.5. The subjective prior reflects the vague subjective information.

Example 3.5 (Determining hyperparameters)

Consider an i.i.d. sample from a normal distribution $N(\mu, 1)$. The parameter of interest θ is the expected value μ. We guess that the parameter μ is lying symmetrically around 1 and the probability that the value is larger than 5 is around 0.25, i.e.,

$$\text{mode}(\pi) = 1 \quad \text{and} \quad P(\mu > 5) \approx 0.25.$$

We assume a Cauchy distribution $C(1, \gamma)$, which is heavy tailed and symmetrical around 1. The hyperparameter is the scaling parameter γ. The quantile function of $C(x_0, \gamma)$ is given by

$$q(p) = x_0 + \gamma \tan(\pi(p - 0.5)).$$

In this case $x_0 = 1$, $p = 0.75$ and $q(p) = 5$. Because $\tan(\frac{\pi}{4}) = 1$ we get $\gamma = 4$. The proposed subjective prior is $C(1, 4)$; see also Figure 3.5.   □

Another way to implement subjective information into a prior distribution is as following. Sometimes it is easier to have a good guess for the data point than for the parameter distribution. This information can also be explored for determining hyperparameters. For detailed discussion of this method, see Berger (1980, Section 3.5). Here we only present an example.

Example 3.6 (Determining hyperparameters via prediction)
We continue Example 2.11 and consider the Bayes model $\theta \sim \text{Beta}(\alpha_0, \beta_0)$, and $\mathbf{X}|\theta \sim \text{Bin}(n, \theta)$. The marginal distribution of X can be calculated by (2.6), as

$$P(X = k) = \int_0^1 \binom{n}{k} \theta^{\alpha_0 + k - 1}(1 - \theta)^{\beta_0 + n - k - 1} B(\alpha_0, \beta_0)^{-1} d\theta,$$

where $B(a, b)$ is the beta function as the solution of the integral equation

$$B(a, b) = \int_0^1 x^{a-1}(1 - x)^{b-1} dx, \quad a > 0, \ b > 0. \tag{3.2}$$

We obtain the beta-binomial distribution $\text{BetaBin}(n, \alpha_0, \beta_0)$,

$$P(X = k) = \binom{n}{k} \frac{B(\alpha_0 + k, \beta_0 + n - k)}{B(\alpha_0, \beta_0)}. \tag{3.3}$$

These probabilities can be calculated, using the beta function implemented in R, see Figure 3.6 and R Code 2. Let us now illustrate two different cases to determine α_0 and β_0. We begin by assuming equal probability,

$$P(X = k) = \frac{1}{n + 1}. \tag{3.4}$$

Using

$$\binom{n}{k} = \frac{1}{(n + 1)B(k + 1, n - k + 1)},$$

we get

$$P(X = k) = \frac{1}{n + 1} \frac{B(\alpha_0 + k, \beta_0 + n - k)}{B(k + 1, n - k + 1)B(\alpha_0, \beta_0)}. \tag{3.5}$$

For $\alpha_0 = 1$ and $\beta_0 = 1$ we obtain (3.4). Recall that it is the Bayes model in the historical Example 2.9.
Now we assume that the marginal distribution in (3.3) is symmetric and the highest probability is around 0.3. The symmetry $P(X = k) = P(X = n - k)$ is fulfilled for $\alpha_0 = \beta_0$, and we plot $P(X = n - k)$ as a function of α to obtain α_0, see Figure 3.6 and following R Code. The assumed subjective prior is $\text{Beta}(3.8, 3.8)$. $\qquad\qquad\square$

R Code 3.1.2. Determining hyperparameters, Example 3.6.

```
n<-4; a=3.8; b=3.8; p<-rep(0,n+1)
for (i in 1:(n+1)){
    p[i]<-beta(a+i-1,b+n-i+1)/beta(a,b)*choose(n,(i-1))
    }
plot(0:n+1,p,"h",ylab="marginal probability",
    xlab="x",lwd=7, col=grey(0.6))
```

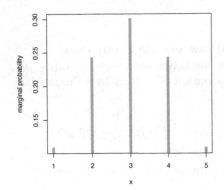 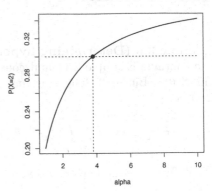

Figure 3.6: Example 3.4. Left: The marginal distribution for X is plotted. Right: The parameter of the prior is determined such that the marginal distribution has the desired properties.

```
aa=seq(1,10,0.01); a0=3.8
plot(aa,beta(aa+2,aa+2)/beta(aa,aa)*choose(n,n/2),"l")
lines(aa,rep(0.3,length(aa)),"l",lty=2)
points(a0,6*beta(a0+2,a0+2)/beta(a0,a0),lwd=3)
segments(a0,6*beta(a0+2,a0+2)/beta(a0,a0),a0,0,lty=2)
```

Another good proposal for a subjective prior is a mixture distribution. Mixture distributions are rich parametric classes. A folk theorem says that every distribution can be approximated by a mixture distribution, which is used in several contexts without a clear statement. We quote here a recent result of Nguyen et al. (2020).

Theorem 3.1 (Approximation by mixture) *Assume $g : \mathbb{R}^p \to \mathbb{R}$ and for all $x \in \mathbb{R}^p$ there exist two positive constants c_1, c_2 such that*

$$|g(x)| \le c_1(1 + \|x\|_2^p)^{-p-c_2}. \tag{3.6}$$

Then for any continuous function $f : \mathbb{R}^p \to \mathbb{R}$, there exists a sequence $h_m : \mathbb{R}^p \to \mathbb{R}$ of mixtures

$$h_m(x) = \sum_{j=1}^m c_j \frac{1}{\sigma_j^p} g(\frac{x - \mu_j}{\sigma_j}), \quad \sigma_j > 0, \ \mu_j \in \mathbb{R}^p, \ c_j > 0, \ \sum_{j=1}^m c_j = 1$$

such that

$$\lim_{m \to \infty} \int |f(x) - h_m(x)| dx = 0.$$

The following classroom example gives some insight in modelling by a mixture distribution.

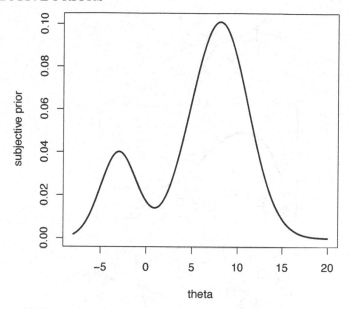

Figure 3.7: Illustration for Example 3.7. The subjective prior is a mixture of normal distributions.

Example 3.7 (Weather)
We are interested in the temperature at noon in February. The parameter of interest is the expectation of the temperature θ. We ask two experts for their knowledge. One expert guesses that the winter is mild with temperature around 8 °C. The other expert has a cold winter in mind. Believing in the climate change we give the first expert the weight 0.8 but higher uncertainty. The subjective prior of the first expert is proposed to be N(8, 10), the subjective prior of the second expert comes out by N(−3, 4). Summarizing, we assume a mixture prior

$$\pi(\theta) = 0.8\,\phi_{(8,10)}(\theta) + 0.2\,\phi_{(-3,4)}(\theta), \tag{3.7}$$

where $\phi_{(\mu,\sigma^2)}$ denotes the density of N(μ, σ^2), see Figure 3.7. This mixture distribution fulfills the assumptions of the above theorem with $g(.) = \phi_{(0,1)}(.)$. $\square$

In principle whenever it is possible to determine a weight function over the parameter set and we can calculate the weight function at every argument, we can apply a Bayesian analysis; see Figure 3.8. There are no closed expressions for the prior or for the likelihood function needed. This approach is explained in Chapter 9 on Bayesian computations.

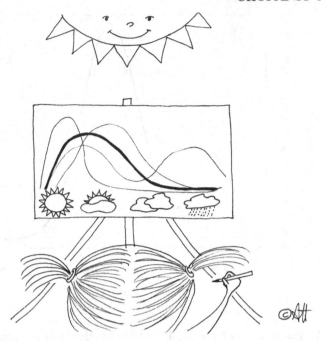

Figure 3.8: Subjective prior for the expected value of hours of sunshine.

In the next section we present the construction of a parametric distribution family, which includes both the prior and the respective posterior. Such a distribution family is called conjugate and it is highly recommended.

3.2 Conjugate Priors

This approach was introduced in Raiffa and Schlaifer (1961). We start with the definition of a conjugate family.

Definition 3.1 (Conjugate family) For a given statistical model $\mathcal{P} = \{P_\theta : \theta \in \Theta\}$ a family $\mathcal{F}$ of probability distributions on Θ is called conjugate iff for every prior $\pi(.) \in \mathcal{F}$ and for every $\mathbf{x}$ the posterior $\pi(.|\mathbf{x}) \in \mathcal{F}$.

The family $\mathcal{F}$ is also called closed under sampling. The prior distribution which is an element of a conjugate family is named **conjugate prior**.

This family is very practical, since only an updating of the hyperparameters is needed to compute the posterior. We illustrate it with the help of the flowers example (Example 2.1).

Example 3.8 (Flowers)

The data (n_1, n_2, n_3) are the number of flowers of respective colours. The sample size $n = n_1 + n_2 + n_3$ is fixed. The data generating distribution a multinomial, $\mathsf{Mult}(n, p_1, p_2, p_3)$:

$$P(n_1, n_2, n_3) = \frac{n!}{n_1! \, n_2! \, n_3!} p_1^{n_1} p_2^{n_2} p_3^{n_3}. \tag{3.8}$$

The unknown parameter $\theta = (p_1, p_2, p_3)$ consists of the probabilities for each colour, so that $p_1 + p_2 + p_3 = 1$. We consider as prior for θ a Dirichlet distribution $\mathsf{Dir}_3(e_1, e_2, e_3)$ with

$$\pi(\theta_1, \theta_2, \theta_3 | e_1, e_2, e_3) \propto \theta_1^{e_1-1} \theta_2^{e_2-1} \theta_3^{e_3-1}. \tag{3.9}$$

The posterior is calculated by

$$\pi(\theta | (n_1, n_2, n_3)) \propto \pi(\theta_1, \theta_2, \theta_3 | e_1, e_2, e_3) P(n_1, n_2, n_3).$$

We obtain

$$\pi(\theta | (n_1, n_2, n_3)) \propto \theta_1^{e_1+n_1-1} \theta_2^{e_2+n_2-1} \theta_3^{e_3+n_3-1},$$

which is the kernel of a Dirichlet distribution with parameters:

$$e_1 + n_1, \ e_2 + n_2, \ e_3 + n_3. \tag{3.10}$$

Thus the set of Dirichlet distributions

$$\mathcal{F} = \{\mathsf{Dir}_3(\alpha_1, \alpha_2, \alpha_3) : \alpha_1 > 0, \alpha_2 > 0, \alpha_3 > 0\}$$

forms a conjugate family. The prior in (3.9) is a conjugate prior. To compute the posterior, we apply the updating rule (3.10). The hyperparameters e_1, e_2, e_3 in (3.9) can be determined by subjective information. We know that violet is the common colour, white occurs also, but pink is an exception. Set $e_0 = e_1 + e_2 + e_3$ and using the subjective expected values of the components,

$$\mathsf{E}\theta_1 = \frac{e_1}{e_0} = 0.2, \quad \mathsf{E}\theta_2 = \frac{e_2}{e_0} = 0.7, \quad \mathsf{E}\theta_3 = \frac{e_3}{e_0} = 0.1,$$

we choose $e_1 = 2, e_2 = 7, e_3 = 1$. In this illustrative example we assume a subjective prior which is also conjugate, i.e.,

$$\theta \sim \mathsf{Dir}_3(2, 7, 1).$$

□

In Example 2.11 the statistical model is a binomial distribution with unknown success probability, the prior and the posterior distribution are beta distributions. Example 2.12 is also a Bayes model with a conjugate prior, where the i.i.d. sample is from a normal distribution, the parameter of interest is the expected value and the conjugate prior is a normal distribution.

In Example 2.13 the i.i.d. sample is from a normal distribution, the parameter of interest is the inverse variance, the conjugate prior is a gamma distribution. The same set up was also considered in Example 2.14. Here the parameter of interest is the variance and the conjugate prior is an inverse-gamma distribution.

In Example 2.16, however, the prior is not conjugate.

The common property of the data generating distributions in Examples 2.11, 3.8 and 2.12 is that the related statistical model belongs to an exponential family. In the following we state a general result for the construction a conjugate family for data generating distributions belonging to an exponential family.

First we define the exponential family.

Definition 3.2 (Exponential family of dimension k) A statistical model $\mathcal{P} = \{P_\theta : \theta \in \Theta\}$ belongs to an exponential family of dimension k, if there exist real-valued functions $\zeta_1, \ldots, \zeta_k$ on Θ, real-valued statistics $T_1, \ldots, T_k$ and a function h on $\mathcal{X}$ such that the probability function has the form

$$p(\mathbf{x}|\theta) = C(\theta) \exp(\sum_{j=1}^{k} \zeta_j(\theta) T_j(\mathbf{x})) h(\mathbf{x}). \tag{3.11}$$

The representation (3.11) is not unique. The exponential family is called **strictly k-dimensional**, iff it is impossible to reduce the number of terms in the sum in (3.11).

Many distributions belong to exponential family. The exponential families have several interesting analytical properties. Especially in statistical inference, estimation and testing theory has been extensively developed for exponential families; see for instance Liero and Zwanzig (2011). For more theoretical results on exponential families we recommend Liese and Miescke (2008), see also Robert (2001) and the literature therein.

The main trick is to reformulate the presentation of the probability function such that it has the form (3.11). Hereby, of most interest are the sufficient statistic $T(\mathbf{x}) = (T_1, \ldots, T_k)$ and the parameter $\zeta(\theta) = (\zeta_1, \ldots, \zeta_k)$. $C(\theta)$ is the normalizing constant. The new parameter $\zeta = \zeta(\theta) = (\zeta_1, \ldots, \zeta_k)$ is called **natural parameter**.

We demonstrate it by three examples. Consider the distributions in Examples 3.8 and 2.12.

Example 3.9 (Multinomial distribution)

The multinomial distribution $\text{Mult}(n, p_1, \ldots, p_k)$ is a discrete distribution with probability function

$$P(n_1, \ldots, n_k) = \frac{n!}{n_1! \ldots n_k!} p_1^{n_1} \ldots p_k^{n_k}, \qquad (3.12)$$

where

$$\sum_{i=1}^{k} n_i = n, \quad \sum_{i=1}^{k} p_i = 1. \qquad (3.13)$$

Using the relation A,

$$z^a = \exp(\ln(z^a)) = \exp(a \ln(z)) \qquad (3.14)$$

we obtain

$$P(n_1, \ldots, n_k) = h(n_1, \ldots, n_k) \exp\left(\sum_{i=1}^{k} \ln(p_i) n_i\right),$$

with

$$h(n_1, \ldots, n_k) = \frac{n!}{n_1! \ldots n_k!}.$$

Hence the multinomial distribution belongs an exponential family. By using (3.13) we can reduce the dimension. Since,

$$\sum_{i=1}^{k} \ln(p_i) n_i = \sum_{i=1}^{k-1} n_i \ln\left(\frac{p_i}{p_k}\right) + n \ln(p_k); \quad p_k = 1 - \sum_{i=1}^{k-1} p_i.$$

the multinomial distribution belongs to a $(k-1)$-dimensional exponential family with

$$T = (n_1, \ldots, n_{k-1}) \text{ and } \zeta = \left(\ln\left(\frac{p_1}{p_k}\right), \ldots, \ln\left(\frac{p_{k-1}}{p_k}\right)\right).$$

$\square$

Example 3.10 (Dirichlet distribution)

For $\mathbf{x} = (x_1, \ldots, x_k)$ with $\sum_{i=1}^{k} x_i = 1$, the Dirichlet distribution $\text{Dir}_k(\alpha_1, \ldots, \alpha_k)$ has the density

$$f(\mathbf{x}|\alpha_1, \ldots, \alpha_k) = \frac{\Gamma(\alpha_0)}{\Gamma(\alpha_1) \ldots \Gamma(\alpha_k)} x_1^{\alpha_1 - 1} \ldots x_k^{\alpha_k - 1}, \qquad (3.15)$$

where $\alpha_0 = \sum_{i=1}^{k} \alpha_i$. Using the relation (3.14) we rewrite

$$f(\mathbf{x}|\alpha_1, \ldots, \alpha_k) = \frac{\Gamma(\alpha_0)}{\Gamma(\alpha_1) \ldots \Gamma(\alpha_k)} \exp\left(\sum_{i=1}^{k} \ln(x_i)(\alpha_i - 1)\right).$$

Set $h(\mathbf{x}) = x_1^{-1} \ldots x_k^{-1}$. We get

$$f(\mathbf{x}|\alpha_1, \ldots, \alpha_k) = \frac{\Gamma(\alpha_0)}{\Gamma(\alpha_1) \ldots \Gamma(\alpha_k)} h(x) \exp(\sum_{i=1}^{k} \ln(x_i)\alpha_i).$$

The family of Dirichlet distributions forms a k-dimensional exponential family with

$$T(\mathbf{x}) = (\ln(x_1), \ldots, \ln(x_k))$$

and

$$\zeta = (\alpha_1, \ldots, \alpha_k).$$

□

Example 3.11 (Normal distribution)
The normal distribution $\mathsf{N}(\mu, \sigma^2)$ has the density

$$f(x|\mu, \sigma^2) = \frac{1}{\sqrt{2\pi\sigma^2}} \exp\left(-\frac{(x-\mu)^2}{2\sigma^2}\right),$$

which can be decomposed into

$$f(x|\theta) = \frac{1}{\sqrt{2\pi\sigma^2}} \exp\left(-\frac{\mu^2}{2\sigma^2}\right) \exp\left(-\frac{x^2}{2\sigma^2} + \frac{x\mu}{\sigma^2}\right).$$

Thus the normal distribution is a two-parameter exponential family with $\theta = (\mu, \sigma^2)$ where

$$T(x) = (x^2, x) \quad \text{and} \quad \zeta(\theta) = (-\frac{1}{2\sigma^2}, \frac{\mu}{\sigma^2}).$$

□

Theorem 4.12 gives a very useful result on exponential families, namely that for an i.i.d. sample to follow an exponential family, it suffices that X_i follows an exponential family; see Liero and Zwanzig (2011).

Theorem 3.2
If $\mathbf{X} = (X_1, \ldots, X_n)$ is an i.i.d. sample from a distribution of the form (3.11) with functions ζ_j and T_j, $j = 1, \ldots, k$ then the distribution of $\mathbf{X}$ belongs to an exponential family with parameter ζ_j, and statistics

$$T_{(n,j)}(\mathbf{x}) = \sum_{i=1}^{n} T_j(x_i), \quad j = 1, \ldots, k.$$

PROOF: Since the distribution of X_i belongs to an exponential family the probability function of the sample is given by

$$p(\mathbf{x}|\theta) = \prod_{i=1}^{n} C(\theta) \exp(\sum_{j=1}^{k} \zeta_j(\theta)T_j(x_i))h(x_i)$$

$$= C(\theta)^n \exp(\sum_{j=1}^{k} \zeta_j(\theta) \sum_{i=1}^{n} T_j(x_i))\tilde{h}(\mathbf{x}) \qquad (3.16)$$

with $\tilde{h}(\mathbf{x}) = \prod_{i=1}^{n} h(x_i)$. Thus the distribution of $\mathbf{X} = (X_1, \ldots, X_n)$ belongs to an exponential family with the functions ζ_j and the statistics $T_{(n,j)}(\mathbf{x}) = \sum_{i=1}^{n} T_j(x_i)$.

□

We apply this result to a normal i.i.d. sample.

Example 3.12 (Normal distribution)
Consider an i.i.d. sample $X_1, \ldots, X_n$ from $N(\mu, \sigma^2)$. Then the distribution of $\mathbf{X} = (X_1, \ldots, X_n)$ belongs to a two-parameter exponential family with $\theta = (\mu, \sigma^2)$ and

$$T_n(\mathbf{x}) = (\sum_{i=1}^{n} \mathbf{x_i}, \sum_{i=1}^{n} \mathbf{x_i^2})$$

and

$$\zeta(\theta) = (\frac{\mu}{\sigma^2}, -\frac{1}{2\sigma^2}).$$

□

The following theorem tells how we can get the conjugate family for the natural parameter of a statistical model belonging to a k-dimensional exponential family. For convenience we write the probability function as

$$p(\mathbf{x}|\theta) = h(\mathbf{x})\exp(\theta^T T(\mathbf{x}) - \Psi(\theta)), \qquad (3.17)$$

where $\mathbf{x} \in \mathbb{R}^n$ and $\theta \in \Theta \subseteq \mathbb{R}^k$; the function $\Psi(\theta)) = -\ln(C(\theta))$ is called *cumulant generating function*; see Robert (2001).

Theorem 3.3
Assume $\mathbf{X}$ has a distribution of the form (3.17). Then the conjugate family over Θ is given by

$$\mathcal{F} = \{\pi(\theta|\mu, \lambda) = K(\mu, \lambda)\exp(\theta^T\mu - \lambda\Psi(\theta)) : \mu \in \mathbb{R}^k, \ \lambda \in \mathbb{R}, \ \lambda > 0\}.$$
$$(3.18)$$

The posterior belonging to the prior $\pi(\theta|\mu_0, \lambda_0)$ has the parameters

$$\mu = \mu_0 + T(\mathbf{x}) \quad and \quad \lambda = \lambda_0 + 1.$$

PROOF: Suppose $\pi(\theta) \in \mathcal{F}$ such that

$$\pi(\theta) = K(\mu_0, \lambda_0) \exp(\theta^T \mu_0 - \lambda_0 \Psi(\theta)).$$

The posterior is determined by $\pi(\theta) l(\theta|\mathbf{x})$. Thus

$$\pi(\theta|\mathbf{x}) \propto K(\mu_0, \lambda_0) \exp(\theta^T \mu_0 - \lambda_0 \Psi(\theta)) h(\mathbf{x}) \exp(\theta^T T(\mathbf{x}) - \Psi(\theta))$$
$$\propto \exp(\theta^T \mu_0 - \lambda_0 \Psi(\theta)) \exp(\theta^T T(\mathbf{x}) - \Psi(\theta))$$
$$\propto \exp(\theta^T (\mu_0 + T(\mathbf{x})) - (\lambda_0 + 1) \Psi(\theta)).$$

Hence the posterior belongs to $\mathcal{F}$ with parameters $\mu_0 + T(\mathbf{x})$ and $\lambda_0 + 1$.

$\square$

We demonstrate the application of Theorem 3.3 by the following example.

Example 3.13 (Gamma distribution and conjugate prior)
Consider an observation $X \sim \mathsf{Gamma}(\alpha, \beta)$ with known shape parameter $\alpha > 0$. Parameter of interest θ is the rate parameter $\beta > 0$. The density of $\mathsf{Gamma}(\alpha, \theta)$ is

$$f(x|\alpha, \theta) = \frac{\theta^\alpha}{\Gamma(\alpha)} x^{\alpha-1} \exp(-\theta x). \tag{3.19}$$

Hence the statistical model belongs to a 1-dimensional exponential family of form (3.17) with natural parameter θ and

$$T(x) = -x, \quad \Psi(\theta) = -\alpha \ln(\theta), \quad h(x) = \frac{x^{\alpha-1}}{\Gamma(\alpha)}.$$

Applying Theorem 3.3, a conjugate prior is

$$\pi(\theta) \propto \exp(\theta \mu_0 - \lambda_0 \Psi(\theta))$$

and the posterior

$$\pi(\theta|x) \propto \exp(\theta(\mu_0 - x) - (\lambda_0 + 1) \Psi(\theta)).$$

For $\mu_0 < 0$ we can rewrite the kernels as those of gamma distributions. We have

$$\pi(\theta) \propto \theta^{\alpha \lambda_0} \exp(\theta \mu_0),$$
$$\pi(\theta|x) \propto \theta^{\alpha(\lambda_0 + 1)} \exp(\theta(\mu_0 - x)).$$

which are the kernels of $\mathsf{Gamma}(\lambda_0 \alpha - 1, -\mu_0)$ and $\mathsf{Gamma}(\alpha(\lambda_0 + 1) - 1, -\mu_0 + x)$ respectively. Summarizing we notice that if $\theta \sim \mathsf{Gamma}(a_0, b_0)$ with $a_0 > 0$ and $b_0 > 0$ then $\theta|x \sim \mathsf{Gamma}(a_0 + \alpha, b_0 + x)$. $\square$

Generalized linear models also belong to exponential families. Let us get back to the Corona Example 2.5.

Example 3.14 (Logistic regression)
We consider a general logistic regression model. The data set is given by $(y_i, x_{1i}, \ldots, x_{pi})$ with $i = 1, \ldots, n$, where $y_1, \ldots, y_n$ are independent response variables. The success probability depends on the covariates $x_i = (x_{1i}, \ldots, x_{pi})^T$, i.e.,

$$P(y_i = 1|x_i, \theta) = p(x_i).$$

The statistical model is

$$p(y_1, \ldots, y_n|x_i, \theta) = \prod_{i=1}^{n} p(x_i)^{y_i}(1 - p(x_i))^{(1-y_i)}$$

$$= \exp\left(\sum_{i=1}^{n} \ln(p(x_i))y_i + \sum_{i=1}^{n} \ln(1 - p(x_i))(1 - y_i)\right)$$

$$= \exp\left(\sum_{i=1}^{n} \ln\left(\frac{p(x_i)}{1 - p(x_i)}\right)y_i + \sum_{i=1}^{n} \ln(1 - p(x_i))\right). \tag{3.20}$$

The main trick is to find a canonical link function $g(.)$, such that

$$\sum_{i=1}^{n} \ln\left(\frac{p(x_i)}{1 - p(x_i)}\right)y_i = \sum_{i=1}^{n} y_i x_i^T \theta.$$

In this case $g(.)$ is the logistic function

$$g(z) = \frac{\exp(z)}{1 + \exp(z)}, \quad \ln\left(\frac{g(z)}{1 - g(z)}\right) = z.$$

Assuming

$$p(x_i) = g(x_i^T \theta), \tag{3.21}$$

we get

$$p(y_1, \ldots, y_n|x_1, \ldots, x_p, \theta) = \exp\left(\sum_{i=1}^{n} y_i x_i^T \theta - \Psi(\theta|x_1, \ldots, x_p)\right).$$

Hence under Assumption (3.21), the statistical model is a p-dimensional exponential family with natural parameter θ and

$$T(\mathbf{y}) = \sum_{i=1}^{n} y_i x_i, \quad \Psi(\theta|x_1, \ldots, x_p) = -\sum_{i=1}^{n} \ln(1 - p(x_i)) = \sum_{i=1}^{n} \ln(1 + \exp(x_i^T \theta)).$$

Applying Theorem 3.3, a conjugate prior is

$$\pi(\theta|x_1, \ldots, x_p) \propto \exp(\theta^T \mu_0 - \lambda_0 \Psi(\theta|x_1, \ldots, x_p)).$$

$\square$

The following table collects the conjugate priors and corresponding posteriors related to some popular one parameter exponential families with parameter θ, assuming all other parameters known, where the posteriors are computed for a single observation x. The table is partly taken from Robert (2001).

Distribution $p(x\|\theta)$	Prior $\pi(\theta)$	Posterior $\pi(\theta\|x)$
Normal $N(\theta,\sigma^2)$	Normal $N(\mu,\tau^2)$	Normal $N\left(\rho(\sigma^2\mu + \tau^2 x), \rho\sigma^2\tau^2\right)$ $\rho^{-1} = \sigma^2 + \tau^2$
Poisson $\mathsf{Poi}(\theta)$	Gamma $\mathsf{Gamma}(\alpha,\beta)$	Gamma $\mathsf{Gamma}(\alpha + x, \beta + 1)$
Gamma $\mathsf{Gamma}(\nu,\theta)$	Gamma $\mathsf{Gamma}(\alpha,\beta)$	Gamma $\mathsf{Gamma}(\alpha + \nu, \beta + x)$
Binomial $\mathsf{Bin}(n,\theta)$	Beta $\mathsf{Beta}(\alpha,\beta)$	Beta $\mathsf{Beta}(\alpha + x, \beta + n - x)$
Negative Binomial $\mathsf{NB}(m,\theta)$	Beta $\mathsf{Beta}(\alpha,\beta)$	Beta $\mathsf{Beta}(\alpha + m, \beta + x)$
Multinomial $\mathsf{Mult}_k(\theta_1,\ldots,\theta_k)$	Dirichlet $\mathsf{Dir}(\alpha_1,\ldots,\alpha_k)$	Dirichlet $\mathsf{Dir}(\alpha_1 + x_1,\ldots,\alpha_k + x_k)$
Normal $N\left(\mu,\frac{1}{\theta}\right)$	Gamma $\mathsf{Gamma}(\alpha,\beta)$	Gamma $\mathsf{Gamma}(\alpha + \frac{1}{2}, \beta + \frac{(\mu-x)^2}{2})$
Normal $N(\mu,\theta)$	InvGamma $\mathsf{InvGamma}(\alpha,\beta)$	InvGamma $\mathsf{InvGamma}(\alpha + \frac{1}{2}, \beta + \frac{(\mu-x)^2}{2})$

In case we want to include the prior knowledge from different sources we can also construct an averaging prior which is conjugate. The following theorem is an extension of Theorem 3.3 to mixtures.

Theorem 3.4 (Mixture of exponential distributions)
Assume $\mathbf{X}$ *has a distribution of the form*

$$p(\mathbf{x}|\theta) = \exp(\theta^T T(\mathbf{x}) - \Psi(\theta))h(\mathbf{x}).$$

Then the set of mixture distributions on Θ,

$$\mathcal{F}_N = \{\sum_{i=1}^{N} \omega_i \pi(\theta|\mu_i, \lambda_i) : \sum_{i=1}^{N} \omega_i = 1, \omega_i > 0,$$

$$\pi(\theta|\mu_i, \lambda_i) = K(\mu_i, \lambda_i) \exp(\theta^T \mu_i - \lambda_i \Psi(\theta)),$$

$$\mu_i \in \mathbb{R}^k, \ \lambda_i \in \mathbb{R}\}$$

$$(3.22)$$

is a conjugate family. For the prior

$$\pi(\theta) = \sum_{i=1}^{N} \omega_i \pi(\theta|\mu_i, \lambda_i),$$

the posterior is

$$\pi(\theta|\mathbf{x}) = \sum_{i=1}^{N} \omega_i(\mathbf{x})\pi(\theta|\mu_i + T(\mathbf{x}), \lambda_i + 1)$$

with

$$\omega_i(\mathbf{x}) \propto \frac{\omega_i K(\mu_i, \lambda_i)}{K(\mu_i + T(\mathbf{x}), \lambda_i + 1)}.$$

PROOF: Applying Theorem 3.3 we get

$$\pi(\theta|\mathbf{x}) \propto \pi(\theta)\ell(\theta|\mathbf{x})$$

$$\propto \sum_{i=1}^{N} \omega_i K(\mu_i, \lambda_i) \exp(\theta^T \mu_i - \lambda_i \Psi(\theta)) \exp(\theta^T T(\mathbf{x}) - \Psi(\theta))$$

$$\propto \sum_{i=1}^{N} \omega_i K(\mu_i, \lambda_i) \exp(\theta^T (\mu_i + T(\mathbf{x})) - (\lambda_i + 1)\Psi(\theta))$$

$$\propto \sum_{i=1}^{N} \omega_i(\mathbf{x}) K(\mu_i + T(\mathbf{x}), \lambda_i + 1) \exp(\theta^T (\mu_i + T(\mathbf{x})) - (\lambda_i + 1)\Psi(\theta)).$$

$$(3.23)$$

□

We now turn back to the classroom Example 3.7 and Figure 3.7.

Example 3.15 (Weather)

We assume that the temperature measurements are normally distributed. Thus

$$f(x|\theta) = \frac{1}{\sqrt{2\pi}}\frac{1}{\sigma}\exp(-\frac{x^2}{2\sigma^2})\exp\left(\frac{1}{\sigma^2}\theta x - \frac{1}{2\sigma^2}\theta^2\right).$$

We have

$$\Psi(\theta) = \frac{1}{2\sigma^2}\theta^2, \quad T(x) = \frac{1}{\sigma^2}x.$$

From Theorem 3.3 a conjugate family is

$$\mathcal{F} = \{K(\mu,\lambda)\exp(\theta\mu - \lambda\frac{1}{2\sigma^2}\theta^2) : \mu \in \mathbb{R}, \ \lambda \in \mathbb{R}, \ \lambda > 0\}$$

with

$$K(\mu,\lambda) = \frac{1}{\sqrt{2\pi}}\frac{\sqrt{\lambda}}{\sigma}\exp\left(-\frac{\sigma^2}{2\lambda}\mu^2\right).$$

Parameterizing by $\tau^2 = \frac{\sigma^2}{\lambda}$ and $m = \frac{\sigma^2}{\lambda}\mu$, it is the family of normal distributions

$$\mathcal{F} = \{N(m,\tau^2) : m \in \mathbb{R}, \ \tau^2 > 0\}.$$

Recall Example 2.12 with $n = 1$, the posterior belonging to the prior $N(m,\tau^2)$ is $N(m(x),\tau_p^2)$ where

$$m(x) = \rho(x\tau^2 + m\sigma^2), \quad \tau_p^2 = \rho\sigma\tau^2, \quad \rho = (\tau^2 + \sigma^2)^{-1}.$$

The prior in Example 3.7 fulfills the conditions of Theorem 3.4. It is a mixture of normal distributions

$$\pi(\theta) - \omega_1\,\phi_{(m_1,\tau_1^2)}(\theta) + \omega_2\,\phi_{(m_2,\tau_2^2)}(\theta).$$

Thus the posterior is a mixture of the related posteriors

$$\pi(\theta|x) = \omega_1(x)\,\phi_{(m_1(x),\tau_{1,p}^2)}(\theta) + \omega_2(x)\,\phi_{(m_2(x),\tau_{2,p}^2)}(\theta) \qquad (3.24)$$

with

$$m_i(x) = \rho_i(x\tau_i^2 + m_i\sigma^2), \quad \tau_{i,p}^2 = \rho_i\sigma\tau_i^2, \quad \rho_i = (\tau_i^2 + \sigma^2)^{-1}, \quad i = 1,2,$$

and

$$\omega_i(x) \propto \omega_i\frac{\tau_i}{\tau_{i,p}}\exp\left(-\frac{1}{2\tau_i^2}m_i^2 + \frac{1}{2\tau_{i,p}^2}m_i(x)^2\right)$$

with $\omega_1(x) + \omega_2(x) = 1$. Figures 3.9 and 3.10 show the prior and posterior. □

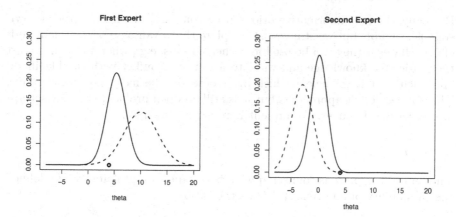

Figure 3.9: Illustration for Example 3.15. Left: The subjective prior (broken line) of expert 1 and the posterior after observing $x = 4$. Right: The prior (broken line) and posterior related to expert 2.

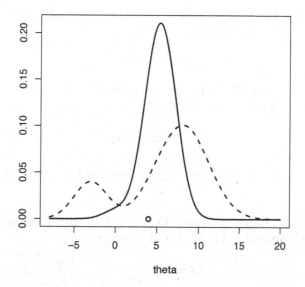

Figure 3.10: Illustration for Example 3.15. The broken line is the prior mixture distribution. The continuous line is the posterior, which is also a mixture, but the posterior weight of the second expert is shifted from 0.2 to 0.035, because the observation $x = 4$ lies in the tail of the second expert's prior.

3.3 Non-informative Priors

Basically, the non–informative priors are recommended when no prior knowledge is available but we still want to explore the advantages of Bayesian modelling. However they can be used for other reasons, e.g., when we do not trust the subjective knowledge and want to avoid any conflict with the likelihood approach. But it is unclear which prior deserves the name non–informative. There are different approaches but it is still an open problem and new set ups are under development. We present here main principles.

3.3.1 Laplace Prior

The name of Laplace is connected with first papers on probability. He defined the probability of an event $A \subset \Omega$ as the ratio

$$\mathsf{P}(A) = \frac{\text{number of cases } \omega \text{ belonging to } A}{\text{number of all } \omega \text{ in } \Omega}.$$

Behind this definition is the imagination that all elements of Ω have the same chance to be drawn.

$$\mathsf{P}(\{\omega\}) = \frac{1}{\text{number of all } \omega \text{ in } \Omega} = \text{const.}$$

It works in case when the number of all ω in Ω is finite. This approach is generalized as constant probability function

$$p(\omega) \propto \text{const.}$$

Note that for unbounded Ω, a constant measure is no longer a probability measure because it cannot be normalized since

$$\int_{-\infty}^{\infty} \text{const} = \infty.$$

Applying this to the Bayesian context we have the following definition.

Definition 3.3 (Laplace prior) Consider a statistical model $\mathcal{P} = \{\mathsf{P}_\theta : \theta \in \Theta\}$. The **Laplace prior** is defined as constant

$$\pi(\theta) \propto \text{const.}$$

Note that, the notational system in statistics is not unique, so that sometimes in literature prior distributions following a Laplace distribution are named Laplace prior, which are of course not constant.

Definition 3.3 is also applied to unbounded Θ. In general, priors which are not probability measures are called **improper priors**.

Suppose a finite set $\Theta = \{\theta_1, \ldots, \theta_m\}$. The *Shannon entropy* is defined by

$$H(\pi) = -\sum_{i=1}^{m} \pi(\theta_i) \log(\pi(\theta_i)). \tag{3.25}$$

It describes how much the probability mass of π is spread out on Θ. The higher the entropy the less informative is the parameter. It holds for all $j = 1, \ldots, m$ that

$$\frac{d}{dk_j}\left(-\sum_{i=1}^{m} k_i \log(k_i)\right) = -k_j \frac{1}{k_j} - \log(k_j) = -1 - \log(k_j) \doteq 0,$$

i.e., the measure with maximal entropy has constant weight on all elements of Θ. Thus the Laplace prior fulfills the idea of no information. For illustration, see Figure 3.11.

Example 3.16 (Lion's appetite)
Consider Example 2.4. The lion has only three different stages: hungry, moderate, lethargic. The Laplace non-informative prior gives every stage the weight $\frac{1}{3}$. □

Consider the Bayes billiard table again.

Example 3.17 (Binomial distribution)
In Example 3.6 we considered a binomial model $X|\theta \sim \mathrm{Bin}(n, \theta)$ with $x \in \mathcal{X} = \{0, 1, \ldots, n\}$ and a beta distributed prior for $\theta \in [0, 1]$. In Example 3.6 we assumed for the marginal distribution of X that $P(\{x\}) = \frac{1}{n+1}$. The related beta prior is the uniform distribution $U[0, 1]$ which is the Laplace non-informative prior. □

Example 3.18 (Normal sample and Laplace prior)
In Example 2.12 we had $\mathbf{X} = (X_1, \ldots, X_n)$ i.i.d from $N(\theta, \sigma^2)$ with known variance σ^2. The parameter space $\Theta = \mathbb{R}$ is unbounded and the Laplace prior $\pi(\theta) \propto \mathrm{const}$ is improper. But the posterior

$$\pi(\theta|\mathbf{x}) \propto \ell(\mu|\mathbf{x}) \propto \exp\left(-\frac{n}{2\sigma^2}(\mu - \bar{x})^2\right)$$

is a normal distribution $N(\bar{x}, \frac{1}{n}\sigma^2)$ which is proper. Furthermore the Laplace non-informative prior is a limiting case of a normal prior with increasing prior variance. The prior variance can be interpreted as a measure of uncertainty, where high variance indicates our higher uncertainty, see Figure 3.11. □

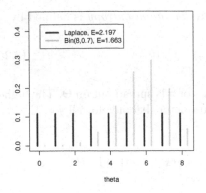
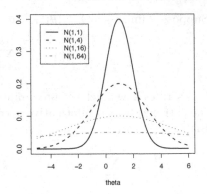

Figure 3.11: Laplace non-informative prior. Left: Illustration of spread and entropy. E stands for the entropy measure (3.25). Right: In Example 3.18 the Laplace prior can be considered as limit of normal priors under increasing variance.

Furthermore the Bayes analysis with Laplace non-informative priors conforms to the likelihood approach as long as we do not change the parameter. But Laplace non-informative priors have a big disadvantage: **Laplace priors depend on parametrization.**
Assume $\Theta \subset \mathbb{R}^p$. Consider the transformation of the prior for reparametrization. Suppose a bijective function $g : \theta \to \eta = g(\theta) \in G \subset \mathbb{R}^p$ and $h : \eta \in G \subset \mathbb{R}^p \to \theta = h(\eta) \in H \subset \mathbb{R}^p$ where $\eta_j = g_j(\theta)$ and $\theta_i = h_i(\eta)$, $i, j = 1, \ldots, p$. Then the *Jacobian matrix* is defined as

$$J = \left[\frac{\partial h}{\partial \eta_1}, \ldots, \frac{\partial h}{\partial \eta_p} \right] = \left[\frac{\partial h_i(\eta)}{\partial \eta_j} \right]_{i=1,\ldots,p, j=1,\ldots,p}. \qquad (3.26)$$

Applying the rules of variable transformation for distributions, we have

$$\pi_\eta(\eta) = \pi_\theta(h(\eta))|J|, \qquad (3.27)$$

where $|J| = \det(J)$ is the Jacobian determinant. A constant prior with respect to θ does not deliver a constant prior with respect to η, so that the Laplace non-informative prior does not fulfill (3.27). We illustrate it in the following example.

Example 3.19 (Binomial distribution and odds ratio)
Consider Example 2.9, where $X|\theta \sim \text{Bin}(n, \theta)$ and $\theta \sim \text{U}[0, 1]$. The parameter θ is the success probability of a binomial distribution. An alternative parameter η is the odds ratio, the ratio between the probabilities of success and failure.

$$\eta = \frac{\theta}{1 - \theta}, \quad \theta = \frac{\eta}{1 + \eta}.$$

The Jacobian is

$$J = \frac{d}{d\eta}\left(\frac{\eta}{1+\eta}\right) = \frac{1}{(1+\eta)^2}.$$

Applying (3.27) we obtain, under $\theta \sim \mathsf{U}[0,1]$, the prior over $[0,\infty)$, as

$$\pi(\eta) = \frac{1}{(1+\eta)^2}.$$

Note that, this distribution does not belong to a conjugate family.   □

3.3.2 Jeffreys Prior

In Jeffreys (1946) Harold Jeffreys proposed non–informative priors which are invariant to re–parametrization. His main argument was that if we are, first and foremost, interested in information on the data generating distribution P_θ, and less in the parameter θ itself, then the distance between distributions matters.

For two probability measures P, Q, the *Kullback–Leibler divergence* $\mathsf{K}(\mathsf{P}|\mathsf{Q})$ measures the information gained if P is used instead of Q. Let P, Q be continuous distributions with densities p, q, respectively, then

$$\mathsf{K}(\mathsf{P}|\mathsf{Q}) = \int p(x)\ln(\frac{p(x)}{q(x)})dx. \tag{3.28}$$

The Kullback–Leibler divergence is not symmetric. For a symmetric metric we use

$$\mathsf{I}_2(\mathsf{P},\mathsf{Q}) = \mathsf{K}(\mathsf{P}|\mathsf{Q}) + \mathsf{K}(\mathsf{Q}|\mathsf{P}) = \int (p(x) - q(x))\ln(\frac{p(x)}{q(x)})dx.$$

Jeffreys argument goes as following. For $\mathsf{P} = \mathsf{P}_\theta$ and $\mathsf{Q} = \mathsf{P}_{\theta'}$ we obtain

$$\mathsf{I}_2(\mathsf{P}_\theta, \mathsf{P}_{\theta'}) = \int (p(x|\theta) - p(x|\theta'))\ln(\frac{p(x|\theta)}{p(x|\theta')})dx.$$

Let $p'(x|\theta)$ denote the p-dimensional vector of first partial derivatives with respect to θ. Then, using

$$p(x|\theta) - p(x|\theta') \approx p'(x|\theta)^T(\theta - \theta'),$$

and

$$\ln(\frac{p(x|\theta)}{p(x|\theta')}) = \ln(p(x|\theta)) - \ln(p(x|\theta')) \approx \frac{1}{p(x|\theta)}p'(x|\theta)^T(\theta - \theta'),$$

we obtain

$$\mathsf{I}_2(\mathsf{P}_\theta, \mathsf{P}_{\theta'}) \approx (\theta - \theta')^T \int \frac{1}{p(x|\theta)}p'(x|\theta)p'(x|\theta)^T dx\, (\theta - \theta').$$

From the definition of the *score function*

$$V(\theta|x) = \frac{\partial \ln p(x|\theta)}{\partial \theta} = \frac{1}{p(x|\theta)} p'(x|\theta)$$

and the *Fisher information* matrix is (see Liero and Zwanzig, 2011)

$$I(\theta) = \mathsf{Cov}_\theta V(\theta|x). \tag{3.29}$$

We have

$$
\begin{aligned}
\int \frac{1}{p(x|\theta)} p'(x|\theta) p'(x|\theta)^T dx &= \int \frac{1}{p(x|\theta)^2} p'(x|\theta) p'(x|\theta)^T p(x|\theta) dx \\
&= \mathsf{E}_\theta (V(\theta|x) V(\theta|x)^T) \\
&= \mathsf{Cov}_\theta (V(\theta|x)) \\
&= I(\theta)
\end{aligned}
$$

so that

$$\mathsf{l}_2(\mathsf{P}_\theta, \mathsf{P}_{\theta'}) \approx (\theta - \theta')^T I(\theta)(\theta - \theta'). \tag{3.30}$$

Thus the distance between the probability measures generates a weighted distance in the parameter space. Changing the parametrization and using

$$\theta - \theta' \approx \mathsf{J}(\eta - \eta')$$

where J is the Jacobian matrix defined in (3.26), we have

$$
\begin{aligned}
\mathsf{l}_2(\mathsf{P}_\theta, \mathsf{P}_{\theta'}) &\approx (\theta - \theta')^T I(\theta)(\theta - \theta') \\
&\approx (\eta - \eta')^T \mathsf{J}^T I(\theta) \mathsf{J}(\eta - \eta').
\end{aligned}
$$

Further, by the chain rule, the score function $\widetilde{V}(\eta|x))$ with respect to η is

$$\widetilde{V}(\eta|x)) = V(\theta|x))^T \mathsf{J}$$

which gives

$$I(\eta) = \mathsf{J}^T I(\theta) \mathsf{J} \tag{3.31}$$

and we see that different parametrizations for the same probability distributions deliver the same distance

$$\mathsf{l}_2(\mathsf{P}_\theta, \mathsf{P}_{\theta'}) = \mathsf{l}_2(\mathsf{P}_\eta, \mathsf{P}_{\eta'}).$$

The relation (3.31) is Jeffreys main argument for an invariant prior.

Definition 3.4 (Jeffreys prior) Consider a statistical model $\mathcal{P} = \{\mathsf{P}_\theta : \theta \in \Theta\}$ with Fisher information matrix $I(\theta)$.
The **Jeffreys prior** is defined as

$$\pi(\theta) \propto \det(I(\theta))^{\frac{1}{2}}.$$

Being *invariant*, Jeffreys prior fulfills the relation (3.27). Using (3.31) and (3.27) we have

$$\pi(\eta) \propto \det(I(\eta))^{\frac{1}{2}}$$
$$\propto \det(J^T I(\theta) J)^{\frac{1}{2}}$$
$$\propto \det(I(\theta))^{\frac{1}{2}} \det(J)$$
$$\propto \pi(\theta) \det(J)$$

Although, Jeffreys first goal was to find an invariant prior, his prior is also *non–informative* in the sense that the prior has no influence because the posterior distribution based on Jeffreys prior coincides approximately with the likelihood function. Let us explain it in more detail.

Assume that we have an i.i.d. sample $\mathbf{x} = (x_1, \ldots, x_n)$ from a regular statistical model, which means the Fisher information matrix exists and it holds that (see Liero and Zwanzig, 2011, Theorem 3.6)

$$I(\theta) = \mathrm{Cov}_\theta V(\theta|x) \text{ and } I_n(\theta) = \mathrm{Cov}_\theta V(\theta|\mathbf{x}) = -\mathsf{E}_\theta J(\theta|\mathbf{x}) = nI(\theta),$$
$$(3.32)$$

where $V(\theta|\mathbf{x})$ is the vector of first derivatives of the log likelihood function and $J(\theta|\mathbf{x})$ is the matrix of second derivatives of the log likelihood function. Let $\widehat{\theta}$ be the maximum likelihood estimator with $V(\widehat{\theta}|\mathbf{x}) = 0$. Applying the Taylor expansion about $\widehat{\theta}$ we have

$$\ln p(\mathbf{x}|\theta) \approx \ln p(\mathbf{x}|\widehat{\theta}) + (\theta - \widehat{\theta})^T V(\widehat{\theta}|\mathbf{x}) + \frac{1}{2}(\theta - \widehat{\theta})^T J(\widehat{\theta}|\mathbf{x})(\theta - \widehat{\theta})$$
$$\approx \ln p(\mathbf{x}|\widehat{\theta}) + \frac{1}{2}(\theta - \widehat{\theta})^T J(\widehat{\theta}|\mathbf{x})(\theta - \widehat{\theta}).$$

Approximating $J(\widehat{\theta}|\mathbf{x})$ by $-nI(\theta)$ we get

$$\ln p(\mathbf{x}|\theta) \approx \ln p(\mathbf{x}|\widehat{\theta}) - \frac{n}{2}(\theta - \widehat{\theta})^T I(\theta)(\theta - \widehat{\theta}). \qquad (3.33)$$

Thus we obtain an approximation of the likelihood function $\ell(\theta|\mathbf{x}) = p(\mathbf{x}|\theta)$ by

$$\ell(\theta|\mathbf{x}) \approx \ell(\widehat{\theta}|\mathbf{x}) \exp\left(-\frac{n}{2}(\theta - \widehat{\theta})^T I(\theta)(\theta - \widehat{\theta})\right)$$
$$\propto \exp\left(-\frac{n}{2}(\theta - \widehat{\theta})^T I(\theta)(\theta - \widehat{\theta})\right). \qquad (3.34)$$

Calculating the posterior with the Jeffreys prior and (3.34) we get

$$\pi(\theta|\mathbf{x}) \propto \pi(\theta)\ell(\theta|\mathbf{x})$$
$$\propto \det(I(\theta))^{\frac{1}{2}}\ell(\theta|\mathbf{x})$$
$$\propto \det(I(\theta))^{\frac{1}{2}} \exp\left(-\frac{n}{2}(\theta - \widehat{\theta})^T I(\theta)(\theta - \widehat{\theta})\right),$$

which is the kernel of $N(\theta, \frac{1}{n}I(\theta)^{-1})$. Thus Jeffreys prior is precisely the right weight needed to obtain the same result as in asymptotic inference theory, where it is shown that (see e.g., van der Vaart, 1998)

$$\sqrt{n}(\hat{\theta} - \theta) \xrightarrow{\mathcal{D}} N(0, I(\theta)^{-1}).$$

The first example we want to calculate Jeffreys prior for, and to compare it with other approaches, is the binomial model.

Example 3.20 (Binomial distribution)
For $X|\theta \sim \mathsf{Bin}(n, \theta)$, we have

$$\ln(p(x|\theta)) = \ln\left(\binom{n}{x}\right) + x\ln(\theta) + (n - x)\ln(1 - \theta)$$

with

$$V(\theta|x) = \frac{x}{\theta} - \frac{n - x}{1 - \theta}$$

$$J(\theta|x) = -\frac{x}{\theta^2} - \frac{n - x}{(1 - \theta)^2}.$$

Using $\mathsf{E}X = n\theta$, we get the Fisher information

$$I(\theta) = -\mathsf{E}J(\theta|x) = \frac{n\theta}{\theta^2} + \frac{n - n\theta}{(1 - \theta)^2} = \frac{n}{\theta(1 - \theta)},$$

and finally the Jeffreys prior

$$\pi(\theta) \propto \theta^{-\frac{1}{2}}(1 - \theta)^{-\frac{1}{2}}.$$

This is the kernel of $\mathsf{Beta}(\frac{1}{2}, \frac{1}{2})$, which is symmetric and gives more weight to small and large parameters, but less weight to parameters around $\frac{1}{2}$; see Figure 3.12. □

Example 3.21 (Location model)
We assume $\mathcal{X} \in \mathbb{R}^p$ and that the data generating distribution P_θ belongs to a location family. In the continuous case the density has the structure

$$p(x|\theta) = f(x - \theta), \qquad (3.35)$$

with known density function $f(x) \geq 0$, $\int_{\mathcal{X}} f(x)dx = 1$ and finite positive definite Fisher information matrix,

$$I(f) = \int \frac{1}{f(x)}f'(x)f'(x)^T dx,$$

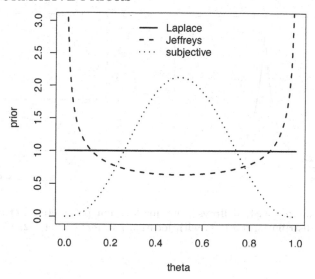

Figure 3.12: Illustration for Examples 3.6, 3.17 and 3.20. Jeffreys prior is Beta($\frac{1}{2}, \frac{1}{2}$), Laplace non-informative prior is Beta$(1, 1)$ and the subjective prior is Beta$(3.8, 3.8)$, as determined in Example 3.6.

where $f'(x)$ is the p-dimensional vector of first derivatives. The unknown parameter $\theta \in \mathbb{R}^p$ is the location parameter and has the same dimension as $\mathcal{X}$. The Fisher information matrix is calculated by

$$I(\theta) = \int \frac{1}{f(x-\theta)} f'(x-\theta) f'(x-\theta)^T dx.$$

Changing the variable of integration by $z = x - \theta$ with $dz = dx$ we have that the Fisher information matrix is independent on θ,

$$I(\theta) = I(f).$$

In the location model Jeffreys prior coincides with Laplace non-informative prior,

$$\pi(\theta) \propto \text{const.}$$

For illustration, see Figure 3.13, which is related to the Laplace distribution La$(\theta, 1)$. Recall that in general the density of a Laplace distribution La(μ, σ) is given by

$$f(x|\mu, \sigma) = \frac{1}{2\sigma} \exp(-\frac{1}{\sigma}|x - \mu|). \qquad (3.36)$$

$\square$

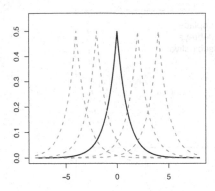 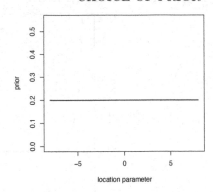

Figure 3.13: Example 3.21. Jeffreys prior for location parameter. Left: Densities of the Laplace distribution $\mathsf{La}(\theta, 1)$ with location parameters $-4, -2, 0, 2, 4$. Right: Jeffreys prior.

We now consider a scale family.

Example 3.22 (Scale model)

We assume $\mathcal{X} \in \mathbb{R}$ and that the data generating distribution P_θ belongs to a scale family, where the unknown parameter $\theta = \sigma \in \mathbb{R}$ is the positive scale parameter. In the continuous case, the density has the structure

$$p(x|\theta) = \frac{1}{\sigma} f\left(\frac{x}{\sigma}\right), \tag{3.37}$$

where $f(x) \geq 0$, $\int f(x)dx = 1$, is a known density with derivative $f'(x) = \frac{d}{dx} f(x)$. Under regularity conditions we have

$$\int \frac{dp(x|\theta)}{d\theta} dx = \frac{d}{d\theta} \int p(x|\theta)dx = \frac{d}{d\theta} 1 = 0,$$

where

$$\frac{dp(x|\theta)}{d\theta} = -\frac{1}{\sigma^2} \left(\frac{x}{\sigma} f'\left(\frac{x}{\sigma}\right) + f\left(\frac{x}{\sigma}\right)\right).$$

Changing the variable of integration $z = \frac{x}{\sigma}$ with $dz = \frac{dx}{\sigma}$, we have

$$\frac{1}{\sigma} \left(\int z f'(z)dz + 1\right) = 0,$$

so that $\int x f'(x)dx = -1$.
The Fisher information matrix is calculated by

$$I(\theta) = \frac{1}{\sigma^4} \int \frac{1}{\frac{1}{\sigma} f\left(\frac{x}{\sigma}\right)} \left(\frac{x}{\sigma} f'\left(\frac{x}{\sigma}\right) + f\left(\frac{x}{\sigma}\right)\right)^2 dx.$$

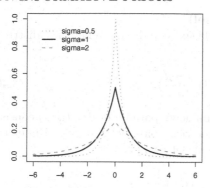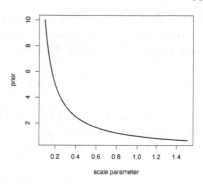

Figure 3.14: Example 3.22. Jeffreys prior for scale parameter. Left: Densities of the Laplace distribution $\mathsf{La}(0, \sigma)$ with different scale parameters. Right: Jeffreys prior.

By the same change of variable, we have

$$I(\theta) = \frac{1}{\sigma^2} \int \frac{1}{f(z)} \left(zf'(z) + f(z)\right)^2 dz$$

$$= \frac{1}{\sigma^2} \left(\int \frac{1}{f(z)} z^2 f'(z)^2 dz + 2 \int zf'(z)dz + \int f(z)dz\right).$$

Applying $\int xf'(x)dx = -1$ and $\int f(x)dx = 1$, we obtain

$$I(\theta) = \frac{1}{\sigma^2} \left(\int \frac{1}{f(z)} z^2 f'(z)^2 dz - 1\right).$$

Hence Jeffreys prior for the scale parameter $\theta = \sigma$ is

$$\pi(\sigma) \propto \frac{1}{\sigma}. \tag{3.38}$$

Jeffreys prior does not coincide with the Laplace non-informative prior, but it is improper, since $\int_0^\infty \frac{1}{x} dx = \infty$. Figure 3.14 illustrates it for density f of the Laplace distribution $\mathsf{La}(0, 1)$. □

The situation changes when we study the location parameter μ and the scale parameter σ simultaneously, i.e., $\theta = (\mu, \sigma)$. In this case, Jeffreys additionally requires independence between μ and σ, so that

$$\pi(\theta) = \pi(\mu)\pi(\sigma). \tag{3.39}$$

Example 3.23 (Location–scale model)
Let us consider a one dimensional sample space and data generating distributions P_θ, $\theta = (\mu, \sigma)$ with density

$$p(x|\theta) = \frac{1}{\sigma} f(\frac{x - \mu}{\sigma}), \qquad (3.40)$$

where $f(x) \geq 0$, $\int_{-\infty}^{\infty} f(x)dx = 1$. The unknown parameter $\theta = (\mu, \sigma)$ consists of the location parameter $\mu \in \mathbb{R}$ and the scale parameter $\sigma > 0$. Under the independence assumption (3.39), the joint Jeffreys prior is

$$\pi(\theta) = \pi(\mu)\pi(\sigma) \propto \text{const}\frac{1}{\sigma} \propto \frac{1}{\sigma}.$$

$\square$

We further illustrate the influence of the independence assumption (3.39) for the normal distribution.

Example 3.24 (Normal distribution)
Consider the location scale model with

$$f(x) = \varphi(x) = \frac{1}{\sqrt{2\pi}} \exp(-\frac{x^2}{2}),$$

i.e., the statistical model is the family of one dimensional normal distributions $N(\mu, \sigma^2)$ where the location parameter μ is the expectation and the scale parameter σ is the standard deviation: $\theta = (\mu, \sigma)$. From

$$\ln(p(x|\theta)) = -\frac{1}{2}\ln(2\pi) - \ln(\sigma) - \frac{1}{2\sigma^2}(x - \mu)^2,$$

we get the score function

$$V(\mu, \sigma|x) = (\frac{1}{\sigma^2}(x - \mu), -\frac{1}{\sigma} + \frac{1}{\sigma^3}(x - \mu)^2)^T,$$

and the components of the Fisher information

$$I(\mu, \sigma)_{11} = \text{Var}\left(\frac{1}{\sigma^2}(X - \mu)\right) = \frac{1}{\sigma^2},$$

$$I(\mu, \sigma)_{12} = \text{Cov}\left(\frac{1}{\sigma^2}(X - \mu), \frac{1}{\sigma^3}(X - \mu)^2\right) = 0,$$

$$I(\mu, \sigma)_{22} = \text{Var}\left(\frac{1}{\sigma^3}(X - \mu)^2\right) = \frac{1}{\sigma^6}(\text{E}(X - \mu)^4 - \sigma^4) = \frac{2}{\sigma^2},$$

so that

$$I(\mu, \sigma) = \begin{pmatrix} \frac{1}{\sigma^2} & 0 \\ 0 & \frac{2}{\sigma^2} \end{pmatrix}.$$

Changing the parametrization to mean and variance $\eta = (\mu, \sigma^2)$, we have the Jacobian

$$J = \begin{pmatrix} 1 & 0 \\ 0 & -\frac{1}{2\sigma} \end{pmatrix}.$$

Applying $I(\mu, \sigma^2) = J^T I(\mu, \sigma) J$, we get

$$I(\mu, \sigma^2) = \begin{pmatrix} \frac{1}{\sigma^2} & 0 \\ 0 & \frac{1}{2\sigma^4} \end{pmatrix}. \tag{3.41}$$

Without the independence condition (3.39), Jeffreys prior is

$$\pi(\mu, \sigma^2) \propto \det(I(\mu, \sigma^2))^{\frac{1}{2}} \propto \frac{1}{\sigma^3}.$$

Under (3.39), we have

$$\pi(\mu, \sigma^2) \propto \frac{1}{\sigma^2}.$$

Using the scale parametrization we have without condition (3.39),

$$\pi(\mu, \sigma) \propto \det(I(\mu, \sigma))^{\frac{1}{2}} \propto \frac{1}{\sigma^2}.$$

This prior is also called *left Haar measure*. Under (3.39) it is

$$\pi(\mu, \sigma) \propto \frac{1}{\sigma},$$

known as *right Haar measure*. □

Example 3.25 (Multinomial distribution)
Consider the multinomial model with k cells. The data $(n_1, \ldots, n_k)$ are the observed frequencies in each cell. The parameter consists of the probabilities p_i of each cell. Since $\sum_{i=1}^{k} p_i = 1$ and $n = \sum_{i=1}^{k} n_i$, we set $\theta = (p_1, \ldots, p_{k-1})$ and $\mathbf{x} = (n_1, \ldots, n_{k-1})$. The probability function is given by

$$p(\mathbf{x}|\theta) \propto \prod_{i=1}^{k-1} p_i^{n_i} (1 - \delta)^{n-s} \tag{3.42}$$

where $\delta = \sum_{i=1}^{k-1} p_i$ and $s = \sum_{i=1}^{k-1} n_i$. The log–likelihood function is

$$\ln(p((n_1, \ldots, n_{k-1})|\theta)) = \sum_{i=1}^{k-1} n_i \ln(p_i) + (n - s) \ln(1 - \delta) + \text{const}$$

with the first and second derivatives for $i, j = 1, \ldots, k-1$

$$\frac{\partial}{\partial p_i} \ln(p(\mathbf{x}|\theta)) = n_i \frac{1}{p_i} - (n-s)\frac{1}{1-\delta}$$

$$\frac{\partial}{\partial p_i p_i} \ln(p(\mathbf{x}|\theta)) = -n_i \frac{1}{p_i^2} - (n-s)\frac{1}{(1-\delta)^2} \qquad (3.43)$$

$$\frac{\partial}{\partial p_i p_j} \ln(p(\mathbf{x}|\theta)) = -(n-s)\frac{1}{(1-\delta)^2}, \quad j \neq i.$$

It holds that $En_i = np_i$ and $E(n-s) = n(1-\delta)$. Using $p_k = 1-\delta$ the Fisher information is

$$I(\theta) = n \begin{pmatrix} \frac{1}{p_1} + \frac{1}{p_k} & \frac{1}{p_k} & \cdots & \frac{1}{p_k} \\ \frac{1}{p_k} & \frac{1}{p_2} + \frac{1}{p_k} & \cdots & \frac{1}{p_k} \\ \vdots & \cdots & \ddots & \frac{1}{p_{k-1}} + \frac{1}{p_k} \end{pmatrix} \qquad (3.44)$$

or, alternatively written as

$$I(\theta) = n \operatorname{diag}(\frac{1}{p_1}, \ldots, \frac{1}{p_{k-1}}) + n\frac{1}{p_k}\mathbf{1}_{k-1}\mathbf{1}_{k-1}^T$$

where $\mathbf{1}_{k-1}$ is the column vector consisting of $k-1$ ones, and $\mathbf{1}_{k-1}\mathbf{1}_{k-1}^T$ is the $(k-1) \times (k-1)$ matrix consisting of ones. Using the rule

$$\det(A + aa^T) = \det(A)(1 + a^T A^{-1} a)$$

we obtain

$$\det(I(\theta)) = n^{k-1} \prod_{i=1}^{k-1} \frac{1}{p_i}(1 + \sum_{i=1}^{k-1} p_i \frac{1}{p_k}) = n^{k-1} \prod_{i=1}^{k} \frac{1}{p_i}.$$

Hence Jeffreys prior for $p_1, \ldots, p_k$ with $\sum_{i=1}^{k} p_i = 1$ is

$$\pi_{\text{Jeff}}(p_1, \ldots, p_k) \propto p_1^{-\frac{1}{2}} \ldots p_k^{-\frac{1}{2}} \qquad (3.45)$$

which is the Dirichlet distribution $\mathrm{Dir}_k(0.5, \ldots, 0.5)$; see (3.15). $\qquad \square$

In the following table, we summarize the Jeffreys priors for some commonly used distributions.

Distribution $p(x\mid\theta)$	Jeffreys prior $\pi(\theta) \propto$	Posterior $\pi(\theta\mid x)$
Normal $N(\theta, \sigma^2)$	const	Normal $N(x, \sigma^2)$
Poisson $\text{Poi}(\theta)$	$\frac{1}{\sqrt{\theta}}$	Gamma $\text{Gamma}(x + 0.5, 1)$
Gamma $\text{Gamma}(\nu, \theta)$	$\frac{1}{\sqrt{\theta}}$	Gamma $\text{Gamma}(\nu, x)$
Binomial $\text{Bin}(n, \theta)$	Beta $\text{Beta}(0.5, 0.5)$	Beta $\text{Beta}(0.5 + x, 0.5 + n - x)$
Negative Binomial $\text{NB}(m, \theta)$	$\theta^{-\frac{1}{2}}(1 - \theta)^{-\frac{1}{2}}$	Beta $\text{Beta}(m, x - 1)$, for $x > 1$
Normal $N(\mu, \theta)$	$\frac{1}{\theta}$	InvGamma $\text{InvGamma}(1, \frac{(\mu-x)^2}{2})$

3.3.3   Reference Priors

Reference priors were introduced by Bernardo in Bernardo (1979). They be-
came more and more popular, and the name *Reference Analysis* is used for a
Bayesian inference based on reference priors, see Bernardo (2005). In Berger
and Bernardo (1992a), an algorithm for the construction of reference priors is
given, also named *Berger–Bernardo method*, see Kass and Wasserman (1996).
A formal definition of a reference prior was first given in Berger et al. (2009).
Bernardo's main contribution was to define the concept of non-information in
a mathematical way. To describe his approach, consider the Kullback–Leibler
divergence

$$K(\pi(.\mid\mathbf{x})\mid\pi) = \int \pi(\theta\mid\mathbf{x}) \ln\left(\frac{\pi(\theta\mid\mathbf{x})}{\pi(\theta)}\right) d\theta, \qquad (3.46)$$

which describes the gain of information using the posterior distribution $\pi(.\mid\mathbf{x})$
instead of the prior π. It is the information coming from the experiment. To
make it independent of the data $\mathbf{x}$, Lindley (1956) proposed the expectation
of the Kullback–Leibler divergence as *expected information*. It depends on the
statistical model $\mathcal{P} = \{\mathsf{P}_\theta : \theta \in \Theta\}$ and on the prior π,

$$I(\mathcal{P}, \pi) = \int p(\mathbf{x})K(\pi(.\mid\mathbf{x})\mid\pi)d\mathbf{x}, \quad \text{where } p(\mathbf{x}) = \int p(\mathbf{x}, \theta)d\theta. \qquad (3.47)$$

The expected information is equal to the *mutual information*, which measures the gain of information using $p(\mathbf{x}, \theta)$ instead of $p(\mathbf{x})\pi(\theta)$. It holds that $p(\mathbf{x}, \theta) = p(\mathbf{x})\pi(\theta|\mathbf{x}) = \pi(\theta)p(\mathbf{x}|\theta)$ and

$$\mathsf{I}(\mathcal{P}, \pi) = \iint p(\mathbf{x}, \theta) \ln\left(\frac{p(\mathbf{x}, \theta)}{p(\mathbf{x})\pi(\theta)}\right) d\theta\, d\mathbf{x}.$$

Now the argument is that a *non-informative* prior should have almost no influence, which in turn means that the information from the experiment should be maximal. Define the *Shannon entropy* of an arbitrary continuous distribution P with density $p(\mathbf{x})$ by

$$\mathsf{H}(\mathsf{P}) = -\int p(\mathbf{x}) \ln(p(\mathbf{x})) d\mathbf{x}, \tag{3.48}$$

recall that in (3.25) the entropy is defined for the discrete case. The expected information can be presented as the difference of the entropy of the prior and the expected entropy of the posterior. By using (3.46) and $\pi(\theta|\mathbf{x})p(\mathbf{x}) = \pi(\theta)p(\mathbf{x}|\theta)$ we get

$$\mathsf{I}(\mathcal{P}, \pi) = \int p(\mathbf{x}) \int \pi(\theta|\mathbf{x}) \ln\left(\frac{\pi(\theta|\mathbf{x})}{\pi(\theta)}\right) d\theta\, d\mathbf{x}$$

$$= -\iint \pi(\theta|\mathbf{x})p(\mathbf{x}) \ln(\pi(\theta)) d\mathbf{x}\, d\theta + \iint p(\mathbf{x})\pi(\theta|\mathbf{x}) \ln(\pi(\theta|\mathbf{x})) d\mathbf{x}\, d\theta$$

$$= -\int \pi(\theta) \ln(\pi(\theta)) \int p(\mathbf{x}|\theta) d\mathbf{x}\, d\theta + \iint p(\mathbf{x})\pi(\theta|\mathbf{x}) \ln(\pi(\theta|\mathbf{x})) d\mathbf{x}\, d\theta$$

$$= -\int \pi(\theta) \ln(\pi(\theta)) d\theta + \int p(\mathbf{x}) \int \pi(\theta|\mathbf{x}) \ln(\pi(\theta|\mathbf{x})) d\theta\, d\mathbf{x}$$

We have

$$\mathsf{I}(\mathcal{P}, \pi) = \mathsf{H}(\pi) - \int p(\mathbf{x}) \mathsf{H}(\pi(.|\mathbf{x})) d\mathbf{x}. \tag{3.49}$$

Thus a prior which generates a large expected information $\mathsf{I}(\mathcal{P}, \pi)$ corresponds to a prior with large entropy and related posterior with small expected entropy. For illustration we consider the formulae in case of a binomial distribution with a beta distribution as prior.

Example 3.26 (Binomial distribution)
We consider $X|\theta \sim \mathrm{Bin}(n, \theta)$ and $\theta \sim \mathrm{Beta}(\alpha, \beta)$. Then the posterior is $\mathrm{Beta}(\alpha + x, \beta + n - x)$, see Example 2.9. The Kullback–Leibler divergence from the prior to the posterior is the divergence between the two beta

distributions, i.e.,

$$K(\mathsf{Beta}(\alpha + x, \beta + n - x)|\mathsf{Beta}(\alpha, \beta))$$

$$=$$

$$\ln\left(\frac{B(\alpha, \beta)}{B(\alpha + x, \beta + n - x)}\right)$$
$$+ x\Psi(\alpha + x) + (n - x)\Psi(\beta + n - x) - n\Psi(\alpha + \beta + n),$$

where Ψ is the digamma function

$$\int_0^1 \frac{1 - x^{\alpha-1}}{1 - x}dx = \Psi(\alpha) - \Psi(1),$$

or alternatively

$$\Psi(x) = \frac{d}{dx}\ln(\Gamma(x)).$$

The marginal distribution of X is calculated in (3.5). For illustration let us consider as candidate class $\mathcal{C}$ of prior distributions on Θ all symmetric beta priors with $\alpha = \beta$. Then the marginal distribution in (3.5) fulfills $p(k) = p(n - k)$ and the expected information can be written as function of α as

$$I(\alpha) = \ln(B(\alpha, \alpha)) - \mathsf{E}_x\ln(B(\alpha + x, \alpha + n - x) + 2\mathsf{E}_x x\Psi(\alpha + x) - n\Psi(2\alpha + n).$$
$$(3.50)$$

In Figure 3.15 the information $I(\alpha)$ is plotted, using the beta and digamma functions defined in R. Let us compare the prior $\mathsf{Beta}(\alpha_p, \alpha_p)$ with $\alpha_p = \arg\min I(\alpha)$ to the prior with maximal entropy. The entropy of $\mathsf{Beta}(\alpha, \beta)$ distribution is given by

$$H(\mathsf{Beta}(\alpha, \beta)) = \ln(B(\alpha, \beta)) - (\alpha-1)\Psi(\alpha) - (\beta-1)\Psi(\beta) + (\alpha+\beta-2)\Psi(\alpha+\beta)$$

$H(\mathsf{Beta}(\alpha, \beta))$ is non-positive for all α, β with maximal value zero attained for $\alpha = 1$, $\beta = 1$, where $B(1, 1) = 1$. The prior $\mathsf{Beta}(1, 1)$ is equal to the uniform distribution $U[0, 1]$, which is also the Laplace non-informative prior; see Example 3.20. $\qquad\square$

Unfortunately the requirement to take the prior which maximizes the expected information $I(\mathcal{P}, \pi)$ does not give tractable results in all cases (see the discussion in Berger and Bernardo, 1992a). Bernardo (1979) proposed to compare the prior information with the theoretical best posterior information, which he described as the limit information for $n \to \infty$ from an experiment with n repeated independent copies of the original experiment $\mathcal{P}^n = \{P_\theta^{\otimes n} : \theta \in \Theta\}$.

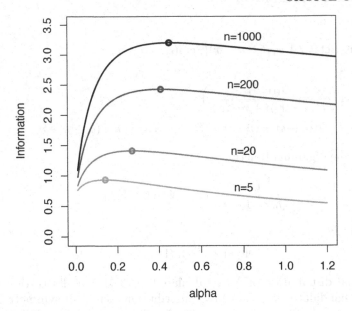

Figure 3.15: Example 3.27. The expected information $\mathsf{I}(\alpha)$ in (3.50) has its maximum at $\alpha = 0.142$ for $n = 5$, $\alpha = 0.272$ for $n = 20$, $\alpha = 0.41$ for $n = 200$ and $\alpha = 0.45$ for $n = 1000$.

Example 3.27 (Binomial distribution)
We continue with Example 3.26. Consider $\mathcal{P}^n = \{\mathsf{Bin}(m, \theta)^{\otimes n} : \theta \in [0, 1]\}$. A sufficient statistic is $t_n(\mathbf{X}) = \sum_{i=1}^{n} X_i$, with $t_n(\mathbf{X}) \sim \mathsf{Bin}(n\,m, \theta)$. Thus we can use the formulae in Example 3.26 with increasing n. In Figure 3.15 the information $\mathsf{I}(\alpha)$ is plotted for different n. We see the symmetric non-informative prior depends on n and approaches $\mathsf{Beta}(0.5, 0.5)$, which is Jeffreys prior; see Example 3.20.   □

In Clarke and Barron (1994) the authors derive an asymptotic expansion of the mutual information $\mathsf{I}(\mathcal{P}, \pi)$ under regularity conditions on the statistical model $\mathcal{P}$. For all positive continuous priors π, supported on a compact subset K, it holds that

$$\mathsf{I}(\mathcal{P}, \pi) = \frac{p}{2} \ln \frac{n}{2\pi e} + \ln c - \mathsf{K}(\pi|\pi^*) + \text{rest}(n), \qquad (3.51)$$

where π^* is Jeffreys prior given by

$$\pi^*(\theta) = \frac{1}{c} \sqrt{\det I(\theta)}, \quad \text{with } c = \int_K \sqrt{\det I(\theta)} d\theta, \quad K \subset \Theta \subset \mathbb{R}^p \text{ compact.}$$

The remainder terms fulfills $\lim_{n \to \infty} \text{rest}(n) = 0$. The regularity conditions on the statistical model $\mathcal{P}$ include the existence and positive–definiteness of the Fisher information $I(\theta)$ and that

$$I(\theta) = \left[\frac{\partial^2}{\partial \theta'_i \partial \theta'_j} \mathsf{K}(\mathsf{P}_\theta | \mathsf{P}_{\theta'}) |_{\theta = \theta'} \right]_{i=1,\ldots,p, j=1,\ldots,p}.$$

The first term in (3.51), $\frac{p}{2} \ln \frac{n}{2\pi e}$, is the entropy of $\mathsf{N}_p(0, I_p)$. The derivation of (3.51) is based on similar background as that of Jeffreys prior, see (3.30). The Kullback–Leibler divergence $\mathsf{K}(\mathsf{P} | \mathsf{Q})$ is non–negative and zero only for $\mathsf{P} = \mathsf{Q}$. Thus under regularity conditions the Jeffreys prior is the reference prior.

The formal definition of Berger et al. (2009) can be presented as follows.

Definition 3.5 (Reference Prior) A function $\pi(\theta)$ depending on the statistical model $\mathcal{P} = \{\mathsf{P}_\theta : \theta \in \Theta\}$ and a class of prior distributions $\mathcal{C}$ on Θ is a reference prior iff

1. For all $\mathbf{x} \in \mathcal{X}$, $\int_\Theta \pi(\theta) p(\mathbf{x} | \theta) d\theta < \infty$.

2. There exists an increasing sequence of compact subsets $\{\Theta_j\}_{j=1,\ldots,\infty}$ such that $\cup_{j=1}^\infty \Theta_j = \Theta$, and the posterior $\pi_j(\theta | \mathbf{x})$ defined on Θ_j fulfills for all $\mathbf{x} \in \mathcal{X}$

$$\lim_{j \to \infty} \mathsf{K}(\pi(.|\mathbf{x}) | \pi_j(.|\mathbf{x})) = 0,$$

   where $\pi(\theta | \mathbf{x}) \propto \pi(\theta) p(\mathbf{x} | \theta)$.

3. For any compact set $\Theta_0 \subseteq \Theta$, and for any $p \in \mathcal{C}$ let π_0 and p_0 be the truncated distributions of π and p defined on Θ_0, such that

$$\lim_{n \to \infty} (\mathsf{I}(\mathcal{P}^n, \pi_0) - \mathsf{I}(\mathcal{P}^n, p_0)) \geq 0.$$

Definition 3.5 is formulated in as general term as possible. Condition 1 includes improper priors, but the related posterior has to be proper. Condition 2 involves a construction of the reference prior as limit of priors on compact subsets; also this condition is a tool for handling improper priors. Condition 3 requires that asymptotically the reference prior should influence the information of the experiment as less as possible.

Definition 3.5 does not provide a construction of reference priors. The following theorem from Berger et al. (2009) provides an explicit expression.

Theorem 3.5
Assume $\mathcal{P} = \{P_\theta : \theta \in \Theta \subseteq \mathbb{R}\}$ and a class of prior distributions $\mathcal{C}$ on Θ. For all $p \in \mathcal{C}$ and all $\theta \in \Theta$, $p(\theta) > 0$ and $\int_\Theta p(\theta)p(\mathbf{x}|\theta)d\theta < \infty$. Let $\mathcal{P}^n = \{P_\theta^{\otimes n} : \theta \in \Theta\}$ and t_n be a sufficient statistic of $\mathcal{P}^n$. Let $\pi^*(\theta)$ be a strictly positive function such that the posterior $\pi^*(\theta|t_n)$ is proper and asymptotically consistent. For any θ_0 in an open subset of Θ, define

$$f_n(\theta) = \exp\left(\int p(t_n|\theta)\ln(\pi^*(\theta|t_n))\,dt_n\right), \quad f(\theta) = \lim_{n\to\infty}\frac{f_n(\theta)}{f_n(\theta_0)}. \quad (3.52)$$

If

1. *for all n, $f_n(\theta)$ is continuous,*

2. *the ratio $\frac{f_n(\theta)}{f_n(\theta_0)}$ is either monotonic in n or bounded by an integrable function and*

3. *conditions 1 and 2 of Definition 3.5 are fulfilled,*

then $f(\theta)$ is a reference prior.

PROOF: See (Berger et al., 2009, Appendix F).

□

The condition that the posterior $\pi^*(\theta|t_n)$ has to be asymptotically consistent is defined in Chapter 5. The following example illustrates an application of Theorem 3.5.

Example 3.28 (Uniform distribution)
Consider an i.i.d. sample from $U[\theta - \frac{1}{2}, \theta + \frac{1}{2}]$, $\theta \in \mathbb{R}$. It is a location model, but it does not fulfill the regularity conditions, because the support of the distribution depends on the parameter. The likelihood function is $\ell(\theta|x_1,\ldots,x_n) = I_{[x_{\max}-0.5, x_{\min}+0.5]}(\theta)$, with sufficient statistic $t = (t_1, t_2) = (x_{\min}, x_{\max})$, where $x_{\min} = \min(x_1,\ldots,x_n)$ and $x_{\max} = \max(x_1,\ldots,x_n)$. Applying Theorem 3.5, we start with $\pi^* = 1$. Then the related posterior $\pi^*(\theta|t) \propto I_{[t_2-0.5, t_1+0.5]}(\theta)$ is the uniform distribution $U[t_2 - 0.5, t_1 + 0.5]$, i.e.,

$$\pi^*(\theta|t) = \frac{1}{1 + t_2 - t_1}I_{[t_2-0.5, t_1+0.5]}(\theta).$$

The statistic $r = t_2 - t_1$ is the range of the distribution which is location-invariant. In this case we have $x_{\max} = u_{\max} - \theta + 0.5$ and $x_{\min} = u_{\min} - \theta + 0.5$ where $u_{\max}$ is the maximum and $u_{\min}$ is the minimum of independent identically $U[0,1]$ distributed random variables. Hence $f_n(\theta) = \exp\left(\int_0^1 p(r)\ln(\frac{1}{1+r})\,dr\right)$ is independent of θ and the ratio $\frac{f_n(\theta)}{f_n(\theta_0)} = 1$. We obtain the reference prior $\pi(\theta) = 1$, which is the Laplace prior. Note that, the prior is improper, but the posterior $U[x_{\max} - 0.5, x_{\min} + 0.5]$ is proper. □

The argument in Example 3.28 can be applied to all location models. In Example 3.21, the regular location model was studied with the result that Jeffreys prior equals Laplace prior.

For cases where the integrals and the limit require complex calculations, Berger et al. (2009) proposed the following algorithm for the numerical computation of the reference prior. The algorithm includes Monte Carlo steps for the integrals in (3.52) and an approximation of the limit, see also Chapter 9.

Algorithm 3.1 Reference prior

1. Starting Values:
   (a) Choose moderate value k (simulate the limit in (3.52)).
   (b) Choose an arbitrary positive function $\pi^*(\theta)$ (take $\pi^*(\theta) = 1$).
   (c) Choose moderate value m for the Monte Carlo steps.
   (d) Choose θ values for which $(\theta, \pi(\theta))$ is required, $\theta^{(1)}, \ldots, \theta^{(M)}$.
   (e) Choose an arbitrary interior point $\theta_0 \in \Theta$.

2. For each $\theta \in \{\theta_0, \theta_1, \ldots, \theta_M\}$:
   (a) For each j with $j = 1, \ldots, m$

      i. Draw independently $\mathbf{x}_{j,1}, \ldots, \mathbf{x}_{j,k}$ from $p(\cdot|\theta)$.
      ii. Compute numerically:

   $$c_j = \int_\Theta \prod_{i=1}^k p(\mathbf{x}_{j,i}|\theta)\pi^*(\theta)d\theta$$

      A. Draw independently $\theta_1, \ldots, \theta_m$ from $\pi^*(\cdot)$.
      B. Calculate $p_h = \prod_{i=1}^k p(\mathbf{x}_{j,i}|\theta_h)$ for $h = 1, \ldots, m$.
      C. Approximate c_j by $\frac{1}{m}\sum_{h=1}^m p_h$.

      iii. Calculate

   $$r_j(\theta) = \ln(\frac{1}{c_j}\prod_{i=1}^k p(\mathbf{x}_{j,i}|\theta)\pi^*(\theta)).$$

   (b) Compute $f_k(\theta) = \exp(\frac{1}{m}\sum_{j=1}^m r_j(\theta))$.

3. Store $(\theta, \pi(\theta))$ with $\pi(\theta) = \frac{f_k(\theta)}{f_k(\theta_0)}$.

The other innovation in Bernardo (1979), besides the information–theoretic foundation of a non-informative prior, is his stepwise procedure for handling nuisance parameters. In the following we present this procedure and give an illustrative example.

Suppose the parameter

$$\theta = (\omega, \lambda)$$

consists of a *parameter of interest* ω and a *nuisance parameter* λ. The main idea is to derive a reference prior for the nuisance parameter given the parameter of interest, $\pi(\lambda|\omega)$, and then to eliminate the nuisance parameter by considering the marginal model,

$$p(\mathbf{x}|\omega) = \int_\Lambda p(\mathbf{x}|\omega, \lambda)\pi(\lambda|\omega)d\lambda. \tag{3.53}$$

Algorithm 3.2 Reference prior with nuisance parameter

1. Split the parameter $\theta = (\omega, \lambda)$ into the parameter of interest ω and the nuisance parameter λ.

2. For each parameter of interest ω, derive the reference prior $\pi(\lambda|\omega)$ related to the statistical model $\{p(\mathbf{x}|\omega, \lambda) : \lambda \in \Lambda\}$.

3. Use $\pi(\lambda|\omega)$ to eliminate the nuisance parameter by

$$p(\mathbf{x}|\omega) = \int_\Lambda p(\mathbf{x}|\omega, \lambda)\pi(\lambda|\omega)\, d\lambda.$$

4. Calculate the reference prior $\pi(\omega)$ related to the statistical model $\{p(\mathbf{x}|\omega) : \omega \in \Omega\}$.

5. Propose as reference prior for the whole parameter

$$\pi(\theta) = \pi(\lambda|\omega)\pi(\omega).$$

Note that, in general an exchange of parameter of interest with the nuisance parameter delivers a different reference prior. Under regularity conditions, this procedure describes a stepwise calculation of Jeffreys prior. In the following example, taken from Polson and Wasserman (1990), we illustrate this method and compare the reference prior with the joint Jeffreys prior.

Example 3.29 (Bivariate Binomial)
Consider the following bivariate binomial model

$$X_1 \sim \mathsf{Bin}(n, p) \text{ and } X_2|X_1 \sim \mathsf{Bin}(X_1, q).$$

The observations are $\mathbf{x} = (x_1, x_2)$. The parameter has two components $\theta = (p, q)$. It holds that

$$p(\mathbf{x}|\theta) = \binom{n}{x_1}p^{x_1}(1-p)^{n-x_1}\binom{x_1}{x_2}q^{x_2}(1-q)^{x_1-x_2}. \tag{3.54}$$

The parameter of interest is $\omega = p$ and the nuisance parameter is $\lambda = q$. We have

$$\{p(\mathbf{x}|\omega, \lambda) : \lambda \in \Lambda\} = \{\mathsf{Bin}(X_1, q) : q \in (0, 1)\}.$$

The binomial model is a regular model, and the reference prior coincides with Jeffreys prior, see Example 3.20. We obtain $\pi(\lambda|\omega)$ as the beta distribution $\mathsf{Beta}(0.5, 0.5)$, which is independent of $\omega = p$. Then the marginal model (3.53) is calculated as

$$p(\mathbf{x}|p)$$

$$= \int_0^1 \binom{n}{x_1} p^{x_1}(1-p)^{n-x_1} \binom{x_1}{x_2} q^{x_2}(1-q)^{x_1-x_2} \frac{1}{B(\frac{1}{2}, \frac{1}{2})} q^{-\frac{1}{2}}(1-q)^{-\frac{1}{2}} dq$$

$$= \binom{n}{x_1} p^{x_1}(1-p)^{n-x_1} \binom{x_1}{x_2} \frac{B(x_2 + \frac{1}{2}, x_1 - x_2 + \frac{1}{2})}{B(\frac{1}{2}, \frac{1}{2})},$$

$$(3.55)$$

where $B(a, b)$ is the beta function. This model is regular. Using

$$\ln(p(\mathbf{x}|p)) = x_1 \ln(\frac{p}{1-p}) + n \ln(1-p) + \text{const}$$

we obtain the Fisher information

$$I(p) = \frac{n}{p(1-p)}.$$

Thus Jeffreys prior is the beta distribution, $\mathsf{Beta}(0.5, 0.5)$. Summarizing, we have the reference prior

$$\pi(\theta) = \pi(p)\pi(q) \propto p^{-\frac{1}{2}}(1-p)^{-\frac{1}{2}} q^{-\frac{1}{2}}(1-q)^{-\frac{1}{2}},$$

where $\pi(p)$ and $\pi(q)$ are the beta distributions, $\mathsf{Beta}(0.5, 0.5)$. Let us now compare the reference prior with the Jeffrey prior for $\theta = (p, q)$. Using (3.54) we have the score function

$$V(\theta|x) = \left(\frac{x_1 - np}{p(1-p)}, \frac{x_2 - x_1 q}{q(1-q)}\right)^T = (V_1(p), V_2(q))^T.$$

and the second derivatives

$$\frac{d}{dp} V_1(p) = \frac{x_1(2p-1) - np^2}{p^2(1-p)^2} \quad \text{and} \quad \frac{d}{dq} V_2(q) = \frac{x_2(2q-1) - x_1 q^2}{q^2(1-q)^2},$$

where $\frac{d}{dq} V_1(p)$ and $\frac{d}{dp} V_2(q)$ are zero. Further,

$$\mathsf{E}X_1 = np \quad \text{and} \quad \mathsf{E}X_2 = \mathsf{E}(\mathsf{E}(X_2|X_1)) = \mathsf{E}(X_1 q) = npq.$$

Applying (3.32) we get

$$I(\theta) = n \begin{pmatrix} \frac{1}{p(1-p)} & 0 \\ 0 & \frac{p}{q(1-q)} \end{pmatrix},$$

and
$$\pi_{\text{Jeff}}(\theta) \propto \det(I(\theta))^{\frac{1}{2}} \propto p^{-\frac{1}{2}}(1-p)^{-\frac{1}{2}}p^{\frac{1}{2}}q^{-\frac{1}{2}}(1-q)^{-\frac{1}{2}},$$

i.e., the Jeffreys prior is

$$\pi_{\text{Jeff}}(\theta) \propto (1-p)^{-\frac{1}{2}}q^{-\frac{1}{2}}(1-q)^{-\frac{1}{2}}.$$

Note that, it is not a product of two beta distributions, but it is still a proper prior.   □

Berger and Bernardo (1992a) proposed an iterative algorithm which is now named *Berger–Bernado method* in the literature. We give the main steps and an illustrative example. The p-dimensional parameter is separated in $m \leq p$ groups. It is recommended to do it in the order of interest. The first group includes the parameters of main interest. The idea is to iteratively eliminate the parameter groups as in the method above. We introduce the following notations.

$$\theta = (\theta_1, \ldots, \theta_p) = (\theta_{(1)}, \ldots, \theta_{(m)}) = (\theta_{[j]}, \theta_{[\sim j]}),$$
$$\theta_{[j]} = (\theta_{(1)}, \ldots, \theta_{(j)}) \text{ and } \theta_{[\sim j]} = (\theta_{(j+1)}, \ldots, \theta_{(m)}) \quad (3.56)$$
$$j = 1, \ldots, m, \quad \theta_{[0]} = 1, \theta_{[\sim 0]} = \theta.$$

Algorithm 3.3 Berger–Bernardo method

1. For $j = m, m-1, \ldots, 1$:

   (a) Suppose the current state is $\pi_{(j+1)}(\theta_{[\sim j]}|\theta_{[j]})$.

   (b) Determine the marginal model

   $$p(\mathbf{x}|\theta_{[j]}) = \int p(\mathbf{x}|\theta)\pi_{(j+1)}(\theta_{[\sim j]}|\theta_{[j]}) \, d\theta_{[\sim j]}.$$

   (c) Determine the reference prior $h_{(j)}(\theta_{(j)}|\theta_{[j-1]})$ related to the model

   $$\{p(\mathbf{x}|\theta_{[j]}) : \theta_{(j)} \in \Theta_j\},$$

   where the parameters $\theta_{[j-1]}$ are considered as given.

   (d) Compute $\pi_{(j)}(\theta_{[\sim(j-1)]}|\theta_{[j-1]})$ by

   $$\pi_{(j)}(\theta_{[\sim(j-1)]}|\theta_{[j-1]}) \propto \pi_{(j+1)}(\theta_{[\sim j]}|\theta_{[j]})h_{(j)}(\theta_{(j)}|\theta_{[j-1]}).$$

2. Take $\pi(\theta) := \pi_{(1)}(\theta_{[\sim 0]}|\theta_{[0]})$, as reference prior.

The following example is an illustration of the Berger–Bernardo method. It is taken from Berger and Bernardo (1992b).

Example 3.30 (Multinomial distribution)

For simplicity we set $k = 4$ and use groups with single parameters. The multinomial distribution $\mathsf{Mult}(n, p_1, \ldots, p_4)$ is a discrete distribution with probability

$$P(n_1, n_2, n_3, n_4) = \frac{n!}{n_1! \ldots n_4!} p_1^{n_1} \ldots p_4^{n_4}, \tag{3.57}$$

where $n_1 + n_2 + n_3 + n_4 = n$ and $p_1 + p_2 + p_3 + p_4 = 1$, thus $p_4 = 1 - \delta$ with $\delta = p_1 + p_2 + p_3$. We have $\mathbf{x} = (n_1, n_2, n_3)$ and $\theta = (p_1, p_2, p_3)$, with $\theta_{(1)} = p_1, \theta_{(2)} = p_2, \theta_{(3)} = p_3$. The first step is to calculate the conditional reference prior for p_3 given p_1, p_2. This means we derive the reference prior for the model

$$\{p(\mathbf{x}|\theta) : \theta_1 = p_1, \theta_2 = p_2, \theta_3 \in (0, 1 - (p_1 + p_2))\},$$

where $p(\mathbf{x}|\theta)$ is given in (3.57). This conditional model is regular. The reference prior is Jeffreys prior, given in (3.45), constrained on $(0, 1 - (p_1 + p_2))$,

$$\pi(p_3|p_1, p_2) \propto p_3^{-\frac{1}{2}}(1 - \delta)^{-\frac{1}{2}}, \quad 0 < p_3 < 1 - (p_1 + p_2).$$

In order to get the constant

$$c = \int_0^{1-(p_1+p_2)} p_3^{-\frac{1}{2}}(1 - \delta)^{-\frac{1}{2}} \, dp_3$$

we apply the integral

$$\int_0^{1-d} (1 - d - x)^a x^b dx = (1 - d)^{a+b+1} B(a+1, b+1), a > -1, b > -1, \tag{3.58}$$

where $B(a + 1, b + 1) = \int_0^1 x^a (1 - x)^b \, dx$ is the beta function. Note that, we obtain the integral (3.58) by changing the variables to $y = \frac{x}{1-d}$ and from the definition of the beta function. The constant $c = B(0.5, 0.5)$ is independent on p_1, p_2. Thus the constrained conditional Jeffreys prior is

$$\pi(p_3|p_1, p_2) = B(0.5, 0.5)^{-1} p_3^{-\frac{1}{2}}(1 - \delta)^{-\frac{1}{2}}.$$

The second step is to determine the marginal model. We have to calculate

$$p(\mathbf{x}|\theta_1, \theta_2) = \int p(\mathbf{x}|\theta)\pi(\theta_3|\theta_1, \theta_2) \, d\theta_3.$$

It holds that

$$p(\mathbf{x}|\theta_1, \theta_2) \propto p_1^{n_1} p_2^{n_2} \int_0^{1-(p_1+p_2)} (1 - \delta)^{n_4 - \frac{1}{2}} p_3^{n_3 - \frac{1}{2}} \, dp_3.$$

Using (3.58) again, the marginal model is

$$p(\mathbf{x}|\theta_1, \theta_2) \propto p_1^{n_1} p_2^{n_2} (1 - (p_1 + p_2))^{(n-n_1-n_2)},$$

which is the multinomial distribution $\mathsf{Mult}(n, p_1, p_2, 1 - (p_1 + p_2))$. We get the conditional reference prior for p_2 given p_1 from (3.45), as

$$\pi(p_2|p_1) \propto p_2^{-\frac{1}{2}} (1 - p_1 - p_2)^{-\frac{1}{2}}, \ 0 < p_2 < 1 - p_1.$$

The marginal model, after eliminating p_2, is

$$p(\mathbf{x}|\theta_1) \propto p_1^{n_1} (1 - p_1)^{n-n_1},$$

which is the binomial distribution, $\mathsf{Bin}(n, p_1)$, and the reference prior is Jeffreys prior

$$\pi(\theta_1) \propto p_1^{-\frac{1}{2}} (1 - p_1)^{-\frac{1}{2}}.$$

This gives

$$\pi(p_1, p_2, p_3) \propto p_1^{-\frac{1}{2}} (1 - p_1)^{-\frac{1}{2}} p_2^{-\frac{1}{2}} (1 - (p_1 + p_2))^{-\frac{1}{2}} p_3^{-\frac{1}{2}} (1 - (p_1 + p_2 + p_3))^{-\frac{1}{2}}.$$

$\square$

3.4 List of Problems

1. Consider the following statistical model on customer satisfaction. In a query, the customer can rate between satisfied (+), disappointed (-), ok (+/-), no answer (0). The parameter θ is the level of satisfaction $0, 1, 2$, where 2 means the customer is satisfied. The probability $P_\theta(x)$ is given in the following table:

x	+	+/-	-	0
$\theta = 2$	0.6	0.1	0	0.3
$\theta = 1$	0.1	0.2	0.1	0.6
$\theta = 0$	0	0.2	0.79	0.01

From earlier studies it is known that the probability of high satisfaction ($\theta = 2$) is 0.2 and the probability of no satisfaction ($\theta = 0$) is 0.3. Consider the result of a single query.

(a) Calculate the posterior distribution, using the information from earlier studies.

(b) Determine the prior with highest entropy and calculate the corresponding posterior distribution.

(c) Determine the prior with $\pi(0) = 2\pi(1)$ and highest entropy.

(d) Determine the prior with $\pi(0) = \pi(1)$ and prior expectation 1.

2. Consider an i.i.d. sample from a Lognormal(μ, σ^2) distribution with density

$$f(x|\theta) = \frac{1}{\sigma x \sqrt{2\pi}} \exp(-\frac{1}{2\sigma^2}(\ln(x) - \mu)^2), \quad -\infty < \mu < \infty, \ \sigma > 0, \ x > 0.$$

Set $\sigma = 1$. The parameter of interest is $\theta = \mu$. We guess that the parameter μ is lying symmetrically around 3. We have only vague knowledge and assume as prior a Cauchy distribution $C(m, \gamma)$. The hyperparameters are m, γ.

(a) Determine the location parameter m of the prior.

(b) Further require that the prior probability $P(\mu > 10) > 0.3$. Determine the hyperparameter γ.

(c) Is the prior conjugate?

(d) Derive Jeffreys prior.

3. Consider an i.i.d. sample $X_1, \ldots, X_n$ from a Pareto distribution $P(\alpha, \mu)$, with

$$f(x \mid \alpha, \mu) = \alpha \frac{\mu^\alpha}{x^{\alpha+1}} I_{[\mu,\infty)}(x), \quad \alpha > 0, \ \mu \in \mathbb{R}$$

(a) Set $\theta = (\alpha, \mu)$. Does the distribution belong to an exponential family?

(b) Set $\alpha = 1$. Does the distribution belong to an exponential family?

(c) Set $\mu = 1$. Does the distribution belong to an exponential family?

(d) Set $\mu = 1$. Derive a conjugate prior for α.

4. Consider an i.i.d sample $X_1, \ldots, X_n$ from a geometric distribution $\mathsf{Geo}(\theta)$,

$$P_\theta(k) = (1 - \theta)^k \theta.$$

(a) Does the sample distribution belong to an exponential family? Determine the sufficient statistic and the natural parameter.

(b) Apply Theorem 3.3 to derive a conjugate family.
   (Hint: $Y = g(X)$, $f_Y(y) = f_X(g^{-1}(y)) \mid \frac{d}{dy} g^{-1}(y) \mid$.)

(c) Which family of distributions is this conjugate family?

(d) Give the conjugate posterior distribution $\pi(\theta \mid x_1, \ldots, x_n)$.

(e) Derive the Fisher information.

(f) Derive Jeffreys prior.

(g) Does Jeffreys prior belong to the conjugate family?

5. Consider an i.i.d. sample $\mathbf{X} = (X_1, \ldots, X_n)$ from Gamma(α, β). The parameter of interest is $\theta = (\alpha, \beta)$.

(a) Is the distribution of the sample a member of an exponential family? Determine the natural parameters and the statistics $T(\mathbf{x})$.

(b) Determine a conjugate prior for θ.

(c) Determine the corresponding posterior.

(d) Is the conjugate family a known distribution family? Is the conjugate family an exponential family?

6. Consider $X|\theta \sim \mathrm{Bin}(n, \theta)$. An alternative parameter η is the odds ratio $\eta = \frac{\theta}{1-\theta}$.

   (a) Consider the prior $\pi(\eta) \propto \eta^{-1}$ for the odds ratio and derive the prior for the success probability θ.

   (b) Is the prior for θ in (a) proper? Is the related posterior proper?

   (c) Set as prior for θ the beta distribution $\mathrm{Beta}(a, b)$. Derive the prior for $\xi = \frac{b}{a}\eta$. Is it a well-known distribution?

7. Suppose $X|\theta \sim \mathrm{Bin}(n, \theta)$.

   (a) Is the distribution of X a member of an exponential family? Determine the natural parameter.

   (b) Determine a conjugate prior for the natural parameter.

   (c) Determine the corresponding posterior.

   (d) Is the conjugate family a known distribution family?

8. Consider the multinomial distribution $\mathrm{Mult}(n, p_1, \ldots, p_3)$ given in (3.12) with $k = 3$. We are interested in the parameters $\eta_1 = \frac{p_1}{p_3}$ and $\eta_2 = \frac{p_2}{p_3}$. Derive Jeffreys prior $\pi_{\mathrm{Jeff}}(\eta)$.

9. The Hardy–Weinberg model states that the genotypes AA, Aa and aa occur with following probabilities:

$$p_\theta(\mathsf{aa}) = \theta^2, \quad p_\theta(\mathsf{Aa}) = 2\theta(1-\theta), \quad p_\theta(\mathsf{AA}) = (1-\theta)^2, \qquad (3.59)$$

   where θ is an unknown parameter in $\Theta = (0, 1)$.

   (a) Does the distribution in (3.59) belong to an exponential family?

   (b) Determine the sufficient statistics and the natural parameter η.

   (c) Derive a conjugate family for η.

   (d) Determine Jeffreys prior for η.

   (e) Does the posterior related to Jeffreys prior belong to that conjugate family?

10. Consider $X \sim \mathrm{N}(\theta, 1)$.

   (a) Recall Jeffreys prior and the related posterior $\pi_{\mathrm{Jeff}}(\theta|x)$.

   (b) Derive Jeffreys prior π_{Jeff}^k and the related posterior $\pi_{\mathrm{Jeff}}^k(\theta|\mathbf{x})$ for $\theta \in [-k, k]$.

   (c) Show that for all $x \in \mathbb{R}$ the Kullback–Leibler divergence converges:

$$\lim_{k \to \infty} \mathsf{K}(\pi_{\mathrm{Jeff}}(.|x) | \pi_{\mathrm{Jeff}}^k(.|x)) = 0.$$

   (d) Determine the reference prior $p_0(\theta)$ by

$$p_0(\theta) = \lim_{k \to \infty} \frac{\pi_{\mathrm{Jeff}}^k(\theta)}{\pi_{\mathrm{Jeff}}^k(\theta_0)}, \quad \theta_0 \in [-k, k].$$

   Does the reference prior depend on the choice of θ_0?

11. Consider $X \sim N(\mu, \sigma^2)$ with density $\varphi_{\mu,\sigma^2}(x)$. Calculate the Shannon entropy $H(N(\mu, \sigma^2)) = -\int \varphi_{\mu,\sigma^2}(x) \ln(\varphi_{\mu,\sigma^2}(x)) dx$.

12. Consider an i.i.d. sample from $X \sim N(\theta, \sigma^2)$ and two priors π_1: $N(0, \sigma_0^2)$ and π_2: $N(\mu_0, \lambda\sigma_0^2)$, $\lambda > 0$.

   (a) Compute the related posteriors $\pi_j(\theta|t(\mathbf{x}))$, $j = 1, 2$, where $t(\mathbf{x})$ is the sufficient statistic.

   (b) Calculate expected information $I(\mathcal{P}^n, \pi_1)$ and $I(\mathcal{P}^n, \pi_2)$.

   (c) Calculate

$$\lim_{n \to \infty} \left(I(\mathcal{P}^n, \pi_1) - I(\mathcal{P}^n, \pi_2) \right).$$

   (d) For which λ and μ_0 is the second prior less informative than the first?

Chapter 4

Decision Theory

This chapter provides an introduction to decision theory, especially the Bayesian decision theory. The main goal is to give a deeper understanding of Bayesian inference. Readers who are mainly interested in methods and applications, can skip this chapter.

We want to explain the connection between Bayesian inference based on the model $\{\mathcal{P}, \pi\}$ and frequentist inference based on the model $\{\mathcal{P}\}$.

Bayes methods are based on the posterior distribution, but nevertheless it makes sense to apply them outside of the Bayes model. Then the question arises: Which optimal properties the Bayes method can have? One answer is: For a special choice of worst case prior π_0, the Bayes method can be minimax optimal.

Otherwise we can also express frequentist methods as Bayesian and derive optimal properties for them.

This chapter is mainly based on Robert (2001, Chapter 2) and Liese and Miescke (2008, Chapter 3).

4.1 Basics of Decision Theory

We begin with the main ingredients. First, we consider the statistical model defined in (2.1),

$$\mathcal{P} = \{\mathsf{P}_\theta : \theta \in \Theta\},$$

where $\theta = (\theta_1, \dots, \theta_p)$. Second, we introduce a **decision space** denoted by $\mathcal{D}$. In case we want to find out the underlying parameter from data $\mathbf{x}$ we set $\mathcal{D} = \Theta \subseteq \mathbb{R}^p$. When we are only interested in θ_1, the first component of θ, then we take $\mathcal{D} = \Theta_{(1)} \subseteq \mathbb{R}$. We may also be interested to predict a future data point $x \in \mathcal{X}_f$. Then we set $\mathcal{D} = \mathcal{X}_f$. For testing problem the decision space is $\mathcal{D} = \{0, 1\}$; for classification problem or model choice, it is $\mathcal{D} = \{1, \dots, k\}$.

An element $d \in \mathcal{D}$ is called **decision**. The main purpose of statistical inference is to make a decision based on the data $\mathbf{x}$. Formulated as **decision rule**, it is defined as

$$\delta : \mathbf{x} \in \mathcal{X} \mapsto \delta(\mathbf{x}) = d \in \mathcal{D}.$$

DOI: 10.1201/9781003221623-4

Figure 4.1: The Glienicker Bridge as Laplace–Bayes bridge.

Depending on the inference problem it can be an estimator, a predictor, a test or a classification rule. The third ingredient is the evaluation of a decision, which is determined by a **loss function**, defined as

$$L : (\theta, d) \in \Theta \times \mathcal{D} \mapsto L(\theta, d) \in \mathbb{R}_+. \qquad (4.1)$$

The loss $L(\theta, d)$ is the penalty, when θ is the true parameter and the decision is d. A high loss means a bad decision. In general a loss can be defined arbitrarily, but it must satisfy some reasonable properties. For instance, for $\Theta = \mathcal{D} = \mathbb{R}$,

$$L(\theta, d) = L(d, \theta)$$
$$L(\theta, d) \text{ is increasing if } |\theta - d| \text{ is increasing,}$$

$$(4.2)$$

see Figure 4.2. Important loss functions are the Laplace or L_1 loss

$$L_1(\theta, d) = |\theta - d|$$

and the Gaussian or L_2 loss

$$L_2(\theta, d) = |\theta - d|^2.$$

We illustrate the meaning of a loss function with help of the toy example on lion's appetite introduced in Example 2.4.

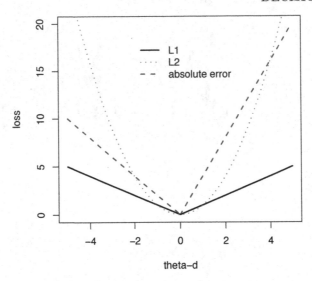

Figure 4.2: Illustration of different loss functions.

Example 4.1 (Lion's appetite) The appetite of a lion has three different stages:

$$\theta \in \Theta = \{\text{hungry, moderate, lethargic}\} = \{\theta_1, \theta_2, \theta_3\}.$$

It would be much more dangerous to classify a hungry lion as lethargic than the other way around. We consider the following losses:

$L(\theta, d)$	$d = \theta_1$	$d = \theta_2$	$d = \theta_3$
θ_1	0	0.8	1
θ_2	0.3	0	0.8
θ_3	0.3	0.1	0

$\square$

Applying the loss on a decision rule gives $L(\theta, \delta(\mathbf{x}))$ which is data dependent. The quality of a decision rule should not depend on the observed data, but on the data generating probability distribution. We have to consider the loss as a random variable and to study the distribution of $L(\theta, \delta(\mathbf{X}))$ under $\mathbf{X} \sim \mathsf{P}_\theta$. In decision theory we are mainly interested in the expectation. We define the **risk** as the expected loss

$$R(\theta, \delta) = \int_{\mathcal{X}} L(\theta, \delta(\mathbf{x})) \, p(\mathbf{x}|\theta) \, d\mathbf{x}. \tag{4.3}$$

The risk in (4.3) is often called **frequentist risk**, because it is based on the statistical model $\mathcal{P}$ only. We consider decision rule δ_1 better than decision rule δ_2 iff

$$R(\theta, \delta_1) \leq R(\theta, \delta_2) \quad \text{for all } \theta \in \Theta.$$

Since $R(\theta, \delta_1)$ should be compared over the entire parameter space, there are risks that are not comparible. We illustrate it with the toy example.

Example 4.2 (Lion's appetite)
Continuing with Example 4.1, we compare the following three decision rules by their risk.

x	0	1	2	3	4
δ_1	θ_3	θ_3	θ_2	θ_2	θ_1
δ_2	θ_3	θ_2	θ_2	θ_1	θ_1
δ_3	θ_1	θ_1	θ_1	θ_1	θ_1

The first and second rule take the observations into account, the higher x the more dangerous the lion is. The third rule always decides that the lion was hungry independent of the observation. We calculate the risk by $R(\theta, \delta) = \sum_{i=1}^{5} L(\theta, \delta(x_i)) \mathsf{P}_\theta(x_i)$, where $\mathsf{P}_\theta(x_i)$ is given in Example 2.4. We obtain

$R(\theta, \delta)$	θ_1	θ_2	θ_3
δ_1	0.73	0.08	0.005
δ_2	0.08	0.07	0.01
δ_3	0	0.3	0.3

As it seems, none of the rules can be preferred; see Figure 4.3.   □

In general we cannot find an optimal decision rule which is better than all the other decision rules. The way out is that we search for the rule for which we cannot find a better one.

Definition 4.1 (Admissible decision) Assume the statistical model $\mathcal{P}$, the decision space $\mathcal{D}$ and the loss function L. A decision rule δ_0 is called **inadmissible** iff there exists a decision rule δ_1 such that

$$\begin{aligned}
R(\theta, \delta_0) &\geq R(\theta, \delta_1) \quad \text{for all } \theta \in \Theta \\
R(\theta_0, \delta_0) &> R(\theta_0, \delta_1) \quad \text{for at least one } \theta_0 \in \Theta.
\end{aligned} \tag{4.4}$$

Otherwise the decision rule δ_0 is called **admissible**.

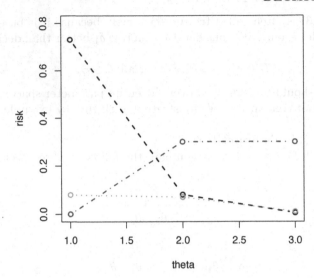

Figure 4.3: Example 4.2. No decision rule has a smaller risk for all parameters. It is impossible to compare them by the frequentist risk.

The admissibility property is weaker than the requirement of minimal risk for all parameters. We have a better chance to find an admissible decision rule than an optimal one. Already from this type of definition it becomes clear that it is easier to prove inadmissibility than the opposite. We will see that most of the proofs in this field are indirect proofs.
We conclude this section with a prominent example on the James–Stein estimator, defined in James and Stein (1960).

Example 4.3 (James–Stein estimator)

Let $X_1, \ldots, X_d$ be independent r.v., $X_i \sim \mathsf{N}(\theta_i, \sigma^2)$, where $\theta = (\theta_1, \ldots, \theta_d) \in \mathbb{R}^d$ is unknown. The variance σ^2 is known, say $\sigma^2 = 1$. Then the unbiased, natural estimator for θ is $T_{\mathrm{nat}}(\mathbf{x}) = (x_1, \ldots, x_d)^T$. The surprising fact here is that for $d > 2$ it is inadmissible. James and Stein (1960) introduced a shrinkage estimator for $d > 2$, as

$$S_{\mathrm{JS}}(\mathbf{x}) = \left(1 - \frac{d-2}{\|\mathbf{x}\|^2}\right)\mathbf{x}. \tag{4.5}$$

For the loss function $L(\theta, d) = \|\theta - d\|^2$ it can be shown that the James–Stein estimator is better. We consider main steps of the proof. Comparing the estimators $T_{\text{nat}}(\mathbf{x}) = \mathbf{x}$ and S_{JS} we get

$$D = R(\theta, T_{\text{nat}}) - R(\theta, S_{\text{JS}})$$

$$= 2 \sum_{j=1}^{d} \mathsf{E}_\theta \left((x_j - \theta_j) \, x_j \frac{d-2}{\|\mathbf{x}\|^2} \right) - \mathsf{E}_\theta \left(\frac{(d-2)^2}{\|\mathbf{x}\|^2} \right). \qquad (4.6)$$

Using repeated integration by parts, it can be shown that

$$\mathsf{E}_\theta \left((x_j - \theta_j) \, x_j \frac{d-2}{\|\mathbf{x}\|^2} \right) = \mathsf{E}_\theta \left(\frac{(d-2)^2}{\|\mathbf{x}\|^2} \right).$$

Thus

$$D = \mathsf{E}_\theta \left(\frac{(d-2)^2}{\|\mathbf{x}\|^2} \right).$$

We need to calculate the expectation of an inverse noncentral χ_d^2 distribution with noncentrality parameter $\lambda = \frac{\|\theta\|^2}{2}$. We use the representation of a noncentral χ_d^2 distribution as a Poisson weighted mixture, with $N \sim \text{Poi}(\lambda)$. Then

$$\mathsf{E}_\theta \left(\frac{1}{\|\mathbf{x}\|^2} \right) = \mathsf{E}_\lambda \left(\mathsf{E}_\theta \left(\frac{1}{\|\mathbf{x}\|^2} | N = k \right) \right).$$

The conditional distribution of $\|\mathbf{x}\|^2$ given $N = k$ is a central χ_{d+2k}^2 distribution. The expectation of an inverse χ_{d+2k}^2 distribution is $\frac{1}{d+2k-2}$. We obtain

$$\mathsf{E}_\theta \left(\frac{1}{\|\mathbf{x}\|^2} \right) = \mathsf{E}_\lambda \left(\frac{1}{d + 2N - 2} \right).$$

Applying the Jensen inequality, it holds that

$$\mathsf{E}_\lambda \left(\frac{1}{d + N - 2} \right) \geq \frac{1}{d + 2\mathsf{E}_\lambda N - 2} = \frac{1}{d + 2\lambda - 2}.$$

Finally we obtain

$$D \geq \frac{(d-2)^2}{d + \|\theta\|^2 - 2} > 0.$$

$\square$

4.2 Bayesian Decision Theory

Now we consider the Bayes model $\{\mathcal{P}, \pi\}$; see Definition 2.3. This avoids the problem that decision rules are not comparable with respect to the risk function, given in (4.3). We define the **integrated risk** as the expectation of frequentist risk with respect to the prior π

$$r(\pi, \delta) = \mathsf{E}^\pi R(\theta, \delta) = \int_\Theta R(\theta, \delta) \pi(\theta) \, d\theta. \qquad (4.7)$$

To illustrate the advantages of this Bayesian approach we continue with our toy example.

Example 4.4 (Lion's appetite)
We continue Example 4.2. In Example 2.4 a subjective prior is set for an adult
lion as $(\pi(\theta_i))_{i=1,2,3} = (0.1, 0.1, 0.8)$. The integrated risk is calculated by

$$r(\pi, \delta) = \sum_{i=1}^{3} R(\theta_i, \delta)\pi(\theta_i).$$

We obtain $r(\pi, \delta_1) = 0.085$, $r(\pi, \delta_2) = 0.023$ and $r(\pi, \delta_3) = 0.27$. The second
rule is better than the first rule. The pessimistic and data-independent deci-
sion rule δ_3 is the worst.   □

Definition 4.2 (Bayes decision rule) Assume the Bayes model $\{\mathcal{P}, \pi\}$,
the decision space $\mathcal{D}$ and the loss function L. A decision rule δ^π is called a
Bayes decision rule iff it minimizes the integrated risk $r(\pi, \delta)$. The value

$$r(\pi) = r(\pi, \delta^\pi) = \inf_{\delta} r(\pi, \delta)$$

is called **Bayes risk**.

The posterior expectation of the loss function is called the **posterior expected loss**, i.e.,

$$\rho(\pi, d|\mathbf{x}) = \mathsf{E}^\pi\left(L(\theta, d)|\mathbf{x}\right) = \int_\Theta L(\theta, d)\pi(\theta|\mathbf{x})\, d\theta. \tag{4.8}$$

The following theorem gives us a method to find the Bayes decision.

Theorem 4.1 *For every* $\mathbf{x} \in \mathcal{X}$, $\delta^\pi(\mathbf{x})$ *is given by*

$$\rho(\pi, \delta^\pi(\mathbf{x})|\mathbf{x}) = \inf_{d \in \mathcal{D}} \rho(\pi, d|\mathbf{x}). \tag{4.9}$$

PROOF: The result follows directly from Fubini's Theorem. Using
$p(\mathbf{x}|\theta)\pi(\theta) = \pi(\theta|\mathbf{x})p(\mathbf{x})$ we have

$$\begin{aligned}
r(\pi, \delta) &= \int_\Theta R(\theta, \delta)\pi(\theta)\, d\theta \\
&= \int_\Theta \int_\mathcal{X} L(\theta, \delta(\mathbf{x}))p(\mathbf{x}|\theta)\, d\mathbf{x}\, \pi(\theta)\, d\theta \\
&= \int_\mathcal{X} \int_\Theta L(\theta, \delta(\mathbf{x}))\pi(\theta|\mathbf{x})\, d\theta\, p(\mathbf{x})\, d\mathbf{x} \\
&= \int_\mathcal{X} \rho(\pi, \delta(\mathbf{x}))p(\mathbf{x})\, d\mathbf{x} \\
&\geq \int_\mathcal{X} \rho(\pi, \delta^\pi(\mathbf{x}))p(\mathbf{x})\, d\mathbf{x} = r(\pi, \delta^\pi).
\end{aligned} \tag{4.10}$$

□

4.3 Common Bayes Decision Rules

In this section we present the Bayes rules for the estimation problem $\mathcal{D} = \Theta$ with the quadratic loss, Laplace loss and also briefly discuss alternative loss function. Exploring the analogy of estimation and prediction we derive Bayes rules for prediction. Furthermore, we present results for $\mathcal{D} = \{0, 1\}$.

To begin with the estimation problem, we consider a Bayes model $\{\mathcal{P}, \pi\}$ with $\Theta \subseteq \mathbb{R}^p$. The goal is to find an "optimal" estimator: $\widehat{\theta} : \mathbf{x} \in \mathcal{X} \mapsto \Theta$. In other words, we set $\mathcal{D} = \Theta$ and search for $\widehat{\theta} = \delta^\pi$.

4.3.1 Quadratic Loss

Consider the *weighted L_2 loss* function,

$$L_W(\theta, d) = \|\theta - d\|^2_{W^{-1}} = (\theta - d)^T W (\theta - d), \qquad (4.11)$$

where W is a $p \times p$ positive definite matrix. For $W = \mathbf{I}_p$ we have $L_W(\theta, d) = L_2(\theta, d)$. For a diagonal weight matrix $W = \operatorname{diag}(w_1, \ldots, w_p)$ with $w_i > 0$, $\theta = (\theta_1, \ldots, \theta_p)$, and $d = (d_1, \ldots, d_p)$, it is

$$L_W(\theta, d) = \sum_{i=1}^{p} w_i (\theta_i - d_i)^2.$$

Theorem 4.2 *For every $\mathbf{x} \in \mathcal{X}$, $\delta^\pi(\mathbf{x})$ with respect to L_W is given by the expectation of the posterior distribution $\pi(\theta|\mathbf{x})$*

$$\delta^\pi(\mathbf{x}) = \mathsf{E}^\pi(\theta|\mathbf{x}). \qquad (4.12)$$

PROOF: The result is a consequence of the projection property of the conditional expectation. To see this, note that

$$\rho(\pi, d|\mathbf{x}) = \int_\Theta L_W(\theta, d)\pi(\theta|\mathbf{x})\, d\theta = \int_\Theta (\theta - d)^T W (\theta - d)\pi(\theta|\mathbf{x})\, d\theta$$

$$= \int_\Theta (\theta \pm \delta^\pi(\mathbf{x}) - d)^T W (\theta \pm \delta^\pi(\mathbf{x}) - d)\pi(\theta|\mathbf{x})\, d\theta$$

$$= \int_\Theta (\theta - \delta^\pi(\mathbf{x}))^T W (\theta - \delta^\pi(\mathbf{x}))\pi(\theta|\mathbf{x})d\theta \qquad (4.13)$$

$$+ 2\int_\Theta (d - \delta^\pi(\mathbf{x}))^T W (\delta^\pi(\mathbf{x}) - \theta)\pi(\theta|\mathbf{x})\, d\theta$$

$$+ \int_\Theta (d - \delta^\pi(\mathbf{x}))^T W (d - \delta^\pi(\mathbf{x}))\pi(\theta|\mathbf{x})\, d\theta.$$

Since $\delta^\pi(\mathbf{x}) = \int_\Theta \theta \, \pi(\theta|\mathbf{x}) \, d\theta$, it holds for the mixture term that

$$\int_\Theta (d - \delta^\pi(x))^T W (\delta^\pi(\mathbf{x}) - \theta) \pi(\theta|\mathbf{x}) \, d\theta$$

$$= (d - \delta^\pi(\mathbf{x}))^T W \int_\Theta (\delta^\pi(\mathbf{x}) - \theta) \pi(\theta|\mathbf{x}) \, d\theta$$

$$= (d - \delta^\pi(\mathbf{x}))^T W \left(\delta^\pi(\mathbf{x}) - \int_\Theta \theta \, \pi(\theta|\mathbf{x}) \, d\theta \right)$$

$$= 0.$$

Hence

$$\rho(\pi, d|\mathbf{x}) = \int_\Theta (\theta - \delta^\pi(\mathbf{x}))^T W (\theta - \delta^\pi(\mathbf{x})) \pi(\theta|\mathbf{x}) \, d\theta$$
$$+ (d - \delta^\pi(\mathbf{x}))^T W (d - \delta^\pi(\mathbf{x}))$$
$$\geq \rho(\pi, \delta^\pi(\mathbf{x})).$$

$\square$

We turn back to the toy data, introduced in Example 2.4.

Example 4.5 (Lion's appetite)
The appetite of a lion has three different stages: hungry, moderate, lethargic. We can order the parameters, in the sense that the stages imply different levels of danger. But we have no distances between the parameters. A quadratic loss is not applicable. $\square$

In case the posterior belongs to a known distribution class, with the first two moments finite and known, we obtain the Bayes decision rule and the Bayes risk without additional problems. We demonstrate it with the help of next three examples.

Example 4.6 (Binomial distribution)
We consider $X|\theta \sim \text{Bin}(n, \theta)$. Recall Examples 3.6 and 3.20, and Figure 3.12. The conjugate prior belongs to the family of beta distributions $\{\text{Beta}(\alpha, \beta): \alpha > 0, \beta > 0\}$; see (2.7). The expected value μ and variance σ^2 of $\text{Beta}(\alpha, \beta)$ are

$$\mu = \frac{\alpha}{\alpha + \beta}, \quad \sigma^2 = \frac{\alpha\beta}{(\alpha + \beta)^2(\alpha + \beta + 1)}. \tag{4.14}$$

The posterior is $\text{Beta}(\alpha + x, \beta + n - x)$. For quadratic loss $(\theta - d)^2$ and for an arbitrary beta prior π, the Bayes estimator is

$$\delta^\pi(x) = \frac{\alpha + x}{\alpha + \beta + n}. \tag{4.15}$$

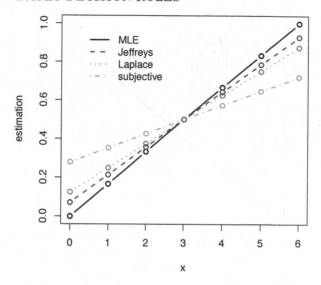

Figure 4.4: Example 4.6. The Bayes estimators are related to the priors $\mathsf{Beta}(0.5, 0.5)$, $\mathsf{Beta}(1, 1)$, $\mathsf{Beta}(3.8, 3.8)$; see also Figure 3.12.

Laplace prior is $\mathsf{Beta}(1, 1)$ and Jeffreys is $\mathsf{Beta}(\frac{1}{2}, \frac{1}{2})$, for which the associated Bayes estimators are

$$\delta^{\pi_{\mathrm{Lap}}}(x) = \frac{1+x}{2+n}, \quad \delta^{\pi_{\mathrm{Jeff}}}(x) = \frac{0.5+x}{1+n}.$$

The moment estimator equals the maximum likelihood $\frac{x}{n}$. See Figure 4.4. $\square$

As second example we consider the normal distribution with normal priors as introduced in Example 2.12.

Example 4.7 (Normal i.i.d. sample and normal prior)
We consider an i.i.d. sample $\mathbf{X} = (X_1, \ldots, X_n)$ from $\mathsf{N}(\mu, \sigma^2)$ with known variance σ^2. The unknown parameter is $\theta = \mu$. For the normal prior $\mathsf{N}(\mu_0, \sigma_0^2)$, the posterior, given in (2.8), is

$$\mathsf{N}(\mu_1, \sigma_1^2), \text{ with } \mu_1 = \frac{\bar{x} n \sigma_0^2 + \mu_0 \sigma^2}{n\sigma_0^2 + \sigma^2} \text{ and } \sigma_1^2 = \frac{\sigma_0^2 \sigma^2}{n\sigma_0^2 + \sigma^2}.$$

We obtain as Bayes estimator

$$\delta^\pi(\mathbf{x}) = \frac{\bar{x} n \sigma_0^2 + \mu_0 \sigma^2}{n\sigma_0^2 + \sigma^2}. \tag{4.16}$$

For different prior parameters μ_0 and σ_0^2 estimators are plotted as functions of the sufficient statistic $\bar{x}$ in Figure 4.5. $\square$

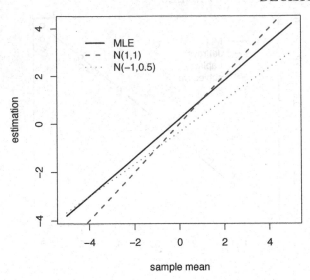

Figure 4.5: Example 4.7. The maximum likelihood estimator and Bayes estimators related to the priors $N(1,1)$ and $N(-1,0.5)$ are plotted as function of the sample mean, where $n = 4$ and $\sigma^2 = 1$.

The last example is related to Example 2.16. Here we have no closed form expression for the Bayes estimator and computer-intensive methods are required.

Example 4.8 (Cauchy i.i.d. sample and normal prior)

We consider an i.i.d. sample $\mathbf{X} = (X_1, \ldots, X_n)$ from a Cauchy distribution $C(\theta, 1)$ with location parameter θ and scale parameter $\gamma = 1$, with density given in (2.13). As prior distribution of θ the normal distribution $N(\mu, \sigma^2)$ is used with density $\varphi(\theta)$. We get for the posterior

$$\pi(\theta|\mathbf{x}) \propto \prod_{i=1}^{n} \frac{1}{1 + (x_i - \theta)^2} \varphi(\theta).$$

The Bayes estimator is defined as

$$\delta^{\pi}(\mathbf{x}) = \frac{\int_{\Theta} \theta h(\theta) \varphi(\theta) \, d\theta}{\int_{\Theta} h(\theta) \, \varphi(\theta) \, d\theta} = \frac{m_1(\mathbf{x})}{m(\mathbf{x})}, \quad \text{with } h(\theta) = \prod_{i=1}^{n} \frac{1}{1 + (x_i - \theta)^2}.$$

Both integrals have to be calculated numerically, for instance by independent Monte Carlo. This can be done as following. Generate independent $\theta^{(1)}, \ldots, \theta^{(N)}$ from $N(\mu, \sigma^2)$ and approximate $m(\mathbf{x})$ by

$$\widehat{m}(\mathbf{x}) = \frac{1}{N} \sum_{j=1}^{N} h(\theta^{(j)}).$$

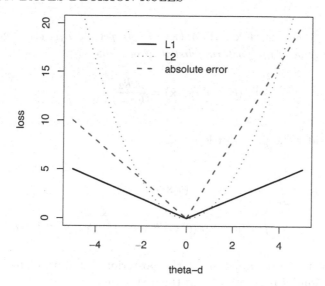

Figure 4.6: Illustration of different loss functions. The absolute error loss defined in (4.17) is plotted with $k_1 = 2, k_2 = 4$.

For the other integral, generate a new sequence $\theta_{(1)}, \ldots, \theta_{(N)}$ from $N(\mu, \sigma^2)$ and approximate $m_1(\mathbf{x})$ by

$$\widehat{m}_1(\mathbf{x}) = \frac{1}{N} \sum_{j=1}^{N} \theta^{(j)} h(\theta^{(j)}).$$

$\square$

4.3.2 Absolute Error Loss

In this subsection we consider the L_1 loss and its asymmetric version. We consider the Bayes model $\{\mathcal{P}, \pi\}$ with $\Theta \subseteq \mathbb{R}$ and $\mathcal{D} = \Theta$, and define the *absolute error loss* by

$$L_{(k_1, k_2)}(\theta, d) = \begin{cases} k_2(\theta - d) & \text{if } d < \theta \\ k_1(d - \theta) & \text{if } d \geq \theta \end{cases}, \quad \text{for } k_1 > 0, k_2 > 0. \quad (4.17)$$

The absolute error loss is convex. For $k_1 > k_2$ overestimation of θ is punished more severely. For $k_1 = k_2 = 1$ it is the L_1 loss. The L_1 loss increases slower than the L_2 loss, hence it is more important in robust statistics, where the influence of outliers needs to be reduced. For illustration see Figure 4.6.

Theorem 4.3 *For every* $\mathbf{x} \in \mathcal{X}$, $\delta^\pi(\mathbf{x})$ *with respect to* $L_{(k_1,k_2)}$ *is given by the* $\frac{k_2}{k_1+k_2}$ *fractile of the posterior distribution* $\pi(\theta|\mathbf{x})$

$$P^\pi(\theta \leq \delta^\pi(\mathbf{x})|\mathbf{x}) = \frac{k_2}{k_1 + k_2}. \qquad (4.18)$$

PROOF: Consider the posterior loss

$$\rho(\pi, d|\mathbf{x}) = \int_\Theta L_{(k_1,k_2)}(\theta, d)\pi(\theta|\mathbf{x})\, d\theta$$

$$= k_1 \int_{-\infty}^d (d - \theta)\pi(\theta|\mathbf{x})\, d\theta + k_2 \int_d^\infty (\theta - d)\pi(\theta|\mathbf{x})\, d\theta.$$

Denote the distribution function of the posterior distribution by $F^\pi(\theta|\mathbf{x})$. Using integration by parts we get for the first term

$$\int_{-\infty}^d (d - \theta)\pi(\theta|\mathbf{x})\, d\theta = (d - \theta)F^\pi(\theta|\mathbf{x})\,|_{-\infty}^d + \int_{-\infty}^d F^\pi(\theta|\mathbf{x})\, d\theta$$

$$= \int_{-\infty}^d F^\pi(\theta|\mathbf{x})\, d\theta.$$

Using $\frac{d}{dx}(-(1 - F(\mathbf{x})) = f(\mathbf{x})$ and integration by parts, we obtain for the second term

$$\int_d^\infty (\theta - d)\pi(\theta|\mathbf{x})\, d\theta = (\theta - d)(-(1 - F^\pi(\theta|\mathbf{x}))\,|_d^\infty + \int_d^\infty (1 - F^\pi(\theta|\mathbf{x}))\, d\theta$$

$$= \int_d^\infty (1 - F^\pi(\theta|\mathbf{x}))\, d\theta.$$

Hence

$$\rho(\pi, d|\mathbf{x}) = k_1 \int_{-\infty}^d F^\pi(\theta|\mathbf{x})\, d\theta + k_2 \int_d^\infty (1 - F^\pi(\theta|\mathbf{x}))\, d\theta.$$

The optimal decision should fulfill (4.9), which implies that the derivative of the posterior loss with respect to d should be zero,

$$\rho'(\pi, d|\mathbf{x}) = 0.$$

We have

$$\rho'(\pi, d|\mathbf{x}) = k_1 F^\pi(d|\mathbf{x}) - k_2(1 - F^\pi(d|\mathbf{x})) = 0$$

and obtain

$$F^\pi(\delta^\pi(\mathbf{x})|\mathbf{x}) = \frac{k_2}{k_1 + k_2}.$$

$\square$

Note that, for the L_1 loss, we have $k_1 = k_2$ and the Bayes estimate is the median of the posterior distribution. Let us consider an example.

Example 4.9 (Binomial distribution)
Continuation of Example 4.6. We consider $X|\theta \sim \text{Bin}(n,\theta)$ and $\theta \sim \text{Beta}(\alpha,\beta)$. The posterior is $\text{Beta}(\alpha+x,\beta+n-x)$. For the L_1 loss, $L(\theta,d) = |\theta-d|$, the Bayes estimator is the median of $\text{Beta}(\alpha+x,\beta+n-x)$. Unfortunately the median of the beta distribution $\text{Beta}(\alpha,\beta)$ is the inverse of an incomplete beta function. We use the following approximation for the median of $\text{Beta}(\alpha,\beta)$

$$\text{med} \approx \frac{\alpha-\frac{1}{3}}{\alpha+\beta-\frac{2}{3}}, \quad \alpha > 1, \beta > 1. \tag{4.19}$$

Hence

$$\delta^\pi(x) \approx \frac{\alpha+x-\frac{1}{3}}{\alpha+\beta+n-\frac{2}{3}}. \tag{4.20}$$

Laplace prior is $\text{Beta}(1,1)$ and Jeffreys is $\text{Beta}(\frac{1}{2},\frac{1}{2})$, and the related Bayes estimators are, for $x > 0, x \leq 1$

$$\delta^{\pi_{\text{Lap}}}(x) \approx \frac{x+\frac{2}{3}}{\frac{4}{3}+n}, \quad \delta^{\pi_{\text{Jeff}}}(x) \approx \frac{\frac{1}{6}+x}{\frac{1}{3}+n}.$$

$\square$

To continue with the prediction problem, we now consider a Bayes model $\{\mathcal{P},\pi\}$ with $\Theta \subseteq \mathbb{R}^p$. The goal is to predict, in an optimal way, a future data point $x_f \in \mathcal{X}_f$ generated by a distribution Q_θ with the same parameter θ as the data generating distribution P_θ. We set $\mathcal{D} = \mathcal{X}_f$.

4.3.3 Prediction

We assume the Bayes model $\{\mathcal{P},\pi\}$ which has the posterior $\pi(\theta|\mathbf{x})$. The future data point x_f is generated by a distribution Q_θ, possibly depending on $\mathbf{x}$; setting $z = x_f$, we have $q(z|\theta,\mathbf{x})$. The **prediction error** $L_{\text{pred}}(z,d)$ is the loss we incur by predicting $z = x_f$ by some $d \in \mathcal{D} = \mathcal{X}_f$. Note that, the future observation is a realization of a random variable. The prediction error is not a loss function as given in (4.1). We define the loss function as the expected prediction error

$$L(\theta,d) = \int_{\mathcal{X}_f} L_{\text{pred}}(z,d)q(z|\theta,\mathbf{x})\,dz. \tag{4.21}$$

Then the posterior expected loss, defined in (4.8) as

$$\rho(\pi,d|\mathbf{x}) = \int_\Theta L(\theta,d)\pi(\theta|\mathbf{x})\,d\theta, \tag{4.22}$$

is given by

$$\rho(\pi, d|\mathbf{x}) = \int_\Theta \int_{\mathcal{X}_f} L_{\mathrm{pred}}(z, d) q(z|\theta, \mathbf{x}) \, dz \, \pi(\theta|\mathbf{x}) \, d\theta$$

$$= \int_{\mathcal{X}_f} L_{\mathrm{pred}}(z, d) \int_\Theta q(z|\theta, \mathbf{x}) \pi(\theta|\mathbf{x}) \, d\theta \, dz. \tag{4.23}$$

We define the **predictive distribution** as the conditional distribution of the future data point x_f given the data $\mathbf{x}$;

$$\pi(x_f|\mathbf{x}) = \int_\Theta q(x_f|\theta, \mathbf{x}) \pi(\theta|\mathbf{x}) \, d\theta. \tag{4.24}$$

Hence

$$\rho(\pi, d|\mathbf{x}) = \int_{\mathcal{X}_f} L_{\mathrm{pred}}(z, d) \pi(z|\mathbf{x}) \, dz. \tag{4.25}$$

Applying Theorem 4.1 we obtain the Bayes rule by minimizing the expected posterior loss for each $\mathbf{x} \in \mathcal{X}$. We call this Bayes rule **Bayes predictor**.

$$\rho(\pi, \delta^\pi(\mathbf{x})|\mathbf{x}) = \inf_{d \in \mathcal{D}} \rho(\pi, d|\mathbf{x})$$

$$= \inf_{d \in \mathcal{D}} \int_{\mathcal{X}_f} L_{\mathrm{pred}}(z, d) \pi(z|\mathbf{x}) \, dz. \tag{4.26}$$

It is the same minimizing problem as for determining the Bayes estimator, where the predictive distribution takes over the role of the posterior. Applying above results on Bayes estimators we obtain following theorem.

Theorem 4.4

1. For $L_{pred}(z, d) = (z - d)^T W(z - d)$, W positive definite a the Bayes predictor is the expectation of the predictive distribution, $\mathsf{E}(z|\mathbf{x})$.

2. Set $\mathcal{X}_f = \mathbb{R}$ and $L_{pred}(z, d) = |z - d|$. The Bayes predictor is the median of the predictive distribution, $\mathrm{Median}(z|\mathbf{x})$.

We illustrate the prediction by the following example.

Example 4.10 (Bayes predictor)

We consider an i.i.d. sample $\mathbf{X} = (X_1, \dots, X_n)$ from $\mathsf{N}(\mu, \sigma^2)$ with known variance σ^2. The unknown parameter is $\theta = \mu$. For the normal prior $\mathsf{N}(\mu_0, \sigma_0^2)$, the posterior, given in (2.8), is

$$\mathsf{N}(\mu_1, \sigma_1^2), \text{ with } \mu_1 = \frac{\bar{x} n \sigma_0^2 + \mu_0 \sigma^2}{n \sigma_0^2 + \sigma^2} \text{ and } \sigma_1^2 = \frac{\sigma_0^2 \sigma^2}{n \sigma_0^2 + \sigma^2}.$$

We want to predict

$$x_f = \bar{x} + a\theta + \varepsilon, \quad \varepsilon \sim \mathsf{N}(0, \sigma_f^2),$$

where σ_f^2 and a are known and ε is independent of $\mathbf{X}$ and θ. Thus we have

$$x_f|(\theta, \mathbf{x}) \sim \mathsf{N}(\bar{\mathbf{x}} + a\theta, \sigma_f^2) \text{ and } \theta|\mathbf{x} \sim \mathsf{N}(\mu_1, \sigma_1^2).$$

Applying known results for normal distribution, i.e., Theorem 6.1, we obtain the predictive distribution

$$x_f|\mathbf{x} \sim \mathsf{N}(\bar{\mathbf{x}} + a\mu_1, a^2\sigma_1^2 + \sigma_f^2).$$

This distribution is symmetric about the expectation. In both cases of Theorem 4.4 we obtain the same Bayes predictor

$$\widehat{x}_f = \bar{\mathbf{x}} + a\mu_1.$$

It is just the prediction rule obtained by plugging in the estimator. For the Bayes predictor we plug in the Bayes estimator.   □

Now we consider the testing problem.

4.3.4 The 0–1 Loss

We assume the Bayes model $\{\mathcal{P}, \pi\}$. We are interested in a testing problem with respect to the statistical model which is split into two disjunct submodels

$$\mathcal{P} = \{\mathsf{P}_\theta : \theta \in \Theta \subseteq \mathbb{R}^p\} = \{\mathsf{P}_\theta : \theta \in \Theta_0\} \cup \{\mathsf{P}_\theta : \theta \in \Theta_1\}, \quad (4.27)$$

where

$$\Theta = \Theta_0 \cup \Theta_1, \quad \Theta_0 \cap \Theta_1 = \varnothing.$$

We set $\mathcal{P}_j = \{\mathsf{P}_\theta : \theta \in \Theta_j\}$ for $j = 0, 1$. The decision space has only two elements $\mathcal{D} = \{0, 1\}$, where $d = j$ stands for the decision that the model $\mathcal{P}_j$ is the true data generating distribution; see also the explanation in Chapter 2. We define the 0–1 **loss** by

$$L_{(0-1)}(\theta, d) = \begin{cases} 0 & \text{if } d = 0 \quad \theta \in \Theta_0 \\ 0 & \text{if } d = 1 \quad \theta \in \Theta_1 \\ 1 & \text{if } d = 0 \quad \theta \in \Theta_1 \\ 1 & \text{if } d = 1 \quad \theta \in \Theta_0 \end{cases} = \begin{cases} d & \text{if } \theta \in \Theta_0 \\ 1 - d & \text{if } \theta \in \Theta_1 \end{cases}. \quad (4.28)$$

The frequentist risk is

$$R(\theta, \delta) = \int_{\mathcal{X}} L_{(0-1)}(\theta, \delta(\mathbf{x}))p(\mathbf{x}|\theta)\, d\mathbf{x} = \begin{cases} \mathsf{P}_\theta(\delta(\mathbf{x}) = 1) & \text{if } \theta \in \Theta_0 \\ \mathsf{P}_\theta(\delta(\mathbf{x}) = 0) & \text{if } \theta \in \Theta_1 \end{cases}.$$

Recall from the theory of hypotheses testing, $H_0: \mathcal{P}_0$ versus $H_1: \mathcal{P}_1$ (see Liero and Zwanzig, 2011, Chapter 5) the decision rule is a test. The error of first type occurs when we reject H_0, but the sample comes from $\mathcal{P}_0$. The error of second type occurs when we do not reject H_0, but the sample comes from $\mathcal{P}_1$. The frequentist risk is

$$R(\theta, \delta) = \begin{cases} \text{P(Error of Type I)} & \text{if } \theta \in \Theta_0 \\ \text{P(Error of Type II)} & \text{if } \theta \in \Theta_1 \end{cases}.$$

Theorem 4.5 *For every* $\mathbf{x} \in \mathcal{X}$, $\delta^\pi(\mathbf{x})$ *with respect to* $L_{(0-1)}$ *is given by*

$$\delta^\pi(\mathbf{x}) = \begin{cases} 1 & \text{if } \mathsf{P}^\pi(\Theta_0|\mathbf{x}) < \mathsf{P}^\pi(\Theta_1|\mathbf{x}) \\ 0 & \text{if } \mathsf{P}^\pi(\Theta_0|\mathbf{x}) \geq \mathsf{P}^\pi(\Theta_1|\mathbf{x}) \end{cases} \qquad (4.29)$$

PROOF: Consider the posterior loss. We have

$$\begin{aligned} \rho(\pi, d|\mathbf{x}) &= \int_\Theta L_{(0-1)}(\theta, d)\pi(\theta|\mathbf{x})\, d\theta \\ &= \int_{\Theta_0} d\,\pi(\theta|\mathbf{x})\, d\theta + \int_{\Theta_1} (1-d)\,\pi(\theta|\mathbf{x})\, d\theta. \\ &= d\,\mathsf{P}^\pi(\Theta_0|\mathbf{x}) + (1-d)\,\mathsf{P}^\pi(\Theta_1|\mathbf{x}) \\ &= d\,(\mathsf{P}^\pi(\Theta_0|\mathbf{x}) - \mathsf{P}^\pi(\Theta_1|\mathbf{x})) + \mathsf{P}^\pi(\Theta_1|\mathbf{x}) \\ &\geq \mathsf{P}^\pi(\Theta_1|\mathbf{x}) + \delta^\pi(\mathbf{x})(\mathsf{P}^\pi(\Theta_0|\mathbf{x}) - \mathsf{P}^\pi(\Theta_1|\mathbf{x})) \\ &= \rho(\pi, \delta^\pi(\mathbf{x})|\mathbf{x}). \end{aligned}$$

The inequality holds, because $\delta^\pi(\mathbf{x}) = 0$ iff $\mathsf{P}^\pi(\Theta_0|\mathbf{x}) - \mathsf{P}^\pi(\Theta_1|\mathbf{x})$ is non–negative.

$\square$

This optimal rule is very intuitive. The posterior $\pi(\theta|\mathbf{x})$ is the weight function on the parameter set Θ after the experiment with result $\mathbf{x}$. We decide for the subset with highest weight, see Figure 4.7.

It works well in all cases where data $\mathbf{x}$ exists such that $0 < \mathsf{P}^\pi(\Theta_0|\mathbf{x}) < 1$. But for simple hypothesis, $\Theta_0 = \{\theta_0\}$, and for continuous posterior distributions it holds that $\mathsf{P}^\pi(\Theta_0|\mathbf{x}) = 0$ for all $\mathbf{x} \in \mathcal{X}$. Then the Bayes rule is $\delta^\pi(\mathbf{x}) = 1$ independent of the result of the experiment. This renders it as an unpracticable method. In Chapter 8 we consider Bayes tests in more detail and offer Bayes solutions for testing simple hypotheses.

Example 4.11 (Normal i.i.d. sample and inverse-gamma prior)

We consider Example 2.14. We have an i.i.d sample from $N(0, \theta)$. The unknown parameter is the variance $\theta = \sigma^2$. We consider as prior distribution of

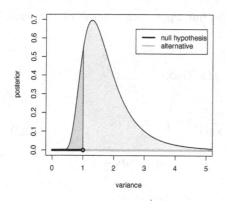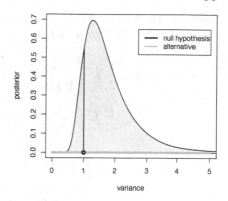

Figure 4.7: Example 4.11. The posterior $\pi(\theta|\mathbf{x})$ is plotted, with prior $\mathsf{InvGamma}(6,9)$. Left: $\Theta_0 = (0,1]$ and $\Theta_1 = (1,\infty)$. The posterior probability of Θ_1 is higher. The Bayes decision is $\delta^\pi(\mathbf{x}) = 1$. Right: $\Theta_0 = \{1\}$ and $\Theta_1 = (0,1)\cup(1,\infty)$. The posterior probability of Θ_0 is zero. The Bayes decision is $\delta^\pi(\mathbf{x}) = 1$.

θ an inverse-gamma distribution $\mathsf{InvGamma}(\alpha,\beta)$, defined in (2.11). The posterior distribution is the inverse-gamma distribution with shape parameter $\alpha + \frac{n}{2}$ and scale parameter $\frac{1}{2}\sum_{i=1}^n x_i^2 + \beta$. We consider the testing problem $H_0\colon \sigma^2 \leq 1$ versus $H_1\colon \sigma^2 > 1$. The situation changes for the testing problem $H_0\colon \sigma^2 = 1$ versus $H_1\colon \sigma^2 \neq 1$, since in this case we will always reject H_0. Figure 4.7 illustrates both situations. □

4.3.5 Intrinsic Losses

Here we briefly discuss two more loss functions, which measure the distance between distributions. They are of interest especially for model choice, when we want to find a distribution which delivers a good fit to the data, but we are less interested in the parameters themselves. We can use the *Kullback–Leibler divergence*

$$L_{\mathrm{KL}}(\theta, d) = \mathsf{K}(\mathsf{P}_\theta|\mathsf{P}_d) = \int_{\mathcal{X}} p(\mathbf{x}|\theta) \ln\left(\frac{p(\mathbf{x}|\theta)}{p(\mathbf{x}|d)}\right) d\mathbf{x} \qquad (4.30)$$

and the *Hellinger distance*

$$L_{\mathrm{H}}(\theta, d) = \mathsf{H}^2(\mathsf{P}_\theta, \mathsf{P}_d) = \frac{1}{2}\int_{\mathcal{X}} \left(\sqrt{\frac{p(\mathbf{x}|d)}{p(\mathbf{x}|\theta)}} - 1\right)^2 p(\mathbf{x}|\theta)\, d\mathbf{x}. \qquad (4.31)$$

Note that the loss $L_{KL}(\theta, d)$ is asymmetric, but $L_H(\theta, d)$ is symmetric. We can reformulate the Hellinger distance as

$$
\begin{aligned}
L_H(\theta, d) &= \frac{1}{2} \int_{\mathcal{X}} \left(\sqrt{\frac{p(\mathbf{x}|d)}{p(\mathbf{x}|\theta)}} - 1 \right)^2 p(\mathbf{x}|\theta) \, d\mathbf{x} \\
&= \frac{1}{2} \int_{\mathcal{X}} \left(\frac{p(\mathbf{x}|d)}{p(\mathbf{x}|\theta)} - 2\sqrt{\frac{p(\mathbf{x}|d)}{p(\mathbf{x}|\theta)}} + 1 \right) p(\mathbf{x}|\theta) \, d\mathbf{x} \\
&= \frac{1}{2} \left(\int_{\mathcal{X}} p(\mathbf{x}|d) \, d\mathbf{x} - 2 \int_{\mathcal{X}} \sqrt{p(\mathbf{x}|\theta)p(\mathbf{x}|d)} \, d\mathbf{x} + 1 \right) \\
&= 1 - \int_{\mathcal{X}} \sqrt{p(\mathbf{x}|\theta)p(\mathbf{x}|d)} \, d\mathbf{x} = 1 - H_{\frac{1}{2}}(P_\theta, P_d).
\end{aligned}
\tag{4.32}
$$

where $H_{\frac{1}{2}}(P_\theta, P_d)$ is the *Hellinger transform*, generally defined as

$$
H_{\frac{1}{2}}(P, Q) = \int_{\mathcal{X}} \sqrt{p(\mathbf{x})q(\mathbf{x})} \, d\mathbf{x}.
\tag{4.33}
$$

We compare the losses under normal distribution.

Example 4.12 (Normal distribution)
Consider the normal distribution $N(\theta, 1)$ with density

$$
\varphi_\theta(x) = \frac{1}{\sqrt{2\pi}} \exp(-\frac{1}{2}(x - \theta)^2).
$$

The likelihood ratio is

$$
\begin{aligned}
\frac{\varphi_\theta(x)}{\varphi_d(x)} &= \exp(-\frac{1}{2}(x - \theta)^2 + \frac{1}{2}(x - d)^2) \\
&= \exp\left((x - \theta)(\theta - d) + \frac{1}{2}(\theta - d)^2 \right)
\end{aligned}
$$

and

$$
\begin{aligned}
L_{KL}(\theta, d) &= E_\theta \left(\ln \left(\frac{\varphi_\theta(x)}{\varphi_d(x)} \right) \right) \\
&= E_\theta \left((x - \theta)(\theta - d) + \frac{1}{2}(\theta - d)^2 \right) \\
&= \frac{1}{2}(\theta - d)^2.
\end{aligned}
\tag{4.34}
$$

Under normal distribution, Kullback–Leibler loss and the quadratic loss are equivalent. Consider the Hellinger loss

$$
L_H(\theta, d) = 1 - \int_{\mathcal{X}} \sqrt{p(x|\theta)p(x|d)} \, dx.
$$

We have

$$\int_{\mathcal{X}} \sqrt{p(x|\theta)p(x|d)}\,dx = \frac{1}{\sqrt{2\pi}} \int_{\mathcal{X}} \exp\left(-\frac{1}{4}((x-\theta)^2 + (x-d)^2)\right)\,dx$$

$$= \frac{1}{\sqrt{2\pi}} \int_{\mathcal{X}} \exp(-\frac{1}{2}(x - \frac{\theta+d}{2})^2 - \frac{1}{8}(\theta-d)^2)\,dx$$

$$= \exp(-\frac{1}{8}(\theta-d)^2).$$

(4.35)

Hence

$$L_{\mathrm{H}}(\theta,d) = 1 - \exp(-\frac{1}{8}(\theta-d)^2).$$

The Hellinger loss is a monotone transformation of the quadratic loss. □

4.4 The Minimax Criterion

In order to show optimality results of a Bayes decision rule in a frequentist context, we need to introduce the minimax criterion.

The first step is to expand the set of possible decisions. We define a **randomized decision rule** $\delta^*(\mathbf{x},.)$ as a distribution over the decision space $\mathcal{D}$. A non-randomized decision rule can be considered as a dirac distribution (one point distribution), $\delta^*(\mathbf{x},a) = 1$, for $\mathbf{x} = a$. Thus we are now searching the "optimal" decision in a bigger class. The loss function of a randomized decision rule is defined as the expected loss

$$L(\theta,\delta^*(\mathbf{x},.)) = \int_{\mathcal{D}} L(\theta,a)\delta^*(\mathbf{x},a)\,da.$$

(4.36)

The risk function, as before, is given by

$$R(\theta,\delta^*) = \mathsf{E}_{\theta} L(\theta,\delta^*(\mathbf{x},.)) = \int_{\mathcal{X}} L(\theta,\delta^*(\mathbf{x},.))p(\mathbf{x}|\theta)\,d\mathbf{x}.$$

(4.37)

A famous example in this context is the Neyman–Pearson test, see for example Liero and Zwanzig (2011, Section 5.3.1).

Example 4.13 (Randomized test) Let $\mathcal{P} = \mathcal{P}_0 \cup \mathcal{P}_1$. We are interested in the testing problem: H_0: $\mathcal{P}_0$ versus H_1: $\mathcal{P}_1$. The decision space is $\mathcal{D} = \{0,1\}$. A randomized decision rule is a distribution over $\mathcal{D} = \{0,1\}$, so that

$$\delta^*(\mathbf{x},.) = \mathsf{Ber}(\varphi(\mathbf{x})).$$

In this case the loss is

$$L(\theta,\delta^*(\mathbf{x},.)) = L(\theta,1)\varphi(\mathbf{x}) + L(\theta,0)(1-\varphi(\mathbf{x})).$$

For the risk we obtain

$$R(\theta,\delta^*) = L(\theta,1)\mathsf{E}_{\theta}\varphi(\mathbf{x}) + L(\theta,0)(1-\mathsf{E}_{\theta}\varphi(\mathbf{x})).$$

□

Denote the class of randomized decisions by $\mathfrak{D}^*$. We apply the minimax principle in this larger class $\mathfrak{D}^*$, and search for the best decision in a bad situation.

Definition 4.3 (Minimax) The **minimax risk** $\overline{R}$ associated with a loss function L is the value

$$\overline{R} = \inf_{\delta \in \mathfrak{D}^*} \sup_{\theta \in \Theta} R(\theta, \delta) = \inf_{\delta \in \mathfrak{D}^*} \sup_{\theta \in \Theta} \mathsf{E}_\theta L(\theta, \delta(\mathbf{x}, .)). \qquad (4.38)$$

The **minimax decision rule** is any rule $\delta_0 \in \mathfrak{D}^*$ such that

$$\sup_{\theta \in \Theta} R(\theta, \delta_0) = \overline{R}. \qquad (4.39)$$

Example 4.14 (Lion's appetite)
Continuing with Examples 4.1 and 4.2, a randomized decision is a three point distribution over $\mathcal{D} = \{d_1, d_2, d_3\}$

$$\delta^*(x, .) = \frac{\begin{array}{c|c|c} d_1 & d_2 & d_3 \\ \hline p_1(x) & p_2(x) & p_3(x) \end{array}}{}, \quad p_3(x) = p_1(x) + p_2(x).$$

The non–randomized decision rule δ_1 in Example 4.2 corresponds to

$$\delta_1^*(1, .) = \frac{\begin{array}{c|c|c} d_1 & d_2 & d_3 \\ \hline 0 & 0 & 1 \end{array}}{} \quad \dots \quad \delta_1^*(4, .) = \frac{\begin{array}{c|c|c} d_1 & d_2 & d_3 \\ \hline 1 & 0 & 0 \end{array}}{}$$

The loss of a randomized decision rule is calculated as

$$L(\theta, \delta^*(x, .)) = p_1(x)L(\theta, d_1) + p_2(x)L(\theta, d_2) + p_3(x)L(\theta, d_3),$$

where $L(\theta, d)$ is defined in Example 4.1. We obtain the risk

$$R(\theta, \delta^*) = \mathsf{E}_\theta p_1(x)L(\theta, d_1) + \mathsf{E}_\theta p_2(x)L(\theta, d_2) + \mathsf{E}_\theta p_3(x)L(\theta, d_3).$$

A randomized decision rule $\delta^*(x, .)$ is defined by

x	0	1	2	3	4
$p_1(x)$	0	0.2	0.5	0.8	0.9
$p_2(x)$	0.2	0.3	0.4	0.2	0.1
$p_3(x)$	0.8	0.5	0.1	0	0

Using the model $\mathcal{P}$ given in Example 2.4, we obtain for $R(\theta, \delta^*)$,

θ_1	θ_2	θ_3
0.194	0.263	0.032

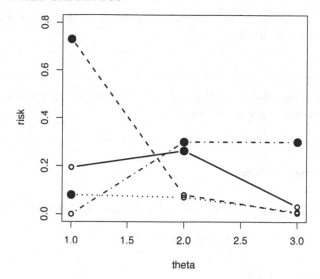

Figure 4.8: Example 4.14. We compare only four rules. $\delta_1, \delta_2, \delta_3$ are the non-randomized rules in Example 4.2. The randomized rule δ^* given in Example 4.14 is plotted as line. The comparison criterion is the maximal risk highlighted by thick points. The decision minimizing the maximal risk is δ_2.

In Figure 4.8 the risk functions are for $\delta_1, \delta_2, \delta_3$ given in Example 4.2 and for δ^*. Comparing these four rules, the minimax rule is δ_2. □

Next we give two statements, which provide an insight into the relationship between the minimax concept and admissibility. As shown in Figure 4.8 the minimax rule is also admissible. Theorem 4.6 gives the corresponding result.

Theorem 4.6
If there exists a unique minimax rule, then it is admissible.

PROOF: Let δ^* be an unique minimax rule. We carry out the proof by contradiction. So let δ^* be not admissible. Then there exists a rule δ_1 such that

$$R(\theta, \delta_1) \leq R(\theta, \delta^*), \quad \text{for all } \theta \in \Theta$$

and there exists a parameter $\theta_0 \in \Theta$ such that

$$R(\theta_0, \widetilde{\delta_1}) < R(\theta_0, \delta^*).$$

This implies

$$\sup_{\theta \in \Theta} R(\theta, \delta_1) \leq \sup_{\theta \in \Theta} R(\theta, \delta^*) = \inf_{\delta \in \mathfrak{D}^*} \sup_{\theta \in \Theta} R(\theta, \delta)$$

and that δ_1 is also minimax, which is a contradiction to the uniqueness of the minimax rule δ^*.

□

The following result is on the opposite direction of the relation between admissibility and minimaxity.

Theorem 4.7
Let δ_0 be an admissible rule. If for all $\theta \in \Theta$ the loss function $L(\theta, .)$ is strictly convex and if for all $\theta \in \Theta$

$$R(\theta, \delta_0) = const,$$

then δ_0 is unique minimax.

PROOF: Let δ_0 be an admissible rule with constant risk, then for an arbitrary $\theta_0 \in \Theta$

$$\sup_{\theta \in \Theta} R(\theta, \delta_0) = R(\theta_0, \delta_0)$$

and there exists no $\tilde{\delta}$ with

$$R(\theta_1, \tilde{\delta}) < R(\theta_1, \delta_0) - R(\theta_0, \delta_0), \text{ for some } \theta_1 \subset \Theta.$$

We carry out the proof by contradiction and state that δ_0 is not unique minimax. Then there exists δ_1 such that

$$\sup_{\theta \in \Theta} R(\theta, \delta_1) = \sup_{\theta \in \Theta} R(\theta, \delta_0) = R(\theta_0, \delta_0).$$

Thus

$$R(\theta_0, \delta_1) \leq \sup_{\theta \in \Theta} R(\theta, \delta_1) = R(\theta_0, \delta_0). \tag{4.40}$$

Let us consider the case

$$R(\theta_0, \delta_1) = R(\theta_0, \delta_0). \tag{4.41}$$

We define a new decision rule

$$\delta_2(\mathbf{x}) = \frac{\delta_1(\mathbf{x}) + \delta_0(\mathbf{x})}{2}.$$

Because of the strong convexity of the loss function we obtain

$$
\begin{aligned}
0 \leq R(\theta_0, \delta_2) &= \int_{\mathcal{X}} L(\theta_0, \delta_2(\mathbf{x})) p(\mathbf{x}|\theta_0) \, d\mathbf{x} \\
&< \int_{\mathcal{X}} \left(\frac{1}{2} L(\theta_0, \delta_1(\mathbf{x})) + \frac{1}{2} L(\theta_0, \delta_0(\mathbf{x})) \right) p(\mathbf{x}|\theta_0) \, d\mathbf{x} \\
&= \frac{1}{2} R(\theta_0, \delta_1) + \frac{1}{2} R(\theta_0, \delta_0) \\
&= R(\theta_0, \delta_0).
\end{aligned}
$$

Thus

$$R(\theta_0, \delta_2) < R(\theta_0, \delta_0),$$

which is a contradiction to the admissibility of δ_0. We can exclude the case
(4.41) and the inequality in (4.40) is strong, which is a contradiction to the
admissibility of δ_0.

$\square$

4.5 Bridges

In this section we show optimality results for Bayesian procedures in the frequentist context. In a sense it is a bridge between the statistician who prefers
the Bayes model and the one who does not like the additional assumption of
a prior distribution. But it is not only some kind of justification of the Bayes
model, rather the bridge results deliver a method to prove optimality results
for a "frequentist" method when it can be rewritten as a Bayes rule.
We start with the statement that Bayes rules are not randomized.

Theorem 4.8 *Assume a Bayes model* $\{\mathcal{P}, \pi\}$. *Then it holds that*

$$\inf_{\delta \in \mathcal{D}} r(\pi, \delta) = \inf_{\delta^* \in \mathcal{D}^*} r(\pi, \delta^*) = r(\pi).$$

PROOF: Consider the posterior risk. For all $\mathbf{x} \in \mathcal{X}$ and all $\delta^* \in \mathcal{D}^*$ we have

$$\rho(\pi, \delta^*(\mathbf{x}, .)|\mathbf{x}) = \int_{\Theta} \int_{\mathcal{D}} L(\theta, a) \delta^*(\mathbf{x}, a) \, da \, \pi(\theta|\mathbf{x}) \, d\theta$$

$$= \int_{\mathcal{D}} \int_{\Theta} L(\theta, a) \pi(\theta|\mathbf{x}) \, d\theta \, \delta^*(\mathbf{x}, a) \, da$$

$$= \int_{\mathcal{D}} \rho(\pi, a|\mathbf{x}) \delta^*(\mathbf{x}, a) \, da$$

$$\geq \inf_{a \in \mathcal{D}} \rho(\pi, a|\mathbf{x}) \int_{\mathcal{D}} \delta^*(\mathbf{x}, a) \, da = \inf_{a \in \mathcal{D}} \rho(\pi, a|\mathbf{x}).$$

Because $r(\pi, \delta^*) = \int_{\mathcal{X}} \rho(\pi, \delta^*(\mathbf{x}, .)|\mathbf{x}) p(\mathbf{x}) d\mathbf{x}$, we further have

$$r(\pi, \delta^*) \geq \inf_{\delta \in \mathcal{D}} r(\pi, \delta)$$

Otherwise $\mathcal{D}^*$ is the larger class, so that

$$\inf_{\delta \in \mathcal{D}} r(\pi, \delta) \geq \inf_{\delta^* \in \mathcal{D}^*} r(\pi, \delta^*).$$

$\square$

Here we have a bridge statement, which says that a Bayes rule can be admissible. The Bayes rule δ^π is optimal within the framework of the Bayes decision
theory, whereas admissibility is an optimality property which does not require
a Bayes model.

Theorem 4.9 *Assume a Bayes model $\{\mathcal{P}, \pi\}$. If*

$$\pi(\theta) > 0, \quad \text{for all } \theta \in \Theta \quad \text{and} \quad r(\pi) = \inf_{\delta \in \mathcal{D}} r(\pi, \delta) < \infty$$

and if

$$R(\theta, \delta) \quad \text{is continuous in } \theta \text{ for all } \delta,$$

then δ^{π} is admissible.

PROOF: The proof is by contradiction. Suppose that δ^{π} is inadmissible. Then there exists a rule δ_1 such that

$$R(\theta, \delta_1) \leq R(\theta, \delta^{\pi}), \quad \text{for all } \theta \in \Theta$$

and

$$R(\theta_1, \delta_1) < R(\theta_1, \delta^{\pi}), \quad \text{for some } \theta_1 \in \Theta.$$

Because $R(., \delta)$ is continuous there exists a neighborhood C, $\theta_1 \in C$ and $\pi(C) > 0$ such that

$$R(\theta, \delta_1) < R(\theta, \delta^{\pi}) \text{ for all } \theta \in C \subset \Theta$$

and

$$\int_C R(\theta, \delta_1) \pi(\theta) \, d\theta < \int_C R(\theta, \delta^{\pi}) \pi(\theta) \, d\theta$$

and for $\overline{C} = \Theta \setminus C$

$$\int_{\overline{C}} R(\theta, \delta_1) \pi(\theta) \, d\theta \leq \int_{\overline{C}} R(\theta, \delta^{\pi}) \pi(\theta) \, d\theta.$$

Hence

$$\begin{aligned}
r(\pi, \delta_1) &= \int_C R(\theta, \delta_1) \pi(\theta) \, d\theta + \int_{\overline{C}} R(\theta, \delta_1) \pi(\theta) \, d\theta \\
&< \int_C R(\theta, \delta^{\pi}) \pi(\theta) \, d\theta + \int_{\overline{C}} R(\theta, \delta^{\pi}) \pi(\theta) \, d\theta \\
&= r(\pi, \delta^{\pi}) = \inf_{\delta \in \mathcal{D}} r(\pi, \delta) < \infty.
\end{aligned}$$

This is a contradiction to the statement that δ^{π} is a Bayes rule.

$\square$

Let us apply this result to the binomial model.

Example 4.15 (Binomial distribution)
We consider $X|\theta \sim \text{Bin}(n, \theta)$ and $\theta \sim \text{Beta}(\alpha, \beta)$. We are interested in Bayes estimation of θ under quadratic loss and apply the results of Example 4.9. The Bayes estimate is

$$\delta^{\pi}(x) = \frac{\alpha + x}{\alpha + \beta + n}. \tag{4.42}$$

It follows that

$$E_\theta \delta^\pi(X) = \frac{\alpha + E_\theta X}{\alpha + \beta + n} = \frac{\alpha + n\theta}{\alpha + \beta + n} \tag{4.43}$$

and

$$\text{Var}_\theta \delta^\pi(X) = \frac{\text{Var}_\theta X}{(\alpha + \beta + n)^2} = \frac{n\theta(1 - \theta)}{(\alpha + \beta + n)^2}. \tag{4.44}$$

We calculate the risk

$$
\begin{aligned}
R(\theta, \delta^\pi) &= \sum_{k=0}^{n} P_\theta(k)(\theta - \delta^\pi(k))^2 \\
&= \text{Var}_\theta \delta^\pi(X) + (\theta - E_\theta \delta^\pi(X))^2 \\
&= \frac{1}{(\alpha + \beta + n)^2} \left(\alpha^2 + \theta(n - 2\alpha^2 - 2\alpha\beta) + \theta^2((\alpha + \beta)^2 - n) \right).
\end{aligned}
\tag{4.45}
$$

The risk $R(\theta, \delta^\pi)$ is a continuous function of θ. Further, the beta distribution is positive over $(0, 1)$. Thus we can apply Theorem 4.9. The Bayes estimator in (4.42) is admissible. Same holds for $\delta^{\pi_{\text{Lap}}}(x) = \frac{1+x}{2+n}$ and $\delta^{\pi_{\text{Jeff}}}(x) = \frac{0.5+x}{1+n}$. $\square$

In the following theorem, instead of continuity we require that the loss is strictly convex and that the Bayes risk is bounded. Then we can go again over the bridge from the Bayes side to the frequentist side.

Theorem 4.10 *Assume a Bayes model $\{\mathcal{P}, \pi\}$, where π can be improper. If for all $\theta \in \Theta$ the loss function $L(\theta, .)$ is strictly convex, and if δ^π is a Bayes rule with finite Bayes risk*

$$r(\pi) = \inf_{\delta \in \mathcal{D}} r(\pi, \delta) < \infty,$$

then δ^π is admissible.

PROOF: Let δ^π be inadmissible. There exists a rule δ_1 such that

$$R(\theta, \delta_1) \leq R(\theta, \delta^\pi), \text{ for all } \theta \in \Theta$$

and

$$R(\theta_1, \delta_1) < R(\theta_1, \delta^\pi), \text{ for some } \theta_1 \in \Theta.$$

We define a new decision rule

$$\delta_2(\mathbf{x}) = \frac{\delta_1(\mathbf{x}) + \delta^\pi(\mathbf{x})}{2}.$$

Because of the strong convexity of the loss function we obtain

$$
\begin{aligned}
R(\theta, \delta_2) &= \int_{\mathcal{X}} L(\theta, \delta_2(\mathbf{x})) p(\mathbf{x}|\theta)\, d\mathbf{x} \\
&< \int_{\mathcal{X}} \left(\frac{1}{2} L(\theta, \delta_1(\mathbf{x})) + \frac{1}{2} L(\theta, \delta^\pi(\mathbf{x})) \right) p(\mathbf{x}|\theta)\, d\mathbf{x} \\
&= \frac{1}{2} R(\theta, \delta_1) + \frac{1}{2} R(\theta, \delta^\pi) \\
&\leq R(\theta, \delta^\pi).
\end{aligned}
$$

Thus

$$
R(\theta, \delta_2) < R(\theta, \delta^\pi) \quad \text{for all } \theta \in \Theta.
$$

For $r(\pi, \delta) = \int R(\theta, \delta)\, \pi(\theta)\, d\theta$ we obtain

$$
r(\pi, \delta_2) < r(\pi, \delta^\pi) = \inf_{\delta \in \mathcal{D}} r(\pi, \delta) < \infty,
$$

which is a contradiction!

$\square$

We apply this theorem in the following example on testing hypotheses.

Example 4.16 (P-value)
Let $X \sim N(\mu, 1)$ and the Laplace prior $\pi(\mu) \propto \text{const}$. Then $\mu|x \sim N(x, 1)$. We consider the testing problem: $H_0 : \mu \leq 0$ versus $H_1 : \mu > 0$. The parameter of interest is defined as

$$
\theta = \begin{cases} 1 & \text{for } \mu \leq 0 \\ 0 & \text{for } \mu > 0 \end{cases}.
$$

Consider a quadratic loss. The Bayes estimate of θ is $p(x) = E(\theta|x)$, where

$$
p(x) = P^\pi(\theta = 1|x) = P^\pi(\mu \leq 0|x) = P^\pi(\mu - x \leq -x|x) = 1 - \Phi(x),
$$

where Φ is the distribution function of $N(0, 1)$. Applying Theorem 4.10 we note that $p(x)$ is admissible. Further, $p(x)$ is the p-value of the test problem above.

$\square$

The p-value example is a good example of using the Bayesian interpretation of a well-known frequentist estimate for showing its admissibility. Now we discuss an example, where it is not possible to apply Theorem 4.10.

Example 4.17 (Normal distribution and Laplace prior)
We suppose $X \sim N(\theta, 1)$ and the Laplace prior $\pi(\theta) \propto \text{const}$. Then $\theta|x \sim N(x, 1)$. Theorem 4.2 states that the Bayes estimate with respect to quadratic loss is $\delta^\pi = E(\theta|x) = x$. We have $R(\theta, \delta^\pi) = \text{Var}(X) = 1$. The Bayes risk is

$$
r(\pi) = \int_{-\infty}^{\infty} R(\theta, \delta^\pi)\, d\theta = \int_{-\infty}^{\infty} 1\, d\theta = \infty,
$$

so that Theorem 4.10 is not applicable.

$\square$

Here is one more bridge, to obtain admissibility of the Bayes rule.

Theorem 4.11   *Assume a Bayes model $\{\mathcal{P}, \pi\}$. If the Bayes rule δ^π is unique, then δ^π is admissible.*

PROOF: Suppose that δ^π is inadmissible. Then there exists a rule δ_1 such that

$$R(\theta, \delta_1) \leq R(\theta, \delta^\pi), \text{ for all } \theta \in \Theta$$

and

$$R(\theta_1, \delta_1) < R(\theta_1, \delta^\pi), \text{ for some } \theta_1 \in \Theta.$$

This implies

$$
\begin{aligned}
r(\pi, \delta_1) &\leq \int_\Theta R(\theta, \delta_1)\pi(\theta)\, d\theta \\
&\leq \int_\Theta R(\theta, \delta^\pi)\pi(\theta)\, d\theta \\
&= r(\pi),
\end{aligned}
$$

i.e., δ_1 is also a Bayes rule, which is a contradiction to the uniqueness of δ^π.

$\square$

In the following, we show a bridge between minimaxity and Bayes optimality.

Theorem 4.12
Assume

$$\inf_{\delta \in \mathcal{D}} \sup_\pi r(\pi, \delta) = \sup_\pi \inf_{\delta \in \mathcal{D}} r(\pi, \delta). \tag{4.46}$$

Then a minimax estimator δ_0 is a Bayes estimator associated with π_0, where

$$\sup_\pi r(\pi) = r(\pi_0). \tag{4.47}$$

The prior π_0 in (4.47) is called **least favourable prior**. The statement above means for a minimax estimator δ_0 that

$$r(\pi_0, \delta) \leq r(\pi_0, \delta_0) = r(\pi_0) \text{ for all } \delta \in \mathcal{D}. \tag{4.48}$$

PROOF: We begin by showing the following general result for $f(x) \geq 0$

$$\sup_x f(x) = \sup_{\{\pi : \int \pi(x)dx=1, \pi(x)\geq 0\}} \int f(x)\pi(x)\, dx. \tag{4.49}$$

We have, for all π, $\int \pi(x)dx = 1$ and $f(x) \geq 0$, so that

$$\int f(x)\pi(x)\, dx \leq \sup_x f(x) \int \pi(x)\, dx = \sup_x f(x).$$

Take a sequence $\{x_n\}$ with $\lim_{n \to \infty} f(x_n) = \sup_x f(x)$ and define a sequence of Dirac measures π_n, $\pi_n(x) = 1$ for $x = x_n$, otherwise $\pi_n(x) = 0$. Then

$$\sup_{\pi} \int f(x)\pi(x)\, dx \geq \int f(x)\pi_n(x) dx = f(x_n)$$

Taking the limit, we obtain the inequality in the other direction

$$\sup_{\pi} \int f(x)\pi(x)\, dx \geq \lim_{n \to \infty} f(x_n) = \sup_x f(x)$$

and (4.49) follows. Now we prove (4.48). It suffices to show that $r(\pi_0, \delta_0) \leq r(\pi_0)$. We have

$$
\begin{aligned}
r(\pi_0, \delta_0) &= \int R(\theta, \delta_0)\pi_0(\theta)\, d\theta \quad \text{(definition of integrated risk)} \\
&\leq \sup_{\theta} R(\theta, \delta_0) \quad \text{(because of (4.49))} \\
&= \inf_{\delta \in \mathcal{D}^*} \sup_{\theta} R(\theta, \delta) \quad (\delta_0 \text{ is minimax}) \\
&= \inf_{\delta \in \mathcal{D}^*} \sup_{\pi} \int R(\theta, \delta)\pi(\theta)\, d\theta \quad \text{(because of (4.49))} \\
&\leq \inf_{\delta \in \mathcal{D}} \sup_{\pi} \int R(\theta, \delta)\pi(\theta)\, d\theta \quad \text{(because of } \mathcal{D} \subset \mathcal{D}^*) \\
&= \inf_{\delta \in \mathcal{D}} \sup_{\pi} r(\pi, \delta) \quad \text{(definition of integrated risk)} \\
&= \sup_{\pi} \inf_{\delta \in \mathcal{D}} r(\pi, \delta) \quad \text{(assumption (4.46))} \\
&= \sup_{\pi} r(\pi).
\end{aligned}
$$

$\square$

We use the toy example of lion's appetite to illustrate the notion of a least favourable prior.

Example 4.18 (Lion's appetite)
Recall Example 4.1. We consider only two stages of lions's appetite θ_0, θ_1, where θ_0 means not hungry and θ_1 means hungry. We set a new and simplified model, instead of that in Example 2.4, by probabilities $P_\theta(x)$ given in the following table:

x	0	1	2	3	4	
θ_0	0.4	0.4	0.1	0.1	0	(4.50)
θ_1	0	0.05	0.05	0.8	0.1	

The prior is a two point distribution over $\{\theta_0, \theta_1\}$. The prior probability, that the lion was hungry, is p. For $x \in \{0, 1, 2, 3, 4\}$, the posterior is calculated as

$$\pi(\theta_0|x) = \frac{k_0(x)}{k_0(x) + k_1(x)}, \quad \pi(\theta_1|x) = 1 - \pi(\theta_0|x)$$

and

$$k_0(x) = (1-p)P_{\theta_0}(x), \quad k_1(x) = pP_{\theta_1}(x).$$

We suppose an asymmetric loss, because it would be much more dangerous to classify a hungry lion as no hungry as compared to the other way around.

$L(\theta, d)$	$d = \theta_0$	$d = \theta_1$
θ_0	0	0.2
θ_1	0.8	0

(4.51)

The posterior loss is defined by

$$\rho(\pi, d|x) = L(\theta_0, d)\pi(\theta_0|x) + L(\theta_1, d)\pi(\theta_1|x).$$

We obtain

$$\rho(\pi, \theta_0|x) = 0.2\,\pi(\theta_1|x) \text{ and } \rho(\pi, \theta_1|x) = 0.8\,\pi(\theta_0|x),$$

and the Bayes decision is given by

$$\delta^\pi(x) = \begin{cases} \theta_0 & \text{if } 0.2\,\pi(\theta_1|x) < 0.8\pi(\theta_0|x) \\ \theta_1 & \text{otherwise} \end{cases} = \begin{cases} \theta_0 & \text{if } p < g(x) \\ \theta_1 & \text{otherwise} \end{cases},$$

(4.52)

where the threshold is

$$g(x) = \frac{0.2\,P_{\theta_0}(x)}{0.8\,P_{\theta_1}(x) + 0.2\,P_{\theta_0}(x)}.$$

Using (4.50) and (4.51) we obtain the following rounded numbers

x	0	1	2	3	4
$g(x)$	1	0.67	0.32	0.2	0

(4.53)

This implies that we have five different Bayes rules δ_j, $j = 1, \ldots, 5$. Set 1 for decision θ_1 and 0 for decision θ_0. Then

			x	0	1	2	3	4
	p	$=$	0	$\delta_1(x)$	0	0	0	0	0
0	$<$ p	$\le$	0.2	$\delta_2(x)$	0	0	0	0	1
0.2	$<$ p	$\le$	0.32	$\delta_3(x)$	0	0	0	1	1
0.32	$<$ p	$\le$	0.67	$\delta_4(x)$	0	0	1	1	1
0.67	$<$ p	$\le$	1	$\delta_5(x)$	0	1	1	1	1

(4.54)

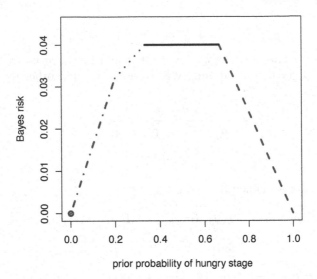

Figure 4.9: Example 4.18. The least favourable prior distributions have p between 0.32 and 0.67.

Using the loss in (4.51) we calculate

$$R(\theta, \delta_j) = \sum_{i=1}^{5} L(\theta, \delta_j(x_i)) \mathsf{P}_\theta(x_i)$$

and

$$r(p) = R(\theta_1, \delta_j)p + R(\theta_0, \delta_j)(1 - p), \text{ for } p \in [g(x_j), g(x_{j-1})] \; j = 1, \dots, 5.$$

The Bayes risk is piecewise linear. The supremum is attained for all $p \in [0.32, 0.67]$. The Bayes risk is plotted in Figure 4.9. For each given prior distribution the Bayes rule is unique. By Theorem 4.11, the Bayes rules in (4.54) are admissible. $\qquad \square$

We now cross the bridge in the other direction. The next theorem gives a sufficient condition for a Bayes rule to be minimax. Note that, the condition is only sufficient!

Theorem 4.13   *Assume a Bayes model* $\{\mathcal{P}, \pi\}$. *If the Bayes rule* δ^π *has constant risk*

$$R(\theta, \delta^\pi) = const, \qquad\qquad (4.55)$$

then δ^π *is minimax.*

PROOF: Suppose that δ^π is not minimax. Then there exists a decision rule δ_1 such that

$$\sup_\theta R(\theta, \delta_1) < \sup_\theta R(\theta, \delta^\pi)$$
$$= R(\theta, \delta^\pi) = \text{const}$$
$$= \int_\Theta R(\theta, \delta^\pi) \pi(\theta) \, d\theta$$
$$= r(\pi, \delta^\pi).$$

Otherwise

$$\sup_\theta R(\theta, \delta_1) \geq \int_\Theta R(\theta, \delta_1) \pi(\theta) \, d\theta = r(\pi, \delta_1).$$

This implies

$$r(\pi, \delta_1) < r(\pi, \delta^\pi)$$

which is a contradiction.

$\square$

Example 4.19 (Lion's appetite)

We continue with Example 4.18. Since the risk is constant for δ_4, we have

$$R(\theta_1, \delta_4) = R(\theta_0, \delta_4) = 0.04$$

Further $r(\pi_0) = 0.04 = \sup_p r(p)$, see Figure 4.9. Hence the rule δ_4 is our favorite,

x	0	1	2	3	4
$\delta_4(x)$	0	0	1	1	1

(4.56)

It is admissible and minimax and associated with a least favourable prior. We are on the safe side and decide that if the lion has eaten more than 1 person the lion was hungry; no surprise! $\square$

Let us now consider a more formal example.

Example 4.20 (Binomial distribution)

We continue with Example 4.15, where $X|\theta \sim \text{Bin}(n, \theta)$, and we consider only the family of conjugate beta priors, $\theta \sim \text{Beta}(\alpha, \beta)$. Then the posterior is also a beta distribution. Under quadratic loss, the Bayes estimator is the expectation of the posterior distribution, given in (4.42). Further the risk function is calculated in (4.45) as

$$R(\theta, \delta^\pi) = \frac{1}{(\alpha + \beta + n)^2} \left(\alpha^2 + \theta(n - 2\alpha^2 - 2\alpha\beta) + \theta^2((\alpha + \beta)^2 - n) \right).$$

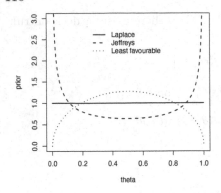
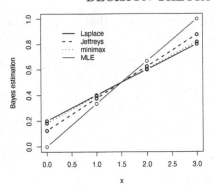

Figure 4.10: Example 4.20. $X \sim \text{Bin}(3, \theta)$. Left: Prior distributions, Laplace: Beta$(1, 1)$, Jeffreys: Beta$(0.5, 0.5)$, least favourable: Beta$(\frac{\sqrt{3}}{2}, \frac{\sqrt{3}}{2})$ are plotted. Right: The Bayes estimators associated with the priors are plotted. For comparison also the maximum likelihood estimator is included. The Bayes estimators overestimate small values and underestimate large values of θ.

Plugging $\alpha = \beta = \frac{\sqrt{n}}{2}$ in (4.45) we get

$$n - 2\alpha^2 - 2\alpha\beta = n - \frac{2n}{4} - \frac{2n}{4} = 0$$

and

$$(\alpha + \beta)^2 - n = (\sqrt{n})^2 - n = 0$$

so that the risk is constant. The least favourable prior belongs to the conjugate family; it is the prior Beta$(\frac{\sqrt{n}}{2}, \frac{\sqrt{n}}{2})$. Applying Theorem 4.13 the estimator

$$\delta(x) = \frac{\frac{\sqrt{n}}{2} + x}{n + \sqrt{n}} \qquad (4.57)$$

is minimax. In Figure 4.10 different priors and Bayes estimates are plotted and compared.   □

4.6   List of Problems

1. Consider a query for analyzing customer satisfaction. The parameter has two levels $\theta \in \{0, 1\}$ where 1 means the customer is satisfied. The costumer can choose between bad, acceptable, and good, coded as $-1, 0, 1$, respectively. The goal is to estimate θ under 0–1 loss. The probabilities

$P_\theta(x) = P(X = x|\theta)$ are given in the following table:

x	-1	0	1
$\theta = 0$	0.5	0.3	0.2
$\theta = 1$	0	0.5	0.5

Assume that the prior probability of $\theta = 1$ is p.

(a) Determine the posterior distribution.

(b) Determine the Bayes estimator of θ.

(c) Calculate the frequentist risk.

(d) Calculate the Bayes risk.

(e) Determine the least favorable prior.

2. Consider the binomial model:

$$X \sim \text{Bin}(n, \theta) \quad \text{and} \quad \theta \sim \text{Beta}(\alpha, \beta)$$

and the asymmetric absolute error loss:

$$L(\theta, d) = \begin{cases} k_2(\theta - d) & \text{if } \theta > d \\ k_1(d - \theta) & \text{if } \theta \leq d \end{cases} \tag{4.58}$$

(a) Derive the Bayes estimator associated with the loss in (4.58).

(b) Set $k_1 = k_2$. Give an approximate expression for the Bayes estimator.

(c) Discuss possible applications of this asymmetric approach. Give an example.

(d) Is it possible to find a prior inside the class of beta distributions $\{\text{Beta}(\alpha, \beta) : \alpha > 1, \beta > 1\}$ such that the Bayes estimator has a constant risk? Why is a constant risk of interest?

3. Consider $X \sim \text{Gamma}(\alpha, \beta)$, where $\alpha > 2$ is known. The parameter of interest is $\theta = \beta$.

(a) Derive Jeffreys prior and the related posterior.

(b) Derive the Bayes estimator $\delta^\pi(x)$ under the L_2 loss function.

(c) Calculate the frequentist risk, the posterior expected loss and the Bayes risk of $\delta^\pi(x)$.

(d) Show that $\delta^\pi(x)$ is admissible.

4. Consider $X \sim \text{Gamma}(\alpha, \beta)$, where $\alpha > 2$ is known. The parameter of interest is $\theta = \frac{1}{\beta}$.

(a) Does $\text{Gamma}(\alpha, \beta)$ belong to an exponential family? Derive the natural parameter and the related sufficient statistics.

(b) Derive the efficient estimator δ^*.

(c) Derive the conjugate prior for θ. Calculate the posterior.

(d) Derive the Bayes estimator $\delta^{\pi}(x)$ under the L_2 loss function.

(e) Calculate the frequentist risk of δ^{π} and of δ^*, the posterior expected loss of δ^{π} and the Bayes risk of δ^{π}.

(f) Is δ^{π} admissible?

(g) Compare the frequentist risk of δ^{π} and of δ^*.

5. Consider $X \sim \text{Mult}(n, \theta_1, \theta_2, \theta_3)$ and

$$\theta = (\theta_1, \theta_2, \theta_3) \in \Theta = \{(\vartheta_1, \vartheta_2, \vartheta_3) : \vartheta_i \in [0,1], \sum_{i=1}^{n} \vartheta_i = 1\}$$

with $\theta \sim \text{Dir}(\alpha_1, \alpha_2, \alpha_3)$. Assume the loss $L(\theta, d) = (\theta - d)^T (\theta - d)$.

(a) Derive the posterior distribution.

(b) Derive the Bayes estimator $\delta^{\pi}(x)$ under the L_2 loss function.

(c) Calculate the frequentist risk of the Bayes estimator.

(d) Show that the Bayes risk is finite.

(e) Determine the least favourable prior.

(f) Calculate the minimax estimator for θ.

Chapter 5

Asymptotic Theory

In this chapter we define and explain the meaning of asymptotic consistency in a Bayes model. The reader not interested in asymptotic theory can jump over this chapter. But we would like to give two important messages on the way. If Bayesian methods should work in the long run then the least we need is that

- the prior does not exclude any data generating parameter, and
- the parameters in $\mathcal{P}$ are identifiable.

The main result we present is Schwartz' Theorem. It is a generalization of Doob's Theorem, which is based on measure theory especially on the martingale method. Schwartz' Theorem has an information–theoretic background. We give the main steps of the proof because of its link to the reference priors. In this textbook we want to avoid measure theory as much as possible, which makes the presentation of results in this chapter essentially complicated. We shall, therefore, try to be as precise as possible. For further reading we recommend Liese and Miescke (2008, Chapter 7, Section 7.5.2) and the original papers Schwartz (1965) and Diaconis and Freedman (1986).

5.1 Consistency

Recall from Chapter 2 that, in a Bayes model $\{\mathcal{P}, \pi\}$, the notion of a true parameter makes no sense. The parameter is a random variable with known distribution π. But given the data $\mathbf{x}$ we can ask for the data generating distribution. Like in decision theory we cross the bridge and consider a Bayes procedure inside the model $\mathcal{P} = \{P_\theta : \theta \in \Theta\}$ and study its consistency.

Having in mind the Bayesian inference principle, that all inference on θ is based on the posterior distribution, we define consistency with respect to the posterior distribution. Note that, we use the capital letter $\mathbf{X}_{(n)}$ for the random variable and $\mathbf{x}_{(n)}$ for its realization, we set $\mathsf{P}^\pi(.)$ for the prior distribution with density $\pi(.)$ and $\mathsf{P}_n^\pi(.|\mathbf{x}_{(n)})$ for the posterior distribution with density $\pi_n(.|\mathbf{x}_{(n)})$.

DOI: 10.1201/9781003221623-5

Definition 5.1 (Strong Consistency) Assume a Bayes model $\{\mathcal{P}^{(n)}, \pi\}$. Assume that $\mathbf{X}_{(n)} \sim \mathsf{P}_{\theta_0}^{(n)}$. Then a sequence of posteriors $\pi_n(\theta|\mathbf{X}_{(n)})$ is called **strongly consistent** at θ_0 iff for every open subset $O \subset \Theta$ with $\theta_0 \in O$ it holds that

$$\mathsf{P}_n^\pi(O|\mathbf{X}_{(n)}) \to 1, \quad \text{as } n \to \infty \text{ with probability 1.} \tag{5.1}$$

This definition is general and also includes dependent data. In case of an i.i.d. sample we have $\mathbf{X}_{(n)} = (X_1, \ldots, X_n) \sim \mathsf{P}_{\theta_0}^{\otimes n}$. The posterior distribution contracts to the data generating parameter θ_0 and becomes the Dirac distribution at the point θ_0. We illustrate the definition for the binomial model.

Example 5.1 (Binomial distribution)
We consider the Bayes model $X|\theta \sim \mathsf{Bin}(n, \theta)$ and $\theta \sim \mathsf{Beta}(\alpha, \beta)$, so that $\theta|X \sim \mathsf{Beta}(\alpha + x, \beta + n - x)$, see Example 2.11. The expectation and the variance of the posterior are

$$\mathsf{E}(\theta|X = x) = \frac{\alpha + x}{\alpha + \beta + n} = \frac{\alpha}{\alpha + \beta + n} + \frac{x}{n}\frac{n}{\alpha + \beta + n}$$

$$\mathsf{Var}(\theta|X = x)\frac{(\alpha + x)(\beta + n - x)}{(\alpha + \beta + n)^2(\alpha + \beta + n + 1)}.$$

Since $0 \le x \le n$, we have

$$\mathsf{E}_{\theta_0}(\mathsf{Var}(\theta|X = x)) \le \frac{(\alpha + n)(\beta + n)}{(\alpha + \beta + n)^2(\alpha + \beta + n + 1)} \to 0, \quad \text{for } n \to \infty.$$

Thus the posterior concentrates more and more around

$$\lim_{n \to \infty} \mathsf{E}_{\theta_0}(\mathsf{E}(\theta|X = x)) = \lim_{n \to \infty} \frac{\alpha}{\alpha + \beta + n} + \lim_{n \to \infty} \mathsf{E}_{\theta_0}\left(\frac{x}{n}\right)\frac{n}{\alpha + \beta + n} = \theta_0.$$

$\square$

Definition 5.1 implies the consistency of Bayes estimators in the frequentist model. First we recall the definition of a Bayes estimator, given in Chapter 4. The loss function

$$L : (\theta, d) \in \Theta \times \Theta \to L(\theta, d) \in \mathbb{R}_+,$$

is the penalty for the choice of d, instead of θ. We assume the following contrast condition, that there exists a constant $c_0 > 0$ for all d, such that

$$\|d - \theta_0\| c_0 \le L(\theta_0, d) - L(\theta_0, \theta_0). \tag{5.2}$$

This condition implies that the loss function $L(\theta_0, .)$, as a function of d, has a unique minimum at θ_0. Furthermore we require the condition that there exists a constant K for all $\mathbf{X}_{(n)} \sim P_{\theta_0}^{(n)}$ such that

$$\int_\Theta L^2(\theta, \theta_0)\, \pi_n(\theta|\mathbf{X}_{(n)})\, d\theta \leq K^2 \quad a.s. \tag{5.3}$$

The posterior expected loss is defined as

$$\rho_n(\pi, d|\mathbf{x}_{(n)}) = \int_\Theta L(\theta, d)\pi_n(\theta|\mathbf{x}_{(n)})\, d\theta.$$

In Theorem 4.1 it is shown that the Bayes estimator $\delta_n^\pi(\mathbf{x}_{(n)})$ fulfills the following condition. For every $\mathbf{x}_{(n)} \in \mathcal{X}^{(n)}$, $\delta_n^\pi(\mathbf{x}_{(n)})$ is given by

$$\rho_n(\pi, \delta_n^\pi(\mathbf{x}_{(n)})|\mathbf{x}_{(n)}) = \inf_d \rho_n(\pi, d|\mathbf{x}_{(n)}).$$

Theorem 5.1
Assume the following conditions:
1. *The loss function fulfills the conditions (5.2) and (5.3).*
2. *For all $\varepsilon > 0$ and all open sets $O \subset \Theta$ with $\theta_0 \in O$ it holds for*

$$B_\varepsilon(\theta_0) = \{\theta; \theta \in O, |L(\theta, d) - L(\theta_0, d)| < \varepsilon, \text{ for all } d\}$$

that
$$P^\pi(B_\varepsilon(\theta_0)) > 0. \tag{5.4}$$

3. *Let $\mathbf{X}_{(n)} \sim P_{\theta_0}^{(n)}$ and the sequence of posteriors $\pi_n(\theta|\mathbf{X}_{(n)})$ be strongly consistent at θ_0.*

Then for $n \to \infty$
$$\delta_n^\pi(\mathbf{X}_{(n)}) \to \theta_0 \quad a.s.$$

PROOF: Set $B = B_\varepsilon(\theta_0)$ and $B^c = \Theta \setminus B$. It holds that

$$\rho_n(\pi, \theta_0|\mathbf{x}_{(n)}) \geq \rho_n(\pi, \delta_n^\pi(\mathbf{x}_{(n)})|\mathbf{x}_{(n)}) = \int_\Theta L(\theta, \delta_n^\pi(\mathbf{x}_{(n)}))\pi_n(\theta|\mathbf{x}_{(\mathbf{n})})\, d\theta$$

$$\geq \int_B L(\theta, \delta_n^\pi(\mathbf{x}_{(n)}))\pi_n(\theta|\mathbf{x}_{(\mathbf{n})})\, d\theta \tag{5.5}$$

$$\geq \int_B (L(\theta_0, \delta_n^\pi(\mathbf{x}_{(n)})) - \varepsilon)\pi_n(\theta|\mathbf{x}_{(\mathbf{n})})\, d\theta$$

$$\geq (L(\theta_0, \delta_n^\pi(\mathbf{x}_{(n)})) - \varepsilon)\, P_n^\pi(B|\mathbf{x}_{(\mathbf{n})}).$$

Otherwise we have

$$\int_B L(\theta, \theta_0)\pi_n(\theta|\mathbf{x}_{(\mathbf{n})})\, d\theta \leq (L(\theta_0, \theta_0) + \varepsilon)P_n^\pi(B|\mathbf{x}_{(\mathbf{n})}).$$

and because of (5.3) and the Cauchy–Schwartz inequality,

$$\left(\int_{B^c} L(\theta, \theta_0)\pi_n(\theta|\mathbf{x}_{(n)})\, d\theta \right)^2$$
$$\leq \int L(\theta, \theta_0)^2 \pi_n(\theta|\mathbf{x}_{(n)})\, d\theta \int I_{B^c}(\theta)\pi_n(\theta|\mathbf{x}_{(n)})\, d\theta$$
$$\leq K^2\, \mathsf{P}_n^\pi(B^c|\mathbf{x}_{(n)}).$$

Summarizing, we get

$$\rho_n(\pi, \theta_0|\mathbf{x}_{(\mathbf{n})}) = \int_B L(\theta, \theta_0)\pi_n(\theta|\mathbf{x}_{(\mathbf{n})})\, d\theta + \int_{B^c} L(\theta, \theta_0)\pi_n(\theta|\mathbf{x}_{(\mathbf{n})})\, d\theta \qquad (5.6)$$
$$\leq (L(\theta_0, \theta_0) + \varepsilon)\mathsf{P}_n^\pi(B|\mathbf{x}_{(\mathbf{n})}) + K\, (\mathsf{P}_n^\pi(B^c|\mathbf{x}_{(\mathbf{n})}))^{\frac{1}{2}}.$$

From (5.5) and (5.6) it follows that

$$(L(\theta_0, \delta^\pi(\mathbf{x}_{(n)})) - \varepsilon)\mathsf{P}_n^\pi(B|\mathbf{x}_{(\mathbf{n})})$$
$$\leq (L(\theta_0, \theta_0) + \varepsilon)\mathsf{P}_n^\pi(B|\mathbf{x}_{(\mathbf{n})}) + K\, (\mathsf{P}_n^\pi(B^c|\mathbf{x}_{(\mathbf{n})}))^{\frac{1}{2}}$$

and

$$L(\theta_0, \delta_n^\pi(\mathbf{x}_{(n)})) \leq 2\varepsilon + L(\theta_0, \theta_0) + K\, \frac{(\mathsf{P}_n^\pi(B^c|\mathbf{x}_{(n)}))^{\frac{1}{2}}}{\mathsf{P}_n^\pi(B|\mathbf{x}_{(n)})}.$$

From (5.4), $B \subset O$ and the consistency of the posterior, it holds that

$$\frac{(\mathsf{P}_n^\pi(B^c|\mathbf{x}_{(n)}))^{\frac{1}{2}}}{\mathsf{P}_n^\pi(B|\mathbf{x}_{(n)})} \to 0, \ a.s.$$

Because of (5.2) we have $L(\theta_0, \theta_0) \leq L(\theta_0, \delta^\pi(\mathbf{x}_{(n)}))$ and for $\varepsilon \to 0$ we obtain

$$0 \leq \|\delta_n^\pi(\mathbf{x}_{(n)}) - \theta_0\|c_0 < L(\theta_0, \delta_n^\pi(\mathbf{x}_{(n)})) - L(\theta_0, \theta_0) \to 0 \ a.s.$$

$$\square$$

The assumption (5.4) includes the requirement that the prior distribution is positive around the data generating parameter θ_0. The following example illustrates what happens when the prior excludes θ_0.

Example 5.2 (Counterexample)
Assume $\mathsf{P}_\theta = \mathsf{N}(\theta, 1)$ and $\theta_0 = 2$. $X_1, \ldots, X_n$ are i.i.d from $\mathsf{N}(2, 1)$. The prior is the uniform distribution over $[-1, 1]$. From Example 2.8, we obtain

$$\pi_n(\theta|\mathbf{x}_{(n)}) \propto \pi(\theta)\ell(\theta|\mathbf{x}_n)\ell(\theta) \propto \exp\left(-\frac{n}{2}(\bar{x} - \theta)^2 \right) I_{[-1,1]}(\theta).$$

Hence the posterior is a truncated normal distribution with $\mu = \bar{x}$ and $\sigma^2 = \frac{1}{n}$ and boundaries $a = -1, b = 1$. The series of posteriors cannot contract around

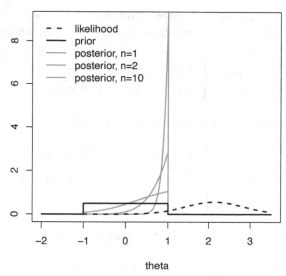

Figure 5.1: Example 5.2. The posterior distributions are concentrated at the border $b = 1$, while the data generating parameter is $\theta_0 = 2$.

$\theta_0 = 2$, see Figure 5.1. We apply a quadratic loss $L(\theta, d) = |\theta - d|^2$. The loss fulfills conditions (5.2)and (5.3). From Theorem 4.2 we get

$$\delta_n^\pi(\mathbf{x}_{(n)}) = \mathsf{E}(\theta|\mathbf{x}_{(n)}) = \bar{x} + \frac{1}{\sqrt{n}} \frac{\varphi(\alpha) - \varphi(\beta)}{\Phi(\beta) - \Phi(\alpha)},$$

where $\alpha = \sqrt{n}(-1 - \bar{x})$, $\beta = \sqrt{n}(1 - \bar{x})$, φ and Φ are the density and distribution functions of $\mathsf{N}(0, 1)$. Note that, this expression cannot be used for calculations, since the posterior support is lying in the tails of a normal distribution that implies the fraction of two nearly zero terms. From the Law of Large Numbers we obtain that $\bar{x} \to 2$ $a.s.$ for $n \to \infty$. Applying the rule of L' Hospitale we get

$$\delta_n^\pi(\mathbf{x}_{(n)}) - \bar{x} \to -1 \quad a.s.$$

□

As a consequence of this example we can formulate a general rule of thumb:

Never exclude any possible parameters from the prior!

Note that, Assumption (5.4) is stronger than that the prior covers all possible parameters. It also requires a continuity of the parametrization. Here we present a counterexample from Bahadur; see Schwartz (1965).

Example 5.3 (Bahadur's counterexample)

Assume $X_1, \ldots, X_n$ are i.i.d from P_θ with $\theta \in \Theta = [1, 2)$, where

$$
\mathsf{P}_\theta = \begin{cases} \mathsf{U}[0, 1] & \text{for} \quad \theta = 1 \\ \mathsf{U}[0, \frac{2}{\theta}] & \text{for} \quad 1 < \theta < 2 \end{cases}
$$

For $y = x_{\max} = \max_{i=1,\ldots,n} x_i$ the likelihood function is

$$
\ell(\theta|\mathbf{x}_{(n)}) = \begin{cases} 1 & \text{for} \quad \theta = 1 \text{ and } y \leq 1 \\ \frac{\theta^n}{2^n} & \text{for} \quad \frac{2}{y} > \theta > 1 \text{ and } y > 1 \\ 0 & \text{else} \end{cases}
$$

and we obtain the maximum likelihood estimator

$$
\widehat{\theta}_{MLE} = \begin{cases} 1 & \text{for} \quad y \leq 1 \\ \frac{2}{y} & \text{for} \quad y > 1. \end{cases}
$$

Consider now the Bayes model with uniform prior $\pi(\theta) = 1$ for $1 \leq \theta < 2$. The Bayes rule under quadratic loss is the expectation of the posterior. We have $\pi(\theta|\mathbf{x}_{(n)}) \propto \ell(\theta|\mathbf{x}_{(n)})$ and

$$
\delta(\mathbf{x}_{(n)}) = \frac{\int_1^2 \theta\, \ell(\theta|\mathbf{x}_{(n)})\, d\theta}{\int_1^2 \ell(\theta|\mathbf{x}_{(n)})\, d\theta}
$$

Applying the integral $\int_a^b z^m dz = \frac{1}{m+1}(b^{m+1} - a^{m+1})$, with $a = 1$ and $b = \frac{2}{y}$ we obtain

$$
\delta(\mathbf{x}_{(n)}) = \begin{cases} \frac{n+1}{n+2} \frac{2^{n+2}-1}{2^{n+1}-1} & \text{for} \quad y \leq 1 \\ \frac{n+1}{n+2} \frac{b^{n+2}-1}{b^{n+1}-1} & \text{for} \quad y > 1, b = \frac{2}{y} \end{cases}.
$$

For $\theta = 1$ we have $\mathsf{P}_\theta(y > 1) = 0$ and $\widehat{\theta}_{MLE} = 1$, but

$$
\lim_{n\to\infty} \delta(\mathbf{x}_{(n)}) = \lim_{n\to\infty} \left(\frac{n+1}{n+2} \frac{2^{n+2}-1}{2^{n+1}-1} \right) = 2.
$$

The maximum likelihood estimator is consistent while the Bayes estimator is not. For illustration see Figure 5.2.   □

In Example 5.1 the limit behavior of the posterior is independent of the prior. This can be proved generally. This property was also one of the key points for the construction of the reference priors; see the discussion in Section 3.3.3.

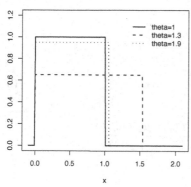 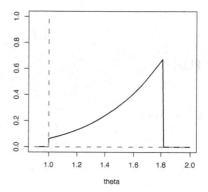

Figure 5.2: Example 5.3. Left: densities for $\theta = 1$ and $\theta = 1.3, 1.9$. Right: likelihood functions with $x_{\max} = 1.1$ and with $x_{\max} = 0.8$ (broken line). The maximum likelihood estimator is $\frac{x_{\max}}{2} \approx 1.82$ in the first case and 1 in the second.

Theorem 5.2 *Assume* $\mathbf{X}_{(n)} \sim \mathsf{P}_{\theta_0}^{(n)}$ *and two different priors* π_1, π_2, *both continuous and positive at* θ_0. *Assume that for* $i = 1, 2$ *the posterior series* $\pi_{i,n}(\theta|X_{(n)})$, *associated with* π_i, *are strongly consistent. Then*

$$\sup_A |\mathsf{P}_n^{\pi_1}(A|\mathbf{X}_{(n)}) - \mathsf{P}_n^{\pi_2}(A|\mathbf{X}_{(n)})| \to 0, \quad \text{for } n \to \infty, \; a.s. \quad (5.7)$$

PROOF: We have for $i = 1, 2$

$$\mathsf{P}_n^{\pi_i}(A|\mathbf{X}_{(n)}) = \frac{\mu_i(A)}{\mu_i(\Theta)} \quad \text{with} \quad \mu_i(A) = \int_A p_n(\mathbf{x}_{(n)}|\theta)\pi_i(\theta) \, d\theta.$$

Let $O_1 \supset O_2 \supset \dots$ be a sequence of open balls with centers θ_0. Because of the consistency of the posterior, for every given $\eta > 1$ there exist m, $n_0 = n_0(m)$ such that for all $n > n_0$

$$\mathsf{P}_n^{\pi_i}(O_m|\mathbf{X}_{(n)}) \geq \frac{1}{\eta}, \; i = 1, 2. \quad (5.8)$$

The continuity of the priors implies the existence of a m_0 such that for all $\theta \in O_{m_0}$

$$\frac{1}{\eta}\alpha \leq \frac{\pi_1(\theta)}{\pi_2(\theta)} \leq \eta\alpha, \quad \text{with} \quad \alpha = \frac{\pi_1(\theta_0)}{\pi_2(\theta_0)}. \quad (5.9)$$

Using (5.9) we obtain

$$\mu_1(O_{m_0}) = \int_{O_{m_0}} p_n(\mathbf{x}_{(n)}|\theta)\frac{\pi_1(\theta)}{\pi_2(\theta)}\pi_2(\theta) \, d\theta \leq \eta\,\alpha\,\mu_2(O_{m_0})$$

and

$$\mu_1(O_{m_0}) \geq \frac{1}{\eta}\,\alpha\,\mu_2(O_{m_0}).$$

Thus

$$\frac{1}{\eta}\,\alpha \leq \frac{\mu_1(O_{m_0})}{\mu_2(O_{m_0})} \leq \eta\,\alpha. \qquad (5.10)$$

We factorize

$$\frac{\mu_1(\Theta)}{\mu_2(\Theta)} = \frac{\mu_1(\Theta)}{\mu_1(O_{m_0})}\,\frac{\mu_1(O_{m_0})}{\mu_2(O_{m_0})}\,\frac{\mu_2(O_{m_0})}{\mu_2(\Theta)}$$

and apply (5.8), (5.10) so that

$$\frac{1}{\eta^2}\,\alpha \leq \frac{\mu_1(\Theta)}{\mu_2(\Theta)} \leq \eta^2\,\alpha. \qquad (5.11)$$

Set $O^c_{m_0} = \Theta \setminus O_{m_0}$. Using (5.11) we can estimate the posteriors

$$
\begin{aligned}
\mathsf{P}_n^{\pi_1}(A|\mathbf{X}_{(n)}) - \mathsf{P}_n^{\pi_2}(A|\mathbf{X}_{(n)}) &= \frac{\mu_1(A)}{\mu_1(\Theta)} - \frac{\mu_2(A)}{\mu_2(\Theta)} \\
&\leq \frac{\mu_1(A \cap O_{m_0})}{\mu_1(\Theta)} - \frac{\mu_2(A \cap O_{m_0})}{\mu_2(\Theta)} + \frac{\mu_1(O^c_{m_0})}{\mu_1(\Theta)} \\
&\leq (\eta^3 - 1)\frac{\mu_2(A \cap O_{m_0})}{\mu_2(\Theta)} + \frac{\mu_1(O^c_{m_0})}{\mu_1(\Theta)} \\
&\leq (\eta^3 - 1)\mathsf{P}_n^{\pi_2}(O_{m_0}|\mathbf{X}_{(n)}) + \mathsf{P}_n^{\pi_1}(O^c_{m_0}|\mathbf{X}_{(n)}).
\end{aligned}
\qquad (5.12)
$$

Since the posteriors are consistent, we have

$$\lim_{n \to \infty} \mathsf{P}_n^{\pi_2}(O_{m_0}|\mathbf{X}_{(n)}) = 1 \; a.s., \quad \lim_{n \to \infty} \mathsf{P}_n^{\pi_1}(O^c_{m_0}|\mathbf{X}_{(n)}) = 0 \; a.s.$$

The constant $\eta > 1$ is arbitrary; letting $\eta \to 1$ we complete the proof.

$\square$

5.2 Schwartz' Theorem

Now we establish the theorem of Lorraine Schwartz, which gives general conditions for posterior consistency.

Recall the Kullback–Leibler divergence $\mathsf{K}(\mathsf{P}_{\theta_0}|\mathsf{P}_\theta)$ in (3.28), $\mathsf{K}(\mathsf{P}|\mathsf{Q}) = \int p(x)\ln(\frac{p(x)}{q(x)})dx$.

The other condition is related to the Hellinger transfom, defined by

$$\mathsf{H}_{\frac{1}{2}}(\mathsf{P}, \mathsf{Q}) = \int \sqrt{p(x)q(x)}\,dx. \qquad (5.13)$$

Theorem 5.3 (Schwartz)   *Assume a Bayes model* $\{\mathcal{P}^{(n)}, \pi\}$ *with* $\mathcal{P}^{(n)} = \{P_\theta^{\otimes n} : \theta \in \Theta\}$. *Let for all* $\varepsilon > 0$,

$$P^\pi(\mathcal{K}_\varepsilon(\theta_0)) > 0, \quad \text{where } \mathcal{K}_\varepsilon(\theta_0) = \{\theta : K(P_{\theta_0}|P_\theta) < \varepsilon\}. \tag{5.14}$$

For every open set $O \subset \Theta$ *with* $\theta_0 \in O$ *and* $O^c = \Theta \backslash O$, *there exist constants* D_0 *and* q_0, $q_0 < 1$, *such that*

$$H_{\frac{1}{2}}(P_{\theta_0}^{\otimes n}, P_{n,O^c}) \leq D_0 \, q_0^n, \tag{5.15}$$

where P_{n,O^c} *is defined by*

$$P_{n,O^c}(A) = \int_A \int_{O^c} p_n(\mathbf{x}_{(n)}|\theta) \frac{\pi(\theta)}{P^\pi(O^c)} \, d\theta \, d\mathbf{x}_{(n)}. \tag{5.16}$$

Then the sequence of posteriors is strongly consistent at θ_0.

PROOF: We only provide a sketch of the proof. For a precise proof based on measure theory we recommend Liese and Miescke (2008, p. 354ff). We have to show that for an arbitrary open subset O with $\theta_0 \in O$

$$P_n^\pi(O^c|\mathbf{X}_{(n)}) \to 0 \quad \text{for } n \to \infty \quad a.s. \tag{5.17}$$

It holds that

$$P_n^\pi(A|\mathbf{x}_{(n)}) = \frac{\mu(A)}{\mu(\Theta)} \quad \text{with } \mu(A) = \int_A p_n(\mathbf{x}_{(n)}|\theta)\, \pi(\theta)\, d\theta. \tag{5.18}$$

We rewrite the fraction as

$$P_n^\pi(O^c|\mathbf{x}_{(n)}) = \frac{\mu(O^c)}{\mu(\Theta)} = \frac{\text{Num}_n(\beta)}{\text{Den}_n(\beta)} \tag{5.19}$$

with

$$\begin{aligned}
\text{Num}_n(\beta) &= \exp(\beta n)\mu(O^c)\, p_n(\mathbf{x}_{(n)}|\theta_0)^{-1}, \\
\text{Den}_n(\beta) &= \exp(\beta n)\mu(\Theta)\, p_n(\mathbf{x}_{(n)}|\theta_0)^{-1}.
\end{aligned} \tag{5.20}$$

The proof is split into two steps. In the first step we show that under assumption (5.15) there exists a β_0 such that

$$\limsup_{n\to\infty} \text{Num}_n(\beta_0) = 0 \quad a.s. \tag{5.21}$$

In the second step we show that under (5.14), for all $\beta > 0$,

$$\liminf_{n\to\infty} \text{Den}_n(\beta) = \infty \quad a.s. \tag{5.22}$$

Setting $\beta = \beta_0$ in (5.22) we obtain (5.17). It remains to show (5.21) and (5.22). We start with (5.21). We have

$$
H_{\frac{1}{2}}(P_{\theta_0}^{\otimes n}, P_{n,O^c}) = \int p_n(\mathbf{x}_{(n)}|\theta_0)^{\frac{1}{2}} \left(\int_{O^c} p_n(\mathbf{x}_{(n)}|\theta) \frac{\pi(\theta)}{P^{\pi}(O^c)} \, d\theta \right)^{\frac{1}{2}} d\mathbf{x}_{(n)}
$$

$$
= \int p_n(\mathbf{x}_{(n)}|\theta_0) g(\mathbf{x}_{(n)})^{\frac{1}{2}} \, d\mathbf{x}_{(n)} = E_{\theta_0} g(\mathbf{X}_{(n)})^{\frac{1}{2}}
$$

$$(5.23)$$

with

$$
g(\mathbf{x}_{(n)}) = \int_{O^c} \frac{p_n(\mathbf{x}_{(n)}|\theta)}{p_n(\mathbf{x}_{(n)}|\theta_0)} \frac{\pi(\theta)}{P^{\pi}(O^c)} \, d\theta. \tag{5.24}
$$

From (5.15) it follows that

$$
E_{\theta_0} g^{\frac{1}{2}} \le D_0 \, q_0^n. \tag{5.25}
$$

Using (5.25) and the inequality for non-negative random variables Z that $P(Z \ge 1) \le EZ$, we obtain for arbitrary $r > 1$

$$
P(r^{2n} g(\mathbf{x}_{(n)}) \ge 1) = P(r^n g(\mathbf{x}_{(n)})^{\frac{1}{2}} \ge 1)
$$

$$
\le r^n E_{\theta_0} g^{\frac{1}{2}} \le D_0 \, (r \, q_0)^n. \tag{5.26}
$$

Choosing r such that $1 < r < \frac{1}{q_0}$ we get

$$
\sum_{n=1}^{\infty} P(r^{2n} g(\mathbf{x}_{(n)}) \ge 1) \le D_0 \sum_{n=1}^{\infty} (r \, q_0)^n < \infty. \tag{5.27}
$$

The Borel–Cantelli Lemma yields

$$
P(r^{2n} g(\mathbf{x}_{(n)}) \ge 1 \text{ infinitely often}) = 0. \tag{5.28}
$$

We set β_0 such that $r = \exp(\beta_0)$, and get from (5.28)

$$
\mathrm{Num}_n(\beta_0) = r^{-n} r^{2n} g(\mathbf{x}_{(n)}) \, P^{\pi}(O^c) \le r^{-n} r^{2n} g(\mathbf{x}_{(n)}) \le r^{-n} \quad a.s.
$$

Since $r > 1$ we get $r^{-n} \to 0$ and (5.21), which completes the first step of the proof. To show (5.22), we set $\beta = 2n\varepsilon$ for arbitrary $\varepsilon > 0$. We have

$$
\liminf_{n \to \infty} \mathrm{Den}_n(\beta) = \liminf_{n \to \infty} \exp(2n\varepsilon) \int_{\Theta} \frac{p_n(\mathbf{x}_{(n)}|\theta)}{p_n(\mathbf{x}_{(n)}|\theta_0)} \pi(\theta) \, d\theta
$$

$$
\ge \liminf_{n \to \infty} \exp(2n\varepsilon) \int_{K_\varepsilon(\theta_0)} \frac{p_n(\mathbf{x}_{(n)}|\theta)}{p_n(\mathbf{x}_{(n)}|\theta_0)} \pi(\theta) \, d\theta
$$

$$
= \liminf_{n \to \infty} \int_{K_\varepsilon(\theta_0)} \exp\left(n2\varepsilon - \ln\left(\frac{p_n(\mathbf{x}_{(n)}|\theta_0)}{p_n(\mathbf{x}_{(n)}|\theta)} \right) \right) \pi(\theta) \, d\theta
$$

$$
= D
$$

$$(5.29)$$

As $P_\theta^{(n)} = P_\theta^{\otimes n}$, we have

$$\ln\left(\frac{p_n(\mathbf{x}_{(n)}|\theta_0)}{p_n(\mathbf{x}_{(n)}|\theta)}\right) = \ln(p_n(\mathbf{x}_{(n)}|\theta_0)) - \ln(p_n(\mathbf{x}_{(n)}|\theta))$$

$$= \sum_{i=1}^{n} \ln(p_n(x_i|\theta_0)) - \sum_{i=1}^{n} \ln(p_n(x_i|\theta)) \qquad (5.30)$$

$$= \sum_{i=1}^{n} \ln\left(\frac{p_n(x_i|\theta_0)}{p_n(x_i|\theta)}\right).$$

Applying Jensen's inequality, that $f(\mathsf{E}X) \leq \mathsf{E}(f(x))$ for convex functions f, and using the factorization

$$\pi(\theta) = \mathsf{P}^\pi(\mathcal{K}_\varepsilon(\theta_0))\pi_\varepsilon(\theta) \text{ with } \pi_\varepsilon(\theta) = \pi(\theta|\mathcal{K}_\varepsilon(\theta_0)) \qquad (5.31)$$

we obtain

$$D \geq \lim_{n\to\infty} \inf \exp\left(n \int_{\mathcal{K}_\varepsilon(\theta_0)} \left(2\varepsilon - \frac{1}{n}\sum_{i=1}^{n} \ln\frac{p_n(x_i|\theta_0)}{p_n(x_i|\theta)}\right) \pi(\theta)\, d\theta\right)$$

$$\geq \lim_{n\to\infty} \inf \exp\left(\mathsf{P}^\pi(\mathcal{K}_\varepsilon(\theta_0))\, n \left(\int_{\mathcal{K}_\varepsilon(\theta_0)} 2\varepsilon\, \pi_\varepsilon(\theta)\, d\theta - \frac{1}{n}\sum_{i=1}^{n} Y_i\right)\right), \qquad (5.32)$$

where

$$Y_i = \int \ln\left(\frac{p(x_i|\theta_0)}{p(x_i|\theta)}\right) \pi_\varepsilon(\theta)\, d\theta.$$

The strong law of large numbers yields $\frac{1}{n}\sum_{i=1}^{n} Y_i \to \mathsf{E}Y_1, a.s.$ for $n \to \infty$. We calculate $\mathsf{E}Y_1$,

$$\mathsf{E}Y_1 = \iint \ln(\frac{p(x_1|\theta_0)}{p(x_1|\theta)})\pi_\varepsilon(\theta)\, d\theta\, p(x_1|\theta_0)\, dx_1$$

$$= \iint \ln(\frac{p(x_1|\theta_0)}{p(x_1|\theta)})p(x_1|\theta_0)\, dx_1\, \pi_\varepsilon(\theta)\, d\theta \qquad (5.33)$$

$$= \int \mathsf{K}(\mathsf{P}_{\theta_0}|\mathsf{P}_\theta)\pi_\varepsilon(\theta)\, d\theta,$$

using Fubini's Theorem.
From (5.14) it follows that

$$\mathsf{E}Y_1 = \int \mathsf{K}(\mathsf{P}_{\theta_0}|\mathsf{P}_\theta)\pi_\varepsilon(\theta)\, d\theta \leq \int_{\mathcal{K}_\varepsilon(\theta_0)} \mathsf{K}(\mathsf{P}_{\theta_0}|\mathsf{P}_\theta)\pi_\varepsilon(\theta)\, d\theta \leq \varepsilon\, \mathsf{P}^{\pi_\varepsilon}(\mathcal{K}_\varepsilon(\theta_0)) \leq \varepsilon.$$

Summarizing, we obtain

$$\lim_{n\to\infty} \inf \operatorname{Den}_n(\beta) \geq \lim_{n\to\infty} \inf \exp\left(n\, \mathsf{P}^\pi(\mathcal{K}_\varepsilon(\theta_0))\, (2\varepsilon - \varepsilon)\right) = \infty$$

which gives (5.22) and completes the proof.

$$\square$$

We complete this chapter with a short discussion of the two main assumptions in Schwartz' Theorem.

The first main assumption (5.14) implies that the prior does not exclude the data generating parameter θ_0; see Example 5.2. Furthermore, a continuity of the prior at θ_0 with respect to the Kullback–Leibler divergence is required.

The second main assumption (5.15) is related to the Hellinger transform (5.13). Recall the relation between Hellinger distance and Hellinger transform in (4.32), so that

$$H^2(\mathsf{P}_{\theta_0}^{\otimes n}, \mathsf{P}_{n,O^c}) = 1 - H_{\frac{1}{2}}(\mathsf{P}_{\theta_0}^{\otimes n}, \mathsf{P}_{n,O^c}) > 1 - D_0\, q_0^n.$$

It describes the ability to differentiate between $\mathsf{P}_{\theta_0}^{\otimes n}$ and P_{n,O^c}, which can be described by the existence of consistent tests. The following statement holds now.

Theorem 5.4   *Assume the model $\mathcal{P}^{(n)} = \{\mathsf{P}_\theta^{\otimes n} : \theta \in \Theta\}$. Let $\theta_0 \in O$, $O \subset \Theta$ be open. Consider the testing problem:*

$$\mathsf{H}_0 : \mathsf{P}_{\theta_0}^{\otimes n} \text{ versus } \mathsf{H}_1 : \{\mathsf{P}_\theta^{\otimes n} : \theta \in \Theta \setminus O\}$$

If there exists a sequence of nonrandomized tests φ_n and positive constants C and β such that

$$\mathsf{E}_{\theta_0}\varphi_n(\mathbf{X}_{(n)}) + \sup_{\theta \in \Theta \setminus O} \mathsf{E}_\theta(1 - \varphi_n(\mathbf{X}_{(n)})) \leq C\exp(-n\beta), \qquad (5.34)$$

then condition (5.15) is fulfilled in the Bayes model $\{\{\mathsf{P}_\theta^{\otimes n} : \theta \in \Theta\}, \pi\}$.

PROOF: Set $C_{1,n} = \{\mathbf{x}_{(n)} : \varphi_n(\mathbf{x}_{(n)}) = 1\}$, the critical region of test φ_n, and $C_{0,n} = \{\mathbf{x}_{(n)} : \varphi_n(\mathbf{x}_{(n)}) = 0\}$ its complement. Recall the definition of P_{n,O^c} in (5.16), and set $\pi_r(\theta)$ for the prior restricted on $\Theta \setminus O$. Then by Schwarz' inequality we have

$$
\begin{aligned}
H_{\frac{1}{2}}(\mathsf{P}_{\theta_0}^{\otimes n}, \mathsf{P}_{n,O^c}) &= \int p(\mathbf{x}_{(n)}|\theta_0)^{\frac{1}{2}} \left(\int p(\mathbf{x}_{(n)}|\theta_0)\pi_r(\theta)\,d\theta \right)^{\frac{1}{2}} d\mathbf{x}_{(n)} \\
&\leq \mathsf{P}_{\theta_0}^{\otimes n}(C_{1,n})^{\frac{1}{2}} \mathsf{P}_{n,O^c}(C_{1,n})^{\frac{1}{2}} + \mathsf{P}_{\theta_0}^{\otimes n}(C_{0,n})^{\frac{1}{2}} \mathsf{P}_{n,O^c}(C_{0,n})^{\frac{1}{2}} \\
&\leq \mathsf{P}_{\theta_0}^{\otimes n}(C_{1,n})^{\frac{1}{2}} + \mathsf{P}_{n,O^c}(C_{0,n})^{\frac{1}{2}}.
\end{aligned}
$$

Since

$$\mathsf{P}_{\theta_0}^{\otimes n}(C_{1,n}) = \mathsf{E}_{\theta_0}\varphi_n(\mathbf{X}_{(n)})$$

and

$$P_{n,O^c}(C_{0,n}) = \int_{C_{0,n}} \int p(\mathbf{x}_{(n)}|\theta)\pi_r(\theta)\, d\theta\, d\mathbf{x}_{(n)}$$

$$\leq \sup_{\theta\in\Theta\backslash O} P_\theta^{\otimes n}(C_{0,n})$$

$$\leq \sup_{\theta\in\Theta\backslash O} E_\theta(1 - \varphi_n(\mathbf{X}_{(n)}))$$

we obtain (5.15) from (5.34).

$\square$

We recommend Liese and Miescke (2008, Section 7.5) for more results.

5.3 List of Problems

1. Consider an i.i.d. sample $\mathbf{X}_{(n)} = (X_1,\ldots,X_n)$ from $\mathsf{Exp}(\theta)$, where θ is the rate parameter. Assume the prior $\theta \sim \mathsf{Gamma}(\alpha,\beta)$.
   (a) Derive the posterior distribution.
   (b) Calculate the Bayes estimator $\delta(\mathbf{x}_{(n)})$ under the L_2 loss.
   (c) Show the consistency of $\delta(\mathbf{x}_{(n)})$.
2. Assume that P_θ belongs to an exponential family.
   (a) Calculate the Kullback–Leibler divergence $K(P_{\theta_0}|P_\theta)$.
   (b) Specify the result (a) for $\mathsf{Gamma}(\alpha,\beta)$.
3. Assume that P_θ is the uniform distribution $U(0,\theta)$.
   (a) Calculate the Kullback–Leibler divergence $K(P_{\theta_0}|P_\theta)$.
   (b) Consider the Kullback–Leibler divergence as a function of θ. Is it continuous at θ_0?
4. Consider an i.i.d. sample $\mathbf{X}_{(n)} = (X_1,\ldots,X_n)$ from $\mathsf{Gamma}(\nu,\theta)$. The parameter ν is known. Consider two different conjugate priors π_1, π_2: $\theta \sim \mathsf{Gamma}(\alpha_1,\beta)$, $\theta \sim \mathsf{Gamma}(\alpha_2,\beta)$. Show that the respective posteriors fulfill (5.7). (Hint: Apply the Pinsker's inequality ,

$$\sup_A |P(A) - Q(A)| \leq \sqrt{\frac{1}{2}K(P|Q)}, \tag{5.35}$$

and show that the Kullback–Leibler divergence converges to zero.)
5. Consider an i.i.d. sample $\mathbf{X}_{(n)} = (X_1,\ldots,X_n)$ from $N(\theta,\sigma^2)$. The parameter σ^2 is known. Suppose $\theta \sim N(\mu,\sigma_0^2)$. Show that the conditions (5.14) and (5.15) of Theorem 5.3 are fulfilled.
6. Consider an i.i.d. sample $\mathbf{X}_{(n)} = (X_1,\ldots,X_n)$ from $N(a+b,1)$. The parameter of interest is $\theta = (a,b)$. Suppose $\theta \sim N_2(0,\mathbf{I}_2)$.
   (a) Show that the Bayes estimator is not consistent.
   (b) Show that the sequence of posteriors is not consistent.
   (c) Show that the condition (5.15) is violated.

Chapter 6

Normal Linear Models

Linear models constitute an important part of statistical inference, with regression and analysis of variance models as their two main components. Searle (1971), later Searle and Gruber (2017), is a prominent classical reference. The article Lindley and Smith (1972) set the stage for Bayesian theory of linear models, although Zellner (1971) is an even older work. Other important references include Box and Tiao (1973), Koch (2007) and Broemeling (2016).

This chapter deals with Bayesian theory of linear models, beginning with a detailed treatment of univariate case, followed by a brief multivariate extension. Our focus will be on Bayesian analysis of linear models under normality assumption, i.e., models of the form

$$\mathcal{P} = \{\mathsf{N}_n(\mathbf{X}\beta, \Sigma(\vartheta)) : \ \beta \in \mathbb{R}^p, \ \Sigma(\vartheta) \succ 0, \ \vartheta \in \mathbb{R}^q\}, \qquad (6.1)$$

where $\mathbf{X}$ is a matrix of known constants and β is the unknown parameter vector. Note that, ϑ is an intrinsic parameter, as variance component of the model. Thus, we shall mostly have $\Sigma(\vartheta) = \sigma^2\Sigma$, so that $\vartheta = \sigma^2$, in which case $\theta = (\beta, \sigma^2)$ and the entire parameter space for the model is $\Theta \subset \mathbb{R}^p \times \mathbb{R}_+$.

We recall that the linearity of such models follows from that of β in the expectation, $\mathbf{X}\beta$. We shall be mainly concerned with two Bayes models based on (6.1), namely,

$$\{\mathcal{P}, \pi_c\} \text{ and } \{\mathcal{P}, \pi_{\text{Jeff}}\},$$

where π_c and π_{Jeff} stand, respectively, for conjugate and Jeffreys priors; see Chapter 3 for details. Under this setting, our aim will be to derive posterior distributions, particularly focusing on closed-form expressions. The conjugate families, as the normal-inverse-gamma distributions or the normal-inverse Wishart distributions, are well studied and computer-intensive methods are not needed. Unfortunately, there is no standard parametrization for these distributions in the literature. We introduce the distributions in a general set up in the running text and sign a gray frame around them for a better reading. Furthermore these explicit posteriors give a chance to test the simulation methods.

DOI: 10.1201/9781003221623-6

6.1   Univariate Linear Models

A univariate linear model can be stated, in matrix form, as

$$\mathbf{y} = \mathbf{X}\beta + \epsilon, \tag{6.2}$$

where $\mathbf{y} \in \mathbb{R}^n$ is the vector of response variables, $\mathbf{X} \in \mathbb{R}^{n \times p}$ is the matrix of known constants, $\beta \in \mathbb{R}^p$ is the vector of unknown parameters, and $\epsilon \in \mathbb{R}^n$ is the vector of unobservable random errors. Expanded in full form, model (6.2) can be expressed as

$$
\begin{pmatrix} y_1 \\ y_2 \\ \vdots \\ y_n \end{pmatrix} =
\begin{pmatrix}
x_{11} & x_{12} & \cdots & x_{1p} \\
x_{21} & x_{22} & \cdots & x_{2p} \\
\vdots & \vdots & \ddots & \vdots \\
x_{n1} & x_{n2} & \cdots & x_{np}
\end{pmatrix}
\begin{pmatrix} \beta_1 \\ \beta_2 \\ \vdots \\ \beta_p \end{pmatrix} +
\begin{pmatrix} \epsilon_1 \\ \epsilon_2 \\ \vdots \\ \epsilon_n \end{pmatrix}, \tag{6.3}
$$

where the first column of $\mathbf{X}$ is often a vector of ones, denoted $\mathbf{1}_n$. Model (6.2) expresses each of n observations in $\mathbf{y}$ as a linear combination of unknown parameters in β with coefficients from $\mathbf{X}$, i.e.,

$$y_i = \mathbf{x}_i^T \beta + \epsilon_i = \sum_{j=1}^{p} x_{ij}\beta_j + \epsilon_i, \tag{6.4}$$

$i = 1, \ldots, n$, where $\mathbf{x}_i \in \mathbb{R}^p$ is the ith row of $\mathbf{X}$. First we illustrate model (6.2) by several examples.

Example 6.1 (Corn plants)
To assess whether corn plants fetch their Phosphorus content from organic or inorganic sources, concentrations for two types of organic (X_1, X_3) and an inorganic X_2 source are measured (in ppm) on $n = 17$ soil samples, along with the Phosphorus content in corn plants as study variable, Y. The data are taken from Snedecor and Cochran (1989). The corresponding linear model can be stated as

$$y_i = \beta_0 + \beta_1 x_{1i} + \beta_2 x_{2i} + \beta_3 x_{3i} + \epsilon_i, \ i = 1, \ldots, n, \tag{6.5}$$

with $\mathbf{y} \in \mathbb{R}^{17}$, $\mathbf{X} \in \mathbb{R}^{17 \times 4}$ (with first column of 1s for intercept) and $\beta = (\beta_0, \ \beta_1, \ \beta_2, \ \beta_3)^T$. It is a univariate multiple linear regression model with three independent variables. Following (6.3), the design matrix and parameter vector, for $n = 17$ and $p = 4$, are

$$
\mathbf{X} =
\begin{pmatrix}
1 & x_{11} & x_{12} & x_{13} \\
1 & x_{21} & x_{22} & x_{23} \\
\vdots & \vdots & \vdots & \vdots \\
1 & x_{17,1} & x_{17,2} & x_{17,3}
\end{pmatrix}, \
\beta =
\begin{pmatrix} \beta_0 \\ \beta_1 \\ \beta_2 \\ \beta_3 \end{pmatrix}.
$$

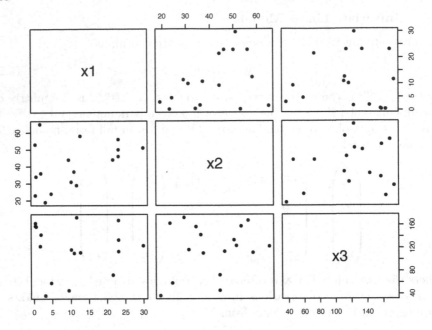

Figure 6.1: Matrix scatter plot of $(X_1,\ X_2,\ X_3)$ for corn plants.

Figure 6.1 depicts matrix scatter plot of independent variables. The model was in fact employed to study two phosphorous contents, i.e., two dependent variables. Here, we use only the first of them, adjourning the treatment of bivariate model as a special case of multivariate linear model in Section 6.4.   □

Example 6.2 (Side effects)
Certain medicines of tuberculosis (TB) are suspected to affect vision, often leading to complete vision loss. Surprisingly, in some cases, the vision returns to normal once the patient stops medicine.

To investigate the issue, a researcher takes a random sample of n patients, who have been on TB medicine for at least six months, and measures each patient's visual acuity for 3 weeks. To account for individual parameters, four concomitant variables are also recorded on each patient: weight, diastolic and systolic blood pressure, and weekly amount of medicine used. As the vision in both eyes deteriorate equally, if it does at all, it is decided to use the average over both eyes as response variable. The linear model is then

$$y_{ij} = \mathbf{x}_i^T \beta + w_i + \epsilon_{ij}, \ j = 1,\ldots,r, \ i = 1,\ldots,a, \tag{6.6}$$

where y_{ij} is the average vision of ith patient in jth week, w_j represents the week effect, and $\mathbf{x}_i^T \beta$ collects four concomitant variables. Model (6.6) consists of all fixed effects components, the first being regression component and second the ANOVA component, resulting into an analysis of covariance (ANOCOVA) model, which is also a special case of model (6.2). The design matrix and parameter vectors, in full form, can be written as

$$
\mathbf{X} = \begin{pmatrix}
x_{11} & x_{12} & x_{13} & x_{14} & 1 & 0 & 0 \\
x_{11} & x_{12} & x_{13} & x_{14} & 0 & 1 & 0 \\
x_{11} & x_{12} & x_{13} & x_{14} & 0 & 0 & 1 \\
x_{21} & x_{22} & x_{23} & x_{24} & 1 & 0 & 0 \\
x_{21} & x_{22} & x_{23} & x_{24} & 0 & 1 & 0 \\
x_{21} & x_{22} & x_{23} & x_{24} & 0 & 0 & 1 \\
x_{31} & x_{32} & x_{33} & x_{34} & 1 & 0 & 0 \\
x_{31} & x_{32} & x_{33} & x_{34} & 0 & 1 & 0 \\
x_{31} & x_{32} & x_{33} & x_{34} & 0 & 0 & 1
\end{pmatrix}, \quad \beta = \begin{pmatrix}
\beta_1 \\
\beta_2 \\
\beta_3 \\
\beta_4 \\
w_1 \\
w_2 \\
w_3
\end{pmatrix}.
$$

The first part of β represents X variables on each patient, where the second part represents the week effect, and these parts correspond, respectively, to first four and last three columns of $\mathbf{X}$.   □

Our main focus is on full rank models, so that $r(\mathbf{X}) = p$ and $\Sigma(\vartheta) = \sigma^2 \Sigma$, where Σ is positive definite and known. Under this set up, the likelihood and log-likelihood functions for model (6.2) with $\theta = (\beta, \sigma^2)$ can be stated as

$$
\ell(\theta|\mathbf{y}) = \frac{1}{(2\pi\sigma^2)^{n/2}} \frac{1}{|\Sigma|^{\frac{1}{2}}} \exp\left(-\frac{1}{2\sigma^2}(\mathbf{y} - \mathbf{X}\beta)^T \Sigma^{-1}(\mathbf{y} - \mathbf{X}\beta)\right) \quad (6.7)
$$

$$
l(\theta|\mathbf{y}) = -\frac{n}{2}\ln(2\pi\sigma^2) - \frac{1}{2}\ln(|\Sigma|) - \frac{1}{2\sigma^2}(\mathbf{y} - \mathbf{X}\beta)^T \Sigma^{-1}(\mathbf{y} - \mathbf{X}\beta), (6.8)
$$

which yield the maximum likelihood estimator, equivalently also the *generalized least-squares estimator* (GLSE), of β,

$$
\widehat{\beta}_\Sigma = (\mathbf{X}^T \Sigma^{-1} \mathbf{X})^{-1} \mathbf{X}^T \Sigma^{-1} \mathbf{y}. \quad (6.9)
$$

Using the linearity of $\widehat{\beta}_\Sigma$ and the properties of multivariate normal distributions, we get

$$
\widehat{\beta}_\Sigma \sim \mathsf{N}_p(\beta, \sigma^2(\mathbf{X}^T \Sigma^{-1} \mathbf{X})^{-1}). \quad (6.10)
$$

For $\Sigma = \mathbf{I}_n$, it reduces to the *ordinary least-squares estimator* (OLSE) of β,

$$
\widehat{\beta} = (\mathbf{X}^T \mathbf{X})^{-1} \mathbf{X}^T \mathbf{y}, \quad \widehat{\beta} \sim \mathsf{N}_p(\beta, \sigma^2(\mathbf{X}^T \mathbf{X})^{-1}). \quad (6.11)
$$

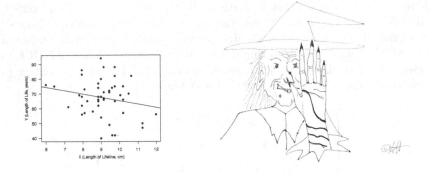

Figure 6.2: Life length data.

For a short introduction into linear models we refer to Liero and Zwanzig (2011, Chapter 6).

Example 6.3 (Life length vs. life line)
Draper and Smith (1966) report an interesting data set, collected to study a popular belief that the length of one's life span can be predicted from the length of *life line* on one's left hand. The data consist of $n = 50$ pairs of observations on the person's age close to death, y, measured in years, and the length of lifeline, X, measured in centimeter; for details and reference to original study, see Draper and Smith (1966, p.105). After removing two outlying observations, the data for $n = 48$ observations yield

$$(\mathbf{X}^T\mathbf{X})^{-1} = \begin{pmatrix} 1.365 & -0.148 \\ -0.148 & 0.016 \end{pmatrix}, \ \mathbf{X}'\mathbf{y} = \begin{pmatrix} 3232 \\ 29282 \end{pmatrix}$$

so that $\widehat{\beta} = (87.89 \ -2.26)$, i.e., the least-squares fitted line predicting life length using life line as predictor, is $\widehat{y}_i = 87.89 - 2.26 x_i$. The fitted line is also shown in Figure 6.2, along with the scatter plot of observed data.
Roughly, the negative slope seems to counter the popular belief. In fact, a test of $H_0 : \beta_1 = 0$ could not be rejected at any reasonable α ($F = 2.09$, p-value $= 0.155$), indicating that the belief is nothing more than a mere perception. Later we analyze the same data under Bayes model.   □

R Code 6.1.3. Life data set, Figure 6.2.

```
library(aprean3)
#"Applied Regression Analysis" by N.R.Draper and H.Smith
life<-dse03v # life data: data(dse03v)
attach(life); age<-life$x; length<-life$y
```

```
plot(length,age)
# outlier age=19, length=13.20
length[45]; age[1] # outlier
# new data without outlier
Age<-c(age[2:44],age[46:50])
Length<-c(length[2:44],length[46:50])
plot(Length,Age)
M<-lm(Age~Length); abline(M); summary(M)
```

6.2 Bayes Linear Models

In this section, we consider Bayes model $\{\mathcal{P}, \pi\}$ with $\mathcal{P}$ as given in (6.1). For convenience, we write

$$\mathcal{P} = \left\{ \mathsf{N}(\mathbf{X}\beta, \Sigma) : \ \beta \in \mathbb{R}^p, \ \Sigma \in \mathbb{R}^{n \times n}, \Sigma \succ 0 \right\}. \tag{6.12}$$

Here, π is a known prior distribution assigned to the parameters. For example, for $\Sigma = \sigma^2 \mathbf{I}$, with σ^2 known, $\theta = \beta$ and a prior is assigned to β; otherwise for $\Sigma = \sigma^2 \mathbf{I}$, with σ^2 unknown, $\theta = (\beta, \sigma^2)$ and a joint prior is assigned to β and σ^2. We shall mainly focus on two types of priors, conjugate and Jeffreys, since in these cases, a closed-form expression of posterior distribution of the parameter of interest can be derived.

In the following subsection we consider only the linear parameter as unknown. We develop methods for handling the linear parameter which will also be applied in the more complex model where the linear parameter β and the covariance parameter are unknown. Especially Lemmas 6.1, 6.2, 6.3, and Theorems 6.1, 6.2 provide a useful toolbox.

6.2.1 Conjugate Prior: Parameter $\theta = \beta$, σ^2 Known

We assume that the matrix Σ in model (6.12) is completely known. The unknown parameter is the linear regression parameter β. It is possible to apply Theorem 3.3 to derive the conjugate family. Here we set as prior the normal distribution $N_p(\gamma, \Gamma)$ and calculate the posterior. We assume that the hyperparameters, γ and Γ, are known. For details on hyperparameters, see Chapter 3.

The Bayes model $\{\mathcal{P}, \pi_c\}$ specializes to

$$\{\{\mathsf{N}_n(\mathbf{X}\beta, \Sigma) : \ \beta \in \mathbb{R}^p\}, \ \mathsf{N}_p(\gamma, \Gamma)\}.$$

We have

$$\mathbf{y}|\beta \sim \mathsf{N}_n(\mathbf{X}\beta, \Sigma) \ \text{ and } \ \beta \sim \mathsf{N}_p(\gamma, \Gamma). \tag{6.13}$$

As the set up is all about joined, marginal and conditional normal distribution, the following two lemmas deliver useful results for simplifying the calculations. Lemma 6.1 states that if the joint distribution is multivariate normal, then all marginal and conditional distributions are also multivariate normal of respective dimensions; see e.g., Mardia et al. (1979).

Lemma 6.1 *Let* $\mathbf{X} \sim N_p(\mu, \Sigma)$, $\Sigma \succ 0$, *be partitioned so that*

$$\mathbf{X} = \begin{pmatrix} \mathbf{X}_1 \\ \mathbf{X}_2 \end{pmatrix} \sim N_p \left(\begin{pmatrix} \mu_1 \\ \mu_2 \end{pmatrix}, \begin{pmatrix} \Sigma_{11} & \Sigma_{12} \\ \Sigma_{21} & \Sigma_{22} \end{pmatrix} \right),$$

where $\mu_1 \in \mathbb{R}^r$, $\Sigma_{11} \in \mathbb{R}^{r \times r}$ *etc. Then*

$$\mathbf{X}_1 \sim N_r(\mu_1, \Sigma_{11})$$
$$\mathbf{X}_2 | \mathbf{X}_1 = \mathbf{x}_1 \sim N_{p-r}(\mu_{2|1}, \Sigma_{2|1}),$$

where

$$\mu_{2|1} = \mu_2 + \Sigma_{21} \Sigma_{11}^{-1} (\mathbf{x}_1 - \mu_1) \text{ and } \Sigma_{2|1} = \Sigma_{22} - \Sigma_{21} \Sigma_{11}^{-1} \Sigma_{12}.$$

The converse of Lemma 6.1 does not hold in general. However, a converse is possible if $\mathsf{E}(\mathbf{X}_2 | \mathbf{X}_1 = \mathbf{x}_1)$ is a linear function of $\mathbf{x}_1$ and $\mathsf{Cov}(\mathbf{X}_2 | \mathbf{X}_1 = \mathbf{x}_1)$ does not even depend on $\mathbf{x}_1$. As this result, unlike Lemma 6.1, is rarely found in statistical literature, we state it below.

Lemma 6.2 *It is given that*

$$\mathbf{X}_1 \sim N_r(\mu_1, \Sigma_{11})$$
$$\mathbf{X}_2 | \mathbf{X}_1 = \mathbf{x}_1 \sim N_{p-r}(\mathbf{A}\mathbf{x}_1 + \mathbf{b}, \Lambda),$$

with $\mathbf{A}$ *and* $\mathbf{b}$ *as a matrix and a vector of constants, respectively, and matrix* Λ *is independent of* $\mathbf{x}_1$. *Then*

$$\mathbf{X} = \begin{pmatrix} \mathbf{X}_1 \\ \mathbf{X}_2 \end{pmatrix} \sim N_p \left(\begin{pmatrix} \mu_1 \\ \mathbf{A}\mu_1 + \mathbf{b} \end{pmatrix}, \begin{pmatrix} \Sigma_{11} & \Sigma_{11}\mathbf{A}^T \\ \mathbf{A}\Sigma_{11} & \Lambda + \mathbf{A}\Sigma_{11}\mathbf{A}^T \end{pmatrix} \right).$$

PROOF: We write the joint density, using the given marginal and conditional normals, as

$$\begin{aligned} f(\mathbf{X}) &= f(\mathbf{X}_1) f(\mathbf{X}_2 | \mathbf{X}_1) \\ &= (2\pi)^{-p/2} |\Sigma_{11}|^{-1/2} |\Lambda|^{-1/2} \exp(-\frac{1}{2}(Q_1 + Q_2), \qquad (6.14) \end{aligned}$$

where

$$Q_1 = (\mathbf{X}_1 - \mu_1)^T \Sigma_{11}^{-1} (\mathbf{X}_1 - \mu_1)$$
$$Q_2 = (\mathbf{X}_2 - \mathbf{A}\mathbf{X}_1 - \mathbf{b})^T \Lambda^{-1} (\mathbf{X}_2 - \mathbf{A}\mathbf{X}_1 - \mathbf{b}).$$

Partitioning this sum, term by term, and collecting quadratic and bilinear, linear, and constant terms together, we have

$$Q_1 + Q_2 = \mathbf{X}^T \mathbf{C}_1^{-1} \mathbf{X} - 2\mathbf{X}^T \mathbf{C}_2 + \mu_1^T \Sigma_{11}^{-1} \mu_1 + \mathbf{b}^T \Lambda^{-1} \mathbf{b},$$

where

$$\mathbf{C}_1^{-1} = \begin{pmatrix} \Sigma_{11}^{-1} + \mathbf{A}^T \Lambda^{-1} \mathbf{A} & -\mathbf{A}^T \Lambda^{-1} \\ -\Lambda^{-1} \mathbf{A} & \Lambda^{-1} \end{pmatrix}, \quad \mathbf{C}_2 = \begin{pmatrix} \Sigma_{11}^{-1} \mu_1 - \mathbf{A}^T \Lambda^{-1} \mathbf{b} \\ \Lambda^{-1} \mathbf{b} \end{pmatrix}.$$

Now, recall that, a random vector $\mathbf{X} \in \mathbb{R}^p$ with $\mathbf{X} \sim \mathsf{N}_p(\mu, \Sigma)$, must have the density:

$$f(\mathbf{X}) \propto \exp\left(-\frac{1}{2} (\mathbf{X}^T \Sigma^{-1} \mathbf{X} - 2\mathbf{X}^T \Sigma^{-1} \mu) \right)$$

Thus, we have to show, that $\mathbf{C}_1 = \Sigma$ and $\mathbf{C}_2 = \Sigma^{-1}\mu$. We use the inverse of a partitioned matrix as given in Seber (2008), as

$$\begin{pmatrix} \mathbf{A}_{11}^{-1} + \mathbf{A}_{11}^{-1} \mathbf{A}_{12} \mathbf{S}^{-1} \mathbf{A}_{21} \mathbf{A}_{11}^{-1} & -\mathbf{A}_{11}^{-1} \mathbf{A}_{12} \mathbf{S}^{-1} \\ -\mathbf{S}^{-1} \mathbf{A}_{21} \mathbf{A}_{11}^{-1} & \mathbf{S}^{-1} \end{pmatrix}^{-1} = \begin{pmatrix} \mathbf{A}_{11} & \mathbf{A}_{12} \\ \mathbf{A}_{21} & \mathbf{A}_{22} \end{pmatrix},$$

where $\mathbf{S} = \mathbf{A}_{22} - \mathbf{A}_{21} \mathbf{A}_{11}^{-1} \mathbf{A}_{12}$ is the Schur complement of $\mathbf{A}_{11}$, assuming $\mathbf{A}_{11} \succ 0$. Applying this to $\mathbf{C}_1$ and comparing the blocks deliver $\Lambda = \mathbf{S}$, $\mathbf{A}_{11} = \Sigma_{11}$ and $\mathbf{A}_{11}^{-1} \mathbf{A}_{12} = \mathbf{A}$. It follows that

$$\mathbf{C}_1 = \begin{pmatrix} \Sigma_{11} & \Sigma_{11} \mathbf{A}^T \\ \mathbf{A}\Sigma_{11} & \Lambda + \mathbf{A}\Sigma_{11} \mathbf{A}^T \end{pmatrix} = \Sigma,$$

as required. Comparison of the linear components yields

$$\mathbf{C}_1 \mathbf{C}_2 = \begin{pmatrix} \Sigma_{11} & \Sigma_{11} \mathbf{A}^T \\ \mathbf{A}\Sigma_{11} & \Lambda + \mathbf{A}\Sigma_{11} \mathbf{A}^T \end{pmatrix} \begin{pmatrix} \Sigma_{11}^{-1} \mu_1 - \mathbf{A}^T \Lambda^{-1} \mathbf{b} \\ \Lambda^{-1} \mathbf{b} \end{pmatrix} = \begin{pmatrix} \mu_1 \\ \mathbf{A}\mu_1 + \mathbf{b} \end{pmatrix},$$

the required mean vector.

□

The Bayes model in (6.13) fulfills the assumptions of Lemma 6.2. The data generating distribution is normal with mean which is linear in β and a covariance which is independent of β. We obtain the following theorem on the joint distribution of β and $\mathbf{y}$.

Theorem 6.1 *For*

$$\beta \sim N_p(\gamma, \Gamma) \quad and \quad \mathbf{y}|\beta \sim N_n(\mathbf{X}\beta, \Sigma)$$

the joint distribution of β and $\mathbf{y}$ is given as

$$\begin{pmatrix} \beta \\ \mathbf{y} \end{pmatrix} \sim N_{n+p} \left(\begin{pmatrix} \gamma \\ \mathbf{X}\gamma \end{pmatrix}, \begin{pmatrix} \Gamma & \Gamma\mathbf{X}^T \\ \mathbf{X}\Gamma & \Sigma + \mathbf{X}\Gamma\mathbf{X}^T \end{pmatrix} \right). \tag{6.15}$$

In the next step we apply Lemma 6.1 to the joint multivariate normal distribution in (6.15) and obtain the following theorem.

Theorem 6.2 *For*

$$\beta \sim N_p(\gamma, \Gamma) \quad and \quad \mathbf{y}|\beta \sim N_n(\mathbf{X}\beta, \Sigma)$$

it holds that

$$\mathbf{y} \sim N_n\left(\mu_{\mathbf{y}}, \Sigma_{\mathbf{y}}\right) \quad and \quad \beta|\mathbf{y} \sim N_p\left(\mu_{\beta|\mathbf{y}}, \Sigma_{\beta|\mathbf{y}}\right),$$

with

$$\mu_{\mathbf{y}} = \mathbf{X}\gamma \tag{6.16}$$

$$\Sigma_{\mathbf{y}} = \Sigma + \mathbf{X}\Gamma\mathbf{X}^T \tag{6.17}$$

$$\mu_{\beta|\mathbf{y}} = \gamma + \Gamma\mathbf{X}^T(\Sigma + \mathbf{X}\Gamma\mathbf{X}^T)^{-1}(\mathbf{y} - \mathbf{X}\gamma) \tag{6.18}$$

$$\Sigma_{\beta|\mathbf{y}} = \Gamma - \Gamma\mathbf{X}^T(\Sigma + \mathbf{X}\Gamma\mathbf{X}^T)^{-1}\mathbf{X}\Gamma. \tag{6.19}$$

Recall that, the posterior moments in (6.18)-(6.19) are expressed in terms of prior covariance matrix Γ, so that Γ need not be invertible, rather the $n \times n$ matrix $\mathbf{X}\Gamma\mathbf{X}^T + \Sigma$ is required to be so, which suffices to keep the posterior non-degenerate.

We also note, however, that the posterior moments often provide a better insight when expressed in terms of precision matrices, Σ^{-1} and Γ^{-1}. The following lemma on special matrix inverse identities will be useful in achieving these objectives.

Lemma 6.3 *Let* $\mathbf{A} \in \mathbb{R}^{m \times m}$ *and* $\mathbf{B} \in \mathbb{R}^{n \times n}$ *be non-singular matrices, and* $\mathbf{C} \in \mathbb{R}^{m \times n}$ *and* $\mathbf{D} \in \mathbb{R}^{n \times m}$ *be any two matrices such that* $\mathbf{A} + \mathbf{CBD}$ *is non-singular. Then*

$$(\mathbf{A} + \mathbf{CBD})^{-1} = \mathbf{A}^{-1} - \mathbf{A}^{-1}\mathbf{C}(\mathbf{B}^{-1} + \mathbf{DA}^{-1}\mathbf{C})^{-1}\mathbf{DA}^{-1}. \qquad (6.20)$$

In particular,

$$\begin{aligned} (\mathbf{I}_m + \mathbf{CC}^T)^{-1} &= \mathbf{I}_m - \mathbf{C}(\mathbf{I}_n + \mathbf{C}^T\mathbf{C})^{-1}\mathbf{C}^T \qquad (6.21) \\ (\mathbf{I}_m + \mathbf{CC}^T)^{-1}\mathbf{C} &= \mathbf{C}(\mathbf{I}_n + \mathbf{C}^T\mathbf{C})^{-1}. \qquad (6.22) \end{aligned}$$

PROOF: The proof follows by showing that $\mathbf{KK}^{-1} = \mathbf{I}$ for $\mathbf{K} = \mathbf{A} + \mathbf{CBD}$. Set

$$\mathbf{M} = \mathbf{B}^{-1} + \mathbf{DA}^{-1}\mathbf{C}.$$

We have

$$\begin{aligned} \mathbf{KK}^{-1} &= (\mathbf{A} + \mathbf{CBD})\left(\mathbf{A}^{-1} - \mathbf{A}^{-1}\mathbf{CM}^{-1}\mathbf{DA}^{-1}\right) \\ &= \mathbf{I} - \mathbf{CM}^{-1}\mathbf{DA}^{-1} + \mathbf{CBDA}^{-1} - \mathbf{CBDA}^{-1}\mathbf{CM}^{-1}\mathbf{DA}^{-1} \\ &= \mathbf{I} - \mathbf{CB}\left(\mathbf{B}^{-1} - \mathbf{M} + \mathbf{DA}^{-1}\mathbf{C}\right)\mathbf{M}^{-1}\mathbf{DA}^{-1} \\ &= \mathbf{I} - \mathbf{CB}\left(\mathbf{M} - \mathbf{M}\right)\mathbf{M}^{-1}\mathbf{DA}^{-1} = \mathbf{I}. \end{aligned}$$

If $\mathbf{A} = \mathbf{I}_m$, $\mathbf{B} = \mathbf{I}_n$, and $\mathbf{D} = \mathbf{C}^T$, then (6.20) reduces to (6.21). Further we use (6.21) to show (6.22):

$$\begin{aligned} (\mathbf{I}_m + \mathbf{CC}^T)^{-1}\mathbf{C} &= \mathbf{C} - \mathbf{C}(\mathbf{I}_n + \mathbf{C}^T\mathbf{C})^{-1}\mathbf{C}^T\mathbf{C} \\ &= \mathbf{C}(\mathbf{I}_n + \mathbf{C}^T\mathbf{C})^{-1}\left(\mathbf{I}_n + \mathbf{C}^T\mathbf{C} - \mathbf{C}^T\mathbf{C}\right) \\ &= \mathbf{C}(\mathbf{I}_n + \mathbf{C}^T\mathbf{C})^{-1}. \end{aligned}$$

$\square$

Following corollary to Theorem 6.2 utilizes the identities in Lemma 6.3 to re-write the posterior moments in terms of Σ^{-1} and Γ^{-1}.

Corollary 6.1 *Assuming* $\Sigma \in \mathbb{R}^{n \times n}$ *and* $\Gamma \in \mathbb{R}^{p \times p}$ *non-singular, the posterior moments in Theorem 6.2 can be re-formulated as*

$$\begin{aligned} \mu_{\beta|\mathbf{y}} &= \Sigma_{\beta|\mathbf{y}}(\mathbf{X}^T\Sigma^{-1}\mathbf{y} + \Gamma^{-1}\gamma) \qquad (6.23) \\ \Sigma_{\beta|\mathbf{y}} &= (\Gamma^{-1} + \mathbf{X}^T\Sigma^{-1}\mathbf{X})^{-1}. \qquad (6.24) \end{aligned}$$

PROOF: We obtain (6.24) by applying (6.20) to (6.19) with $\mathbf{A} = \Gamma$, $\mathbf{C} = \mathbf{D}^T = \mathbf{X}$ and $\mathbf{B} = \Sigma^{-1}$. Now we show (6.23). From (6.18) we have

$$\mu_{\beta|\mathbf{y}} = \gamma + \Gamma\mathbf{X}^T(\mathbf{X}\Gamma\mathbf{X}^T + \Sigma)^{-1}(\mathbf{y} - \mathbf{X}\gamma)$$
$$= m_1 + m_2$$

with

$$m_1 = \gamma - \Gamma\mathbf{X}^T\left(\mathbf{X}\Gamma\mathbf{X}^T + \Sigma\right)^{-1}\mathbf{X}\gamma$$
$$= \left(\Gamma - \Gamma\mathbf{X}^T(\mathbf{X}\Gamma\mathbf{X}^T + \Sigma)^{-1}\mathbf{X}\Gamma\right)\Gamma^{-1}\gamma$$
$$= \Sigma_{\beta|\mathbf{y}}\Gamma^{-1}\gamma$$

and

$$m_2 = \Gamma\mathbf{X}^T\left(\mathbf{X}\Gamma\mathbf{X}^T + \Sigma\right)^{-1}\mathbf{y}$$
$$= \Gamma^{\frac{1}{2}}\Gamma^{\frac{1}{2}}\mathbf{X}^T\Sigma^{-\frac{1}{2}}\left(\Sigma^{-\frac{1}{2}}\mathbf{X}\Gamma\mathbf{X}^T\Sigma^{-\frac{1}{2}} + \mathbf{I}_n\right)^{-1}\Sigma^{-\frac{1}{2}}\mathbf{y}$$
$$= \Gamma^{\frac{1}{2}}\mathbf{C}\left(\mathbf{C}^T\mathbf{C} + \mathbf{I}_n\right)^{-1}\Sigma^{-\frac{1}{2}}\mathbf{y},$$

where $\mathbf{C} = \Gamma^{\frac{1}{2}}\mathbf{X}^T\Sigma^{-\frac{1}{2}}$. Applying (6.22) gives

$$m_2 = \Gamma^{\frac{1}{2}}\left(\mathbf{I}_m + \mathbf{C}\mathbf{C}^T\right)^{-1}\mathbf{C}\Sigma^{-\frac{1}{2}}\mathbf{y}$$
$$= \Gamma^{\frac{1}{2}}\left(\mathbf{I}_m + \Gamma^{\frac{1}{2}}\mathbf{X}^T\Sigma^{-1}\mathbf{X}\Gamma^{\frac{1}{2}}\right)^{-1}\Gamma^{\frac{1}{2}}\mathbf{X}^T\Sigma^{-1}\mathbf{y}$$
$$= \left(\Gamma^{-1} + \mathbf{X}^T\Sigma^{-1}\mathbf{X}\right)^{-1}\mathbf{X}^T\Sigma^{-1}\mathbf{y}$$
$$= \Sigma_{\beta|\mathbf{y}}\mathbf{X}^T\Sigma^{-1}\mathbf{y}.$$

Finally, substituting m_1 and m_2 in $\mu_{\beta|\mathbf{y}}$ gives (6.23).

$\square$

Note that, we have

$$\mathsf{Cov}(\beta) = \Gamma \text{ and } \mathsf{Cov}(\beta|\mathbf{y}) = (\Gamma^{-1} + \mathbf{X}^T\Sigma^{-1}\mathbf{X})^{-1}.$$

It holds that
$$\mathsf{Cov}(\beta) \succeq \mathsf{Cov}(\beta|\mathbf{y})$$

since $\mathbf{X}^T\Sigma^{-1}\mathbf{X} \succeq 0$. This implies that the posterior not only updates the prior, using the information in data, rather also improves it by reducing its variance.

Before we present special cases of (6.23) and (6.24) we show an alternative method of their derivation.

Following the general notion (see Chapter 2), the posterior distribution of $\beta|\mathbf{y}$ can be obtained as

$$\pi(\beta|\mathbf{y}) \propto \pi(\beta)\ell(\beta|\mathbf{y}).$$

For the set up in (6.13), the prior and the likelihood can be written, respectively, as

$$\pi(\beta) = \frac{1}{(2\pi)^{p/2}|\Gamma|^{1/2}} \exp\left(-\frac{1}{2}(\beta - \gamma)^T \Gamma^{-1}(\beta - \gamma)\right)$$

$$\ell(\theta|\mathbf{y}) = \frac{1}{(2\pi)^{n/2}|\Sigma|^{1/2}} \exp\left(-\frac{1}{2}(\mathbf{y} - \mathbf{X}\beta)^T \Sigma^{-1}(\mathbf{y} - \mathbf{X}\beta)\right),$$

where γ, Γ, and Σ are assumed known. Thus

$$\pi(\beta|\mathbf{y}) \propto \exp\left(-\frac{1}{2}(Q_1 + Q_2)\right)$$

with

$$Q_1 = (\beta - \gamma)^T \Gamma^{-1}(\beta - \gamma)$$
$$Q_2 = (\mathbf{y} - \mathbf{X}\beta)^T \Sigma^{-1}(\mathbf{y} - \mathbf{X}\beta).$$

Consider $Q_1 + Q_2$ as function of β and collecting the quadratic, linear, and constant terms together we obtain

$$Q_1 + Q_2 = \beta^T(\Gamma^{-1} + \mathbf{X}^T\Sigma^{-1}\mathbf{X})\beta - 2\beta^T(\Gamma^{-1}\gamma + \mathbf{X}^T\Sigma^{-1})\mathbf{y} + \text{const}_1$$
$$= \beta^T\Gamma_1^{-1}\beta - 2\beta^T\Gamma_1^{-1}\gamma_1 + \text{const}_1$$

with

$$\Gamma_1 = (\Gamma^{-1} + \mathbf{X}^T\Sigma^{-1}\mathbf{X})^{-1}$$
$$\gamma_1 = \Gamma_1(\Gamma^{-1}\gamma + \mathbf{X}^T\Sigma^{-1}\mathbf{y})$$
$$\text{const}_1 = \gamma^T\Gamma^{-1}\gamma + \mathbf{y}^T\Sigma^{-1}\mathbf{y}.$$

Completing the squares by $Q_3 = \gamma_1^T\Gamma_1^{-1}\gamma_1$, we obtain

$$Q_1 + Q_2 = \beta^T\Gamma_1^{-1}\beta - 2\beta^T\Gamma_1^{-1}\gamma_1 + \gamma_1^T\Gamma_1^{-1}\gamma_1 + \text{const}_1 - Q_3$$
$$= (\beta - \gamma_1)^T\Gamma_1^{-1}(\beta - \gamma_1) + \text{const}_2$$

where
$$\text{const}_2 = \gamma^T\Gamma^{-1}\gamma + \mathbf{y}^T\Sigma^{-1}\mathbf{y} - \gamma_1^T\Gamma_1^{-1}\gamma_1.$$

The constant can also be re-written as

$$\text{const}_2 = (\mathbf{y} - \mathbf{X}\gamma_1)^T\Sigma^{-1}(\mathbf{y} - \mathbf{X}\gamma_1) + (\gamma - \gamma_1)^T\Gamma^{-1}(\gamma - \gamma_1).$$

To see this we expand the two quadratic forms above as follows

$$\mathbf{y}^T\Sigma^{-1}\mathbf{y} + \gamma^T\Gamma^{-1}\gamma + Q_4.$$

with

$$Q_4 = -2\gamma_1^T \mathbf{X}^T \Sigma^{-1} \mathbf{y} + \gamma_1^T \mathbf{X}^T \Sigma^{-1} \mathbf{X} \gamma_1 - 2\gamma_1 \Gamma^{-1} \gamma + \gamma_1 \Gamma^{-1} \gamma_1$$
$$= -2\gamma_1^T (\mathbf{X}^T \Sigma^{-1} \mathbf{y} + \Gamma^{-1} \gamma) + \gamma_1^T (\Gamma^{-1} + \mathbf{X}^T \Sigma^{-1} \mathbf{X}) \gamma_1$$
$$= -2\gamma_1^T \Gamma_1^{-1} \gamma_1 + \gamma_1^T \Gamma_1^{-1} \gamma_1$$
$$= -\gamma_1^T \Gamma^{-1} \gamma_1$$

which gives the formula of const_2. Summarizing we obtain for the posterior

$$\pi(\beta|\mathbf{y}) \propto \exp\left(-\frac{1}{2}(\beta - \gamma_1)^T \Gamma_1^{-1}(\beta - \gamma_1)\right).$$

This is the kernel of $\mathsf{N}_p(\gamma_1, \Gamma_1)$. Comparing the formulas for γ_1 and Γ_1 with (6.23) and (6.24) we obtain again

$$\beta|\mathbf{y} \sim \mathsf{N}_p(\mu_{\beta|\mathbf{y}}, \Sigma_{\beta|\mathbf{y}}).$$

For later application we summarize the calculations in the following lemma.

Lemma 6.4 *It holds that*

$$Q = (\beta - \gamma)^T \Gamma^{-1}(\beta - \gamma) + (\mathbf{y} - \mathbf{X}\beta)^T \Sigma^{-1}(\mathbf{y} - \mathbf{X}\beta)$$

can be expressed as

$$Q = (\beta - \gamma_1)^T \Gamma_1^{-1}(\beta - \gamma_1) + (\mathbf{y} - \mathbf{X}\gamma_1)^T \Sigma^{-1}(\mathbf{y} - \mathbf{X}\gamma_1) + (\gamma - \gamma_1)^T \Gamma^{-1}(\gamma - \gamma_1) \tag{6.25}$$

or alternatively as

$$Q = (\beta - \gamma_1)^T \Gamma_1^{-1}(\beta - \gamma_1) + \mathbf{y}^T \Sigma^{-1} \mathbf{y} + \gamma^T \Gamma^{-1} \gamma - \gamma_1^T \Gamma_1^{-1} \gamma_1 \tag{6.26}$$

with

$$\gamma_1 = \Gamma_1(\Gamma^{-1}\gamma + \mathbf{X}^T \Sigma^{-1} \mathbf{y}).$$
$$\Gamma_1 = (\Gamma^{-1} + \mathbf{X}^T \Sigma^{-1} \mathbf{X})^{-1}$$

6.2.1.1 Special Cases

We now explore a few special cases. To begin with, we assume that the errors $\epsilon_1, \ldots, \epsilon_n$ in (6.4) are i.i.d. $\mathsf{N}_1(0, \sigma^2)$ distributed, such that for ϵ in (6.2)

$$\epsilon \sim \mathsf{N}_n(0, \sigma^2 \mathbf{I}_n),$$

which implies $\text{Cov}(\mathbf{y}) = \sigma^2 \mathbf{I}_n$. Note that, structures like $\sigma^2 \mathbf{I}_n$ are called *spherical* because the level sets of the density are spheres. In this case, the posterior distribution is $N(\mu_{\beta|\mathbf{y}}, \Sigma_{\beta|\mathbf{y}})$ with

$$\mu_{\beta|\mathbf{y}} = \left(\Gamma^{-1} + \sigma^{-2}\mathbf{X}^T\mathbf{X}\right)^{-1}\left(\sigma^{-2}\mathbf{X}^T\mathbf{y} + \Gamma^{-1}\gamma\right) \qquad (6.27)$$

$$\Sigma_{\beta|\mathbf{y}} = \left(\Gamma^{-1} + \sigma^{-2}\mathbf{X}^T\mathbf{X}\right)^{-1}. \qquad (6.28)$$

Another special case is that each component of β has independent prior information with same precision, i.e., we assume

$$\beta \sim N_p(\gamma, \tau^2\mathbf{I}_p).$$

Then the moments of the posterior further simplify to

$$\mu_{\beta|\mathbf{y}} = \left(\tau^{-2}\mathbf{I}_p + \sigma^{-2}\mathbf{X}^T\mathbf{X}\right)^{-1}\left(\mathbf{X}^T\mathbf{y} + \tau^{-2}\gamma\right) \qquad (6.29)$$

$$\Sigma_{\beta|\mathbf{y}} = \left(\tau^{-2}\mathbf{I}_p + \sigma^{-2}\mathbf{X}^T\mathbf{X}\right)^{-1}. \qquad (6.30)$$

If we additionally assume that the design matrix $\mathbf{X}$ is orthogonal, such that $\mathbf{X}^T\mathbf{X} = \eta^2\mathbf{I}_p$, then the maximum likelihood estimator is given by

$$\widehat{\beta} = (\mathbf{X}^T\mathbf{X})^{-1}\mathbf{X}^T\mathbf{y} = \eta^{-2}\mathbf{X}^T\mathbf{y}.$$

We can express the posterior moments as

$$\begin{aligned}
\mu_{\beta|\mathbf{y}} &= \left(\tau^{-2}\mathbf{I}_p + \sigma^{-2}\eta^2\mathbf{I}_p\right)^{-1}\left(\sigma^{-2}\mathbf{X}^T\mathbf{y} + \tau^{-2}\gamma\right) \\
&= \frac{1}{\sigma^2/\tau^2 + \eta^2}\left(\eta^2\widehat{\beta} + \gamma\sigma^2/\tau^2\right) \qquad (6.31)
\end{aligned}$$

$$\begin{aligned}
\Sigma_{\beta|\mathbf{y}} &= \left(\tau^{-2}\mathbf{I}_p + \sigma^{-2}\eta^2\mathbf{I}_p\right)^{-1} \\
&= \frac{\sigma^2}{\sigma^2/\tau^2 + \eta^2}\mathbf{I}_p. \qquad (6.32)
\end{aligned}$$

The expectation of the posterior in (6.31) is the convex combination of the maximum likelihood estimator and the prior expectation.

In particular, we notice that the prior precision and the posterior precision coincide if we let $\sigma^2 \to \infty$, keeping τ^2 and η^2 fixed, i.e., if we make the data *sacrifice* its precision completely. Note also that, if we let $\tau^2 \to \infty$, i.e., the prior becomes non-informative, then the posterior precision reduces to $\eta^2\sigma^{-2}\mathbf{I}_p$.

Example 6.4 (Life length vs. life line)
Here, we analyze the data for Example 6.3 under the model

$$y_i = \alpha + \beta x_i + \varepsilon_i, \ i = 1, \dots, 48, \ \varepsilon_i \sim N(0, \sigma^2) \ i.i.d. \ \text{with} \ \sigma^2 = 160, \quad (6.33)$$

using two conjugate priors for $\theta = (\alpha, \beta)$, namely

$$\begin{pmatrix} \alpha \\ \beta \end{pmatrix} \sim N_2 \left(\begin{pmatrix} 0 \\ 0 \end{pmatrix}, \sigma^2 \begin{pmatrix} 100 & 0 \\ 0 & 100 \end{pmatrix} \right)$$

and

$$\begin{pmatrix} \alpha \\ \beta \end{pmatrix} \sim N_2 \left(\begin{pmatrix} 100 \\ 0 \end{pmatrix}, \sigma^2 \begin{pmatrix} 100 & 0 \\ 0 & 100 \end{pmatrix} \right).$$

We apply Corollary 6.1 and obtain, for the first prior, the posterior

$$\begin{pmatrix} \alpha \\ \beta \end{pmatrix} \Big| \mathbf{y} \sim N_2 \left(\begin{pmatrix} 86.69 \\ -2.13 \end{pmatrix}, \begin{pmatrix} 215.40 & -23.30 \\ -23.30 & 2.56 \end{pmatrix} \right)$$

and for the second prior, the posterior

$$\begin{pmatrix} \alpha \\ \beta \end{pmatrix} \Big| \mathbf{y} \sim N_2 \left(\begin{pmatrix} 88.04 \\ -2.27 \end{pmatrix}, \begin{pmatrix} 215.40 & -23.30 \\ -23.30 & 2.56 \end{pmatrix} \right).$$

The respective marginal posteriors of the slope are

$$\beta|\mathbf{y} \sim N(-2.13, 2.56) \quad \text{and} \quad \beta|\mathbf{y} \sim N(-2.27, 2.56).$$

We see that the choice of prior of the intercept has an influence on the posterior expectation of the slope. We observe much reduced variance in the posterior of the slope compared with the prior. The posterior variance of the intercept in model (6.33) is still high.

Because we are mostly interested in the slope, we center the response and the X–variable. The new model is:

$$y_i - \bar{y} = \beta(x_i - \bar{x}) + \varepsilon_i, \ i = 1, \ldots, 48, \ \varepsilon_i \sim N(0, \sigma^2) \ i.i.d., \ \sigma^2 = 160. \quad (6.34)$$

We assume the prior $\beta \sim N(0, \sigma^2 100)$ and obtain $\beta|\mathbf{y} \sim N(-2.257, 2.59)$. One of posteriors of the slope is plotted in Figure 6.3.   □

R Code 6.2.4. Life data, Example 6.4.

```
library(matrixcalc)
sigma<-sqrt(160) # variance known
# alpha~N(a,sa^2); beta~N(b,sb^2) # prior
# First case
a<-0 # prior mean intercept
```

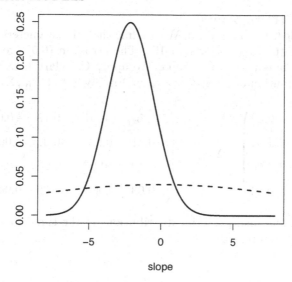

Figure 6.3: Example 6.4 on life length data. The broken line is the prior density, $N(0, \sigma^2 100)$, of the slope, and the continuous curve is the respective posterior density.

```
sa<-sigma*10 # prior sd intercept
b<-0# prior mean slope
sb<-sigma*10 # prior sd slope
Gamma<-matrix(c(sa^2,0,0,sb^2),ncol=2) # prior covariance
solve(Gamma) # inverse matrix
L<-Length;
XX<-matrix(c(48,sum(L),sum(L),sum(L*L)),ncol=2)
P<-XX*(sigma)^(-2)+solve(Gamma)## precision
Gamma1<-solve(P) # posterior covariance
xy<-c(sum(Age),sum(Age*L))
bpost<-Gamma1%*%(sigma^(-2)*xy+solve(Gamma)%*%c(a,b))
bpost # posterior mean of intercept and slope
# centered model
Agec<-Age-mean(Age); y<-Agec
Lengthc<-L-mean(L); x<-Lengthc
# Bayes calculation in centered model
bb<-0 # prior mean
varbb<-100*160 # prior variance
Gamma1c<-(varbb^(-1)+(sigma)^(-2)sum(x*x))^(-1)
Gamma1c # posterior variance
gamma1bb<-Gamma1c*(sigma^(-2)*sum(x*y)+bb*varbb^(-1))
gamma1bb # posterior mean
```

Example 6.5 (Corn plants)

Recall Example 6.1 on corn data. We assume that all parameters are independent and have the same prior $N(0, 10)$. The errors in (6.5) are independent normally distributed with $\sigma^2 = 300$. We apply Corollary 6.1. The resulting posterior distribution of $\beta = (\beta_0, \beta_1, \beta_2, \beta_3)$ is $\beta | \mathbf{y} \sim N_4(\mu_{\beta|\mathbf{y}}, \Sigma_{\beta|\mathbf{y}})$ with

$$
\mu_{\beta|\mathbf{y}} = \begin{pmatrix} 2.22 \\ 1.34 \\ 0.65 \\ 0.24 \end{pmatrix}, \quad \Sigma_{\beta|\mathbf{y}} = \begin{pmatrix} 9.65 & -0.01 & -0.12 & -0.03 \\ -0.01 & 0.23 & -0.07 & 0.004 \\ -0.12 & -0.07 & 0.09 & -0.02 \\ -0.03 & 0.004 & -0.02 & 0.007 \end{pmatrix}.
$$

The prior and marginal posterior distributions of the three components of β are depicted in Figure 6.4.   □

We finish this section by discussing a posterior distribution for a one-way ANOVA model, followed by two related examples. For simplicity, we only focus on fixed effects models, i.e.,

$$
y_{ij} = \mu_i + \epsilon_{ij}, \quad \epsilon_{ij} \sim N(0, \sigma^2) \tag{6.35}
$$

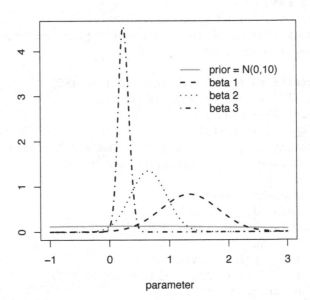

Figure 6.4: Example 6.5. Prior and posterior distributions of $\beta_1, \beta_2, \beta_3$ for corn data with conjugate prior.

where y_{ij} are independent observations measured on jth subject in ith group, $j = 1, \ldots, r_i$, $i = 1, \ldots, a$. It holds that $y_{ij} \sim \mathsf{N}(\mu_i, \sigma^2)$, hence

$$\bar{y}_i \sim \mathsf{N}(\mu_i, \sigma^2/r_i) \quad \text{where} \quad \bar{y}_i = \sum_{j=1}^{r_i} y_{ij}/r_i.$$

We let $\mu_i \sim \mathsf{N}(\gamma_i, \tau^2)$ be the prior, independent for each μ_i and assume γ_i and τ^2, along with σ^2, known. Model (6.35) consists of a independent samples. For known variance σ^2 the statistic $\bar{y}_i$ is sufficient in each sample. Following theorem gives the posterior distribution for each group.

Theorem 6.3 *Given* $\bar{y}_i \sim \mathsf{N}(\mu_i, \sigma^2/r_i)$ *with known* σ^2 *and the prior* $\mu_i \sim \mathsf{N}(\gamma_i, \tau^2)$, *where* $\bar{y}_i = \sum_{j=1}^{r_i} y_{ij}/r_i$, $i = 1, \ldots, a$. *The posterior distribution of* $\mu_i|\bar{y}_i$ *follows for each group* i *as*

$$\mu_i|\bar{y}_i \sim \mathsf{N}(A_i, m_i^2) \tag{6.36}$$

where

$$A_i = \frac{\bar{y}_i \tau^2 + \gamma_i \sigma^2/r_i}{\tau^2 + \sigma^2/r_i} \quad \text{and} \quad m_i^2 = \frac{\tau^2 \sigma^2/r_i}{\tau^2 + \sigma^2/r_i}. \tag{6.37}$$

PROOF: Apply the results of Example 2.12.

$\square$

In the following, we apply Theorem 6.3 on two examples, the data sets of which are taken from Daniel and Cross (2013, Chapter 8).

Example 6.6 (Parkinson)

The aim of the experiment is to assess the effects of weights on postural hand tremor during self-feeding in patients with Parkinson's disease. A random sample of $n = 48$ patients is divided into $a = 3$ groups of $r = 16$ each. The three groups pertain to three different conditions, namely holding a built-up spoon (108 grams), holding a weighted spoon (248 grams), and holding a built-up spoon while wearing a weighted wrist cuff (470 grams). The amplitude of the tremor is measured on each patient (in mm).

Denoting the amplitude measured on jth patient under ith condition as y_{ij}, the model can be stated as

$$y_{ij} = \mu_i + \epsilon_{ij}, \ i = 1, 2, 3, \ j = 1, \ldots, 16.$$

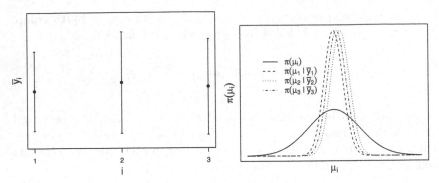

Figure 6.5: Illustration for Example 6.6. Left: Bar plot of the Parkinson data. Right: Prior and posterior distributions of μ_i.

The three sample means, $\bar{y}_i$, are 0.495, 0.575, 0.535, with the corresponding variances as 0.144, 0.234 and 0.208. The left panel of Figure 6.5 shows the basic statistics as bar plots.

For the analysis, we use $\mu_i \sim N(0.5, 0.3)$ as prior and take $\epsilon_{ij} \sim N(0, 0.2)$. This leads to the posterior distributions as $N(0.496, 0.012)$, $N(0.574, 0.012)$ and $N(0.534, 0.012)$, respectively for μ_1, μ_2 and μ_3. The prior and posteriors are plotted in the right panel of Figure 6.5. $\square$

Example 6.7 (Smokers)

The data consist of serum concentration of lipoprotein on a random sample of $r = 7$ individuals classified under each of $a = 4$ groups as non-smokers, light smokers, moderate smokers and heavy smokers. The objective of the study is to see if the average serum concentration differs across four groups.

The model for concentration measured on jth subject in ith group is stated as
$$y_{ij} = \mu_i + \epsilon_{ij}, \ i = 1, \ldots, 4, \ j = 1, \ldots, 7.$$

The sample means and variances of the four groups are, respectively, (10.857, 8.286, 6.143, 3.286) and (2.476, 2.571, 2.810, 3.238). The downward trend of the averages across four groups and almost identical variances are also clear from the bar plot of the data shown in the left panel of Figure 6.6.

Using $\mu_i \sim N(5, 2.5)$ as prior for each group and $\epsilon \sim N(0, 2.5)$, the posterior distributions as obtained as $N(10.125, 0.313)$, $N(7.875, 0.313)$, $N(6, 0.313)$ and $N(3.5, 0.313)$, respectively for μ_1, μ_2, μ_3 and μ_4, and the same are plotted, along with the prior, in the right panel of Figure 6.6. $\square$

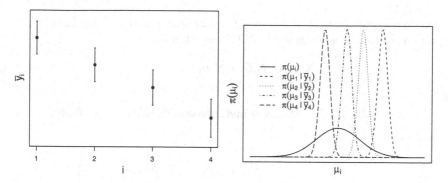

Figure 6.6: Example 6.7. Left: Bar plot of smokers data. Right: Prior and posterior distributions of μ_i.

6.2.2   Conjugate Prior: Parameter $\theta = (\beta, \sigma^2)$

Consider model (6.12) where the covariance matrix is known up to a scaling parameter. We assume the model

$$\mathcal{P} = \left\{ \mathsf{N}_n(\mathbf{X}\beta, \sigma^2\Sigma) : \ \beta \in \mathbb{R}^p, \ \sigma^2 \in \mathbb{R}, \ \sigma^2 > 0 \right\}. \tag{6.38}$$

The unknown parameter θ consists of the linear regression parameter β and the scalar parameter σ^2. The design matrix $\mathbf{X} \in \mathbb{R}^{n \times p}$ and the positive definite matrix $\Sigma \in \mathbb{R}^{n \times n}$ are known. This model belongs to an exponential family of order $p + 1$. It is possible to apply Theorem 3.3 for deriving the conjugate family. Here we take the direct way. First we introduce the distribution family from which we chose the prior and later we show that the posterior also belongs to the same distribution.

We already know that the conjugate prior of β given σ^2 is a normal distribution. Further we know that the conjugate prior of σ^2 in a normal model with known expectation is an inverse–gamma distribution, see Example 2.14. The conjugate family for the joint parameter $\theta = (\beta, \sigma^2)$ is a combination of both. We assume that $\beta|\sigma^2 \sim \mathsf{N}_p(\gamma, \sigma^2\Gamma)$ and $\sigma^2 \sim \mathsf{InvGamma}(a/2, b/2)$. We write the entire set up hierarchically as

$$
\begin{aligned}
\mathbf{y}|\beta, \sigma^2 &\sim \mathsf{N}_n(\mathbf{X}\beta, \sigma^2\Sigma) \\
\beta|\sigma^2 &\sim \mathsf{N}_p(\gamma, \sigma^2\Gamma) \\
\sigma^2 &\sim \mathsf{InvGamma}(a/2, b/2).
\end{aligned}
\tag{6.39}
$$

Note that, β and σ^2 are not independent. The joint distribution of $\theta = (\beta, \sigma^2)$ is called *normal-inverse-gamma* (NIG) distribution . We introduce the distribution in a general setting as follows.

A random vector $\mathbf{X} \in \mathbb{R}^p$ and a positive random scalar $\lambda \in \mathbb{R}_+$ have jointly a *normal-inverse-gamma* (NIG) distribution

$$(\mathbf{X}, \lambda) \sim \mathsf{NIG}(\alpha, \beta, \mu, \Sigma)$$

iff

$$\mathbf{X} | \lambda \sim \mathsf{N}_p(\mu, \lambda \Sigma) \quad \text{and} \quad \lambda \sim \mathsf{InvGamma}(\alpha/2, \beta/2) \qquad (6.40)$$

with density

$$f(\mathbf{X}, \lambda) = c\, \lambda^{-\frac{p+\alpha+2}{2}} \exp\left(-\frac{1}{2\lambda}\left(\beta + (\mathbf{X} - \mu)^T \Sigma^{-1} (\mathbf{X} - \mu) \right) \right). \tag{6.41}$$

The constant c is given by

$$c = \frac{1}{(2\pi)^{p/2} |\Sigma|^{1/2}} \frac{b^{\frac{\alpha}{2}}}{2^{\frac{\alpha}{2}} \Gamma(\frac{\alpha}{2})}. \tag{6.42}$$

The Bayes model $\{\mathcal{P}, \pi_c\}$ specializes to

$$\left\{ \{ \mathsf{N}_n(\mathbf{X}\beta, \sigma^2 \Sigma) : \beta \in \mathbb{R}^p, \sigma^2 \in \mathbb{R}_+ \},\ \mathsf{NIG}(a, b, \gamma, \Gamma) \right\}. \tag{6.43}$$

The following theorem shows that a normal-inverse-gamma prior leads to a normal-inverse-gamma posterior with updated parameters.

Theorem 6.4 *For the Bayes model (6.43), the joint posterior distribution of (β, σ^2) is the normal-inverse-gamma (NIG) distribution,*

$$(\beta, \sigma^2) | \mathbf{y} \sim \mathsf{NIG}(a_1, b_1, \gamma_1, \Gamma_1), \tag{6.44}$$

with

$$
\begin{aligned}
a_1 &= a + n \\
b_1 &= b + (\mathbf{y} - \mathbf{X}\gamma_1)^T \Sigma^{-1}(\mathbf{y} - \mathbf{X}\gamma_1) + (\gamma - \gamma_1)^T \Gamma^{-1}(\gamma - \gamma_1) \\
\gamma_1 &= \Gamma_1 (\mathbf{X}^T \Sigma^{-1} \mathbf{y} + \Gamma^{-1}\gamma) \\
\Gamma_1 &= (\mathbf{X}^T \Sigma^{-1} \mathbf{X} + \Gamma^{-1})^{-1}.
\end{aligned}
\tag{6.45}
$$

PROOF: Consider prior in (6.43). Taking the likelihood function,

$$\ell(\mathbf{y}|\beta,\sigma^2) = \frac{1}{(2\pi)^{n/2}(\sigma^2)^{n/2}|\Sigma|^{1/2}} \exp\left(-\frac{1}{2\sigma^2}(\mathbf{y}-\mathbf{X}\beta)\Sigma^{-1}(\mathbf{y}-\mathbf{X}\beta)\right),$$

into account, the posterior can be written as

$$\pi(\beta,\sigma^2|\mathbf{y}) \propto \frac{1}{(\sigma^2)^{(n+p+a+2)/2}} \exp\left\{-\frac{1}{2\sigma^2}(Q+b)\right\},$$

with

$$Q = (\mathbf{y}-\mathbf{X}\beta)^T\Sigma^{-1}(\mathbf{y}-\mathbf{X}\beta) + (\beta-\gamma)^T\Gamma^{-1}(\beta-\gamma).$$

Applying Lemma 6.4, (6.25) delivers the statement.

□

The posterior moments in Theorem 6.4 include the inverse matrices of Σ and Γ. In case σ^2 is given, two different representations of the posterior moments follow from Theorem 6.2 and Corollary 6.1. The formulas in (6.45) are related to Corollary 6.1. We derive the presentations of the posterior moments related to Theorem 6.4 for $\theta = (\beta,\sigma^2)$ in the following corollary. One advantage of the new formulas is that the inverses of Σ and Γ are not needed.

Corollary 6.2 *For the Bayes model (6.43), the joint posterior distribution of (β,σ^2) is the normal-inverse gamma (NIG) distribution,*

$$(\beta,\sigma^2)|\mathbf{y} \sim \mathsf{NIG}(a_1,b_1,\gamma_1,\Gamma_1), \tag{6.46}$$

with

$$\begin{aligned}
a_1 &= a + n \\
b_1 &= b + (\mathbf{y}-\mathbf{X}\gamma)^T(\Sigma+\mathbf{X}\Gamma\mathbf{X}^T)^{-1}(\mathbf{y}-\mathbf{X}\gamma) \\
\gamma_1 &= \gamma + \Gamma\mathbf{X}^T(\Sigma+\mathbf{X}\Gamma\mathbf{X}^T)^{-1}(\mathbf{y}-\mathbf{X}\gamma) \\
\Gamma_1 &= \Gamma - \Gamma\mathbf{X}^T(\Sigma+\mathbf{X}\Gamma\mathbf{X}^T)^{-1}\mathbf{X}\Gamma.
\end{aligned} \tag{6.47}$$

PROOF: The formulas for γ_1 and Γ_1 in (6.45) are the same as in Corollary 6.1. We can apply Theorem 6.2. It remains to check the expression of b_1. Set

$$\mathbf{M} = \Sigma + \mathbf{X}\Gamma\mathbf{X}^T. \tag{6.48}$$

We have to show that $Q = (\mathbf{y} - \mathbf{X}\gamma)^T \mathbf{M}^{-1}(\mathbf{y} - \mathbf{X}\gamma) = Q_1 + Q_2$, where

$$Q_1 = (\mathbf{y} - \mathbf{X}\gamma_1)^T \Sigma^{-1}(\mathbf{y} - \mathbf{X}\gamma_1), \quad Q_2 = (\gamma - \gamma_1)^T \Gamma^{-1}(\gamma - \gamma_1).$$

We start with Q_2 and apply

$$\gamma_1 = \gamma + \Gamma \mathbf{X}^T \mathbf{M}^{-1}(\mathbf{y} - \mathbf{X}\gamma) \tag{6.49}$$

thus

$$Q_2 = (\mathbf{y} - \mathbf{X}\gamma)^T \mathbf{M}^{-1} \mathbf{X}\Gamma \mathbf{X}^T \mathbf{M}^{-1}(\mathbf{y} - \mathbf{X}\gamma). \tag{6.50}$$

Using (6.49) and (6.48), we get

$$
\begin{aligned}
\mathbf{y} - \mathbf{X}\gamma_1 &= \mathbf{y} - \mathbf{X}\gamma - \mathbf{X}\Gamma \mathbf{X}^T \mathbf{M}^{-1}(\mathbf{y} - \mathbf{X}\gamma) \\
&= (\mathbf{M} - \mathbf{X}\Gamma \mathbf{X}^T)\mathbf{M}^{-1}(\mathbf{y} - \mathbf{X}\gamma) \\
&= \Sigma \mathbf{M}^{-1}(\mathbf{y} - \mathbf{X}\gamma),
\end{aligned}
$$

and

$$Q_1 = (\mathbf{y} - \mathbf{X}\gamma)^T \mathbf{M}^{-1}\Sigma \mathbf{M}^{-1}(\mathbf{y} - \mathbf{X}\gamma).$$

Applying (6.48) we obtain

$$
\begin{aligned}
Q = Q_1 + Q_2 \\
= (\mathbf{y} - \mathbf{X}\gamma)^T \mathbf{M}^{-1}(\Sigma + \mathbf{X}\Gamma \mathbf{X}^T)\mathbf{M}^{-1}(\mathbf{y} - \mathbf{X}\gamma) \\
= (\mathbf{y} - \mathbf{X}\gamma)^T \mathbf{M}^{-1}(\mathbf{y} - \mathbf{X}\gamma),
\end{aligned}
$$

which completes the proof.

$\square$

Theorem 6.4 gives the joint posterior distribution of $\theta = (\beta, \sigma^2)$. If σ^2 is the parameter of interest and β is the nuisance parameter, then the conditional distribution of σ^2 given $\mathbf{y}$ is the most important. By the hierarchical construction of the normal-inverse-gamma distribution $\mathsf{NIG}(a_1, b_1, \gamma_1, \Gamma_1)$ we know that

$$\sigma^2 | \mathbf{y} \sim \mathsf{InvGamma}(a_1/2, b_1/2).$$

In case β is the parameter of interest and σ^2 is the nuisance parameter we are interested in the marginal posterior distribution of $\beta | \mathbf{y}$. The following general lemma gives the marginal distributions of a normal-inverse-gamma distribution. Before we state it, we introduce the multivariate t-distribution.

A random vector $\mathbf{X} \in \mathbb{R}^p$ follows a *multivariate* t-*distribution*, denoted

$$\mathbf{X} \sim t_p(\nu, \mu, \Sigma),$$

if it has the density

$$\frac{\Gamma\left(\frac{\nu+p}{2}\right)}{\Gamma\left(\frac{\nu}{2}\right)\nu^{p/2}\pi^{p/2}|\Sigma|^{1/2}} \left(1 + \frac{1}{\nu}(\mathbf{X}-\mu)^T\Sigma^{-1}(\mathbf{X}-\mu)\right)^{-(\nu+p)/2}, \quad (6.51)$$

where ν denotes the degrees of freedom, $\mu \in \mathbb{R}^p$ is the location parameter and $\Sigma \in \mathbb{R}^{p \times p}$ is the positive definite scale matrix.

Lemma 6.5 *Assume* $(\mathbf{X}, \lambda) \sim \mathsf{NIG}(\alpha, \beta, \mu, \Sigma)$. *Then*

$$\mathbf{X} \sim t_p(\alpha, \mu, \frac{\beta}{\alpha}\Sigma) \ \text{and} \ \lambda \sim \mathsf{InvGamma}(\alpha/2, \beta/2).$$

PROOF: To derive $f(\mathbf{X})$, we need to integrate λ out of $f(\mathbf{X}) = \int\limits_0^\infty f(\mathbf{X}, \lambda)d\lambda$,

where

$$f(\mathbf{X}, \lambda) \propto \lambda^{-(p+\alpha+2)/2} \exp\left(-\frac{1}{2\lambda}(Q+\beta)\right) \ \text{with} \ Q = (\mathbf{X}-\mu)^T\Sigma^{-1}(\mathbf{X}-\mu).$$

We use the substitution $\frac{Q+\beta}{2\lambda} = x$ such that $d\lambda = -\frac{Q+\beta}{2x^2}dx$ and get

$$f(\mathbf{X}) \ \propto \ (Q+\beta)^{-(p+\alpha)/2} \int\limits_0^\infty x^{(p+\alpha)/2-1} \exp(-x)dx.$$

Recall that, for any real $a > 0$, the gamma function is defined as $\Gamma(a) = \int_0^\infty x^{a-1}\exp(-x)dx$; see e.g., Mathai and Haubold (2008). Thus, the integral in $f(\mathbf{X})$ is a gamma function with $a = (p+\alpha)/2$, and we can write

$$f(\mathbf{X}) \propto (Q+\beta)^{-(p+\alpha)/2} \propto (1 + \frac{1}{\beta}(\mathbf{X}-\mu)^T\Sigma^{-1}(\mathbf{X}-\mu))^{-(p+\alpha)/2}. \quad (6.52)$$

Comparing this kernel with (6.51) delivers the statement. The marginal distribution of λ is given by the hierarchical set up of NIG.

$\square$

For a better overview, we summarize once more all results related to the conjugate prior set up in Theorem 6.5.

Theorem 6.5 *Given*

$$\mathbf{y}|\beta,\sigma^2 \sim \mathsf{N}_n(\mathbf{X}\beta, \sigma^2\Sigma) \tag{6.53}$$
$$\beta|\sigma^2 \sim \mathsf{N}_p(\gamma, \sigma^2\Gamma) \tag{6.54}$$
$$\sigma^2 \sim \mathsf{InvGamma}(a/2, b/2), \tag{6.55}$$

then

$$\mathbf{y} \sim \mathsf{t}_n(a, \mathbf{m}, \frac{b}{a}\mathbf{M}) \tag{6.56}$$
$$\beta|\mathbf{y}, \sigma^2 \sim \mathsf{N}_p(\gamma_1, \sigma^2\Gamma_1) \tag{6.57}$$
$$\beta|\mathbf{y} \sim \mathsf{t}_p(a_1, \gamma_1, \frac{b_1}{a_1}\Gamma_1) \tag{6.58}$$
$$\sigma^2|\mathbf{y} \sim \mathsf{InvGamma}(a_1/2, b_1/2). \tag{6.59}$$

with

$$\mathbf{m} = \mathbf{X}\gamma \tag{6.60}$$
$$\mathbf{M} = \Sigma + \mathbf{X}\Gamma\mathbf{X}^T \tag{6.61}$$
$$\gamma_1 = \Gamma_1(\mathbf{X}^T\Sigma^{-1}\mathbf{y} + \Gamma^{-1}\gamma) \tag{6.62}$$
$$= \gamma + \Gamma\mathbf{X}^T\mathbf{M}^{-1}(\mathbf{y} - \mathbf{X}\gamma) \tag{6.63}$$
$$\Gamma_1 = (\Gamma^{-1} + \mathbf{X}^T\Sigma^{-1}\mathbf{X})^{-1} \tag{6.64}$$
$$= \Gamma - \Gamma\mathbf{X}^T\mathbf{M}^{-1}\mathbf{X}\Gamma. \tag{6.65}$$
$$a_1 = a + n \tag{6.66}$$
$$b_1 = b + (\mathbf{y} - \mathbf{X}\gamma_1)^T\Sigma^{-1}(\mathbf{y} - \mathbf{X}\gamma_1) + (\gamma - \gamma_1)^T\Gamma^{-1}(\gamma - \gamma_1) \tag{6.67}$$
$$= b + (\mathbf{y} - \mathbf{X}\gamma_1)^T\mathbf{M}^{-1}(\mathbf{y} - \mathbf{X}\gamma_1). \tag{6.68}$$
$$= b + \mathbf{y}^T\Sigma^{-1}\mathbf{y} + \gamma^T\Gamma^{-1}\gamma - \gamma_1^T\Gamma_1^{-1}\gamma_1. \tag{6.69}$$

PROOF: The equivalent formulas (6.62) and (6.63) are given in Theorem 6.4 and Corollary 6.2. The same is true for (6.64) and (6.65). The three equivalent presentations of b_1 are given in Theorem 6.4, Corollary 6.2 and Lemma 6.4. The t–distribution of β in (6.58) is a consequence of Lemma 6.5. It remains to show (6.56). From Theorem 6.1 it follows for given σ^2 that

$$\begin{pmatrix} \beta \\ \mathbf{y} \end{pmatrix} | \sigma^2 \sim \mathsf{N}_{n+p}\left(\begin{pmatrix} \gamma \\ \mathbf{m} \end{pmatrix}, \begin{pmatrix} \sigma^2\Gamma & \sigma^2\Gamma\mathbf{X}^T \\ \sigma^2\mathbf{X}\Gamma & \sigma^2\mathbf{M} \end{pmatrix} \right).$$

This includes
$$\mathbf{y}|\sigma^2 \sim N_n(\mathbf{m}, \sigma^2 \mathbf{M}),$$
which, together with (6.55), implies
$$(\mathbf{y}, \sigma^2) \sim NIG(a, b, \mathbf{m}, \mathbf{M}).$$

Lemma 6.5 delivers (6.56).

□

Before we introduce non-informative priors, we illustrate some of the expressions in Theorem 6.5 by simple examples.

Example 6.8 (Simple linear regression) We assume
$$y_i = \beta x_i + \varepsilon_i, \quad i = 1, \ldots, n, \quad \varepsilon_i \sim N(0, \sigma^2), \quad i.i.d.$$
with $\theta = (\beta, \sigma^2)$. The prior $NIG(a, b, \gamma, \Gamma)$ means
$$\beta|\sigma^2 \sim N(\gamma, \sigma^2 \Gamma) \quad \text{and} \quad \sigma^2 \sim InvGamma(a/2, b/2).$$

The hyperparameter γ includes a first guess for β; the variance $\sigma^2 \Gamma$ describes the precision of the guess. Here Γ is a scalar, we set $\Gamma = \lambda$. It is the prior information about the ratio of the prior variance σ^2 to the error variance. The hyperparameter a can be interpreted as a type of sample size on which the prior information on σ^2 is based. The guess for σ^2 is given by $\frac{b}{a-2}$. The formulas (6.62), (6.64) and (6.69) become

$$\gamma_1 = \left(\sum_{i=1}^{n} x_i^2 + \lambda^{-1}\right)^{-1} \left(\sum_{i=1}^{n} x_i y_i + \lambda^{-1}\gamma\right)$$

$$\Gamma_1 = \lambda_1 = \left(\sum_{i=1}^{n} x_i^2 + \lambda^{-1}\right)^{-1}$$

$$b_1 = b + \sum_{i=1}^{n} y_i^2 + \lambda^{-1}\gamma^2 - \left(\sum_{i=1}^{n} x_i^2 + \lambda^{-1}\right)^{-1} \left(\sum_{i=1}^{n} x_i y_i + \lambda^{-1}\gamma\right)^2.$$

□

In the following example we consider again the simple linear regression model, but now we include the intercept as parameter. Also in this case the formulas can be written explicitly.

Example 6.9 (Simple linear regression with intercept)
We assume
$$y_i = \alpha + \beta x_i + \varepsilon_i, \quad i = 1, \ldots, n, \quad \varepsilon_i \sim N(0, \sigma^2), \quad i.i.d.$$

with $\theta = (\alpha, \beta, \sigma^2)$. The prior is $\mathsf{NIG}(a, b, \gamma, \Gamma)$. We assume that intercept and slope have conditionally independent priors

$$\alpha|\sigma^2 \sim \mathsf{N}(\gamma_a, \sigma^2 \lambda_1) \text{ and } \beta|\sigma^2 \sim \mathsf{N}(\gamma_b, \sigma^2 \lambda_2),$$

with

$$\sigma^2 \sim \mathsf{InvGamma}(a/2, b/2).$$

Then $\Sigma = \mathbf{I}_2$,

$$\gamma = \begin{pmatrix} \gamma_a \\ \gamma_b \end{pmatrix} \text{ and } \Gamma = \begin{pmatrix} \lambda_1 & 0 \\ 0 & \lambda_2 \end{pmatrix}.$$

We apply (6.62) and (6.64). It holds that

$$\mathbf{X}^T\mathbf{X} + \Gamma^{-1} = \begin{pmatrix} n + \lambda_1^{-1} & \sum_{i=1}^{n} x_i \\ \sum_{i=1}^{n} x_i & \sum_{i=1}^{n} x_i^2 + \lambda_2^{-1} \end{pmatrix}.$$

Using the inversion formula

$$\begin{pmatrix} a & c \\ c & b \end{pmatrix} = \frac{1}{ab - c^2} \begin{pmatrix} b & -c \\ -c & a \end{pmatrix}$$

and

$$\sum_{i=1}^{n} x_i^2 = s_{xx} + n\bar{x}^2, \text{ with } s_{xx} = \sum_{i=1}^{n}(x_i - \bar{x})^2$$

we obtain

$$\Gamma_1 = \frac{1}{d} \begin{pmatrix} s_{xx} + n\bar{x}^2 + \lambda_2^{-1} & -n\bar{x} \\ -n\bar{x} & n + \lambda_1^{-1} \end{pmatrix}$$

with

$$d = (n + \lambda_1^{-1})(s_{xx} + \lambda_2^{-1}) + \lambda_1^{-1} n \bar{x}^2. \tag{6.70}$$

Further

$$\mathbf{X}^T\mathbf{y} + \Gamma^{-1}\gamma = \begin{pmatrix} \sum_{i=1}^{n} y_i + \lambda_1^{-1}\gamma_a \\ \sum_{i=1}^{n} x_i y_i + \lambda_2^{-1}\gamma_b \end{pmatrix}$$

and

$$\sum_{i=1}^{n} x_i y_i = s_{xy} + n\bar{x}\bar{y}, \text{ with } s_{xy} = \sum_{i=1}^{n}(x_i - \bar{x})(y_i - \bar{y})$$

so that

$$\gamma_{1,a} = \frac{1}{d}\left(ns_{xx}\widehat{\alpha} + n\lambda_2^{-1}\widehat{\alpha}_{prior} + \lambda_1^{-1}\gamma_a + \lambda_2^{-1}\right)$$
$$\gamma_{1,b} = \frac{1}{d}\left((n + \lambda_1^{-1})(s_{xx}\widehat{\beta} + \lambda_2^{-1}\gamma_b) + n\bar{x}\lambda_1^{-1}(\bar{y} - \gamma_a)\right) \tag{6.71}$$

with

$$\begin{aligned} \widehat{\alpha} &= \bar{y} - \widehat{\beta}\bar{x} \\ \widehat{\alpha}_{prior} &= \bar{y} - \gamma_b\bar{x} \\ \widehat{\beta} &= \frac{s_{xy}}{s_{xx}}. \end{aligned}$$

(6.72)

(6.73)

□

Example 6.10 (Life length vs. life line)

We continue with Example 6.3 using the centered model (6.34)

$$y_i - \bar{y} = \beta(x_i - \bar{x}) + \varepsilon_i, \ i = 1, \ldots, 48, \ \varepsilon_i \sim \mathsf{N}(0, \sigma^2), \ i.i.d.$$

with unknown slope β and unknown variance σ^2, thus $\theta = (\beta, \sigma^2)$. We assume the conjugate prior $\mathsf{NIG}(10, 1280, 0, 100)$, i.e., $\mathsf{E}\sigma^2 = 160$, since

$$\beta|\sigma^2 \sim \mathsf{N}(0, 100\sigma^2), \ \ \sigma^2 \sim \mathsf{InvGamma}(5, 640), \ , \ \beta \sim \mathsf{t}_1(10, 0, 12800).$$

We apply (6.62), (6.64), (6.69) and obtain the posteriors

$$\beta|\mathbf{y} \sim \mathsf{t}_1(58, -2.257, 2.29), \ \ \sigma^2|\mathbf{y} \sim \mathsf{InvGamma}(29, 4096).$$

In Figure 6.7 the posteriors together with the prior are plotted. □

R Code 6.2.5. Life data, Example 6.10.

```
y<-Agec; x<-Lengthc
# prior NIG(a,b,gamma,Gamma)
a<-10; b<-2*640
gamma<-0; Gamma<-100
## posterior NIG(a1,b1,gamma1,Gamma1)
a1<-a+n
Gamma1<-(Gamma^(-1)+sum(x^2))^(-1)
gamma1<-Gamma1*(sum(x*y)+Gamma^(-1)*gamma)
b1<-b+sum(y*y)+gamma^2*Gamma^(-1)-gamma1^2*Gamma1^(-1)
b1/a1*Gamma1
```

6.2.3 Jeffreys Prior

In Chapter 3 the background of Jeffreys prior as non-informative prior is explained. We assume the model

$$\mathcal{P} = \left\{\mathsf{N}_n(\mathbf{X}\beta, \sigma^2\Sigma) : \theta = (\beta, \sigma^2) \in \mathbb{R}^p \times \mathbb{R}_+\right\}.$$

(6.74)

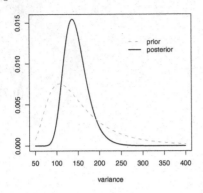
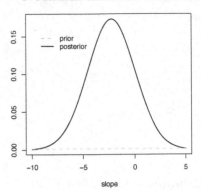

Figure 6.7: Example 6.10 on life length data. Left: The broken line is the prior InvGamma$(5, 640)$ of the error variance, and the continuous line is the posterior InvGamma$(29, 4096)$. Right: The broken line is the prior density $t_1(10, 0, 12800)$ of the slope, and the continuous line is the posterior density $t_1(58, -2.26, 2.29)$.

Under this assumption Jeffreys prior and the reference prior coincide, see (3.51) and the discussion in Subsection 3.3.3. In Definition 3.4 Jeffreys prior is defined as the square root of the determinant of Fisher information. We begin by computing Fisher information.

Theorem 6.6 *The Fisher information matrix for model (6.74) is*

$$I(\theta) = \begin{pmatrix} \frac{1}{\sigma^2} \mathbf{X}^T \Sigma^{-1} \mathbf{X} & 0 \\ 0 & \frac{n}{2\sigma^4} \end{pmatrix}. \tag{6.75}$$

PROOF: From (3.32), the Fisher information $I(\theta)$ is given by

$$I(\theta) = -\mathsf{E}_\theta J(\theta|x),$$

where $J(\theta|x)$ is the matrix of the second derivatives, Hessian, of the log-likelihood function. We have

$$\mathbf{y}|\theta \sim \mathsf{N}_n(\mathbf{X}\beta, \sigma^2 \Sigma), \tag{6.76}$$

with $\theta = (\beta, \sigma^2)$, so that the log-likelihood function is

$$l(\theta|\mathbf{y}) = -\frac{n}{2}\ln(\sigma^2\pi) - \frac{1}{2}\ln(|\Sigma|) - \frac{1}{2\sigma^2}(\mathbf{y} - \mathbf{X}\beta)^T \Sigma^{-1} (\mathbf{y} - \mathbf{X}\beta).$$

Differentiating with respect to β and σ^2, the gradient function is

$$\partial l/\partial \theta = (\partial l/\partial \beta, \ \partial l/\partial \sigma^2)^T$$

with

$$\frac{\partial l}{\partial \beta} = \frac{1}{\sigma^2}\mathbf{X}^T\Sigma^{-1}(\mathbf{y} - \mathbf{X}\beta)$$

$$\frac{\partial l}{\partial \sigma^2} = -\frac{n}{2\sigma^2} + \frac{1}{2(\sigma^2)^2}(\mathbf{y} - \mathbf{X}\beta)^T\Sigma^{-1}(\mathbf{y} - \mathbf{X}\beta).$$

Differentiating again,

$$\frac{\partial^2 l}{\partial\beta\beta^T} = -\frac{1}{\sigma^2}\mathbf{X}^T\Sigma^{-1}\mathbf{X}$$

$$\frac{\partial^2 l}{\partial(\sigma^2)^2} = \frac{n}{2(\sigma^2)^2} - \frac{1}{(\sigma^2)^3}(\mathbf{y} - \mathbf{X}\beta)^T\Sigma^{-1}(\mathbf{y} - \mathbf{X}\beta)$$

$$\frac{\partial^2 l}{\partial\beta\partial\sigma^2} = -\frac{1}{(\sigma^2)^2}\mathbf{X}^T\Sigma^{-1}(\mathbf{y} - \mathbf{X}\beta)$$

so that

$$J(\theta|\mathbf{y}) = \begin{pmatrix} -\frac{1}{\sigma^2}\mathbf{X}^T\Sigma^{-1}\mathbf{X} & -\frac{1}{\sigma^4}\mathbf{X}^T\Sigma^{-1}(\mathbf{y} - \mathbf{X}\beta) \\ \frac{1}{\sigma^4}(\mathbf{y} - \mathbf{X}\beta)^T\mathbf{X}^T\Sigma^{-1} & \frac{n}{2\sigma^4} - \frac{1}{\sigma^6}(\mathbf{y} - \mathbf{X}\beta)^T\Sigma^{-1}(\mathbf{y} - \mathbf{X}\beta) \end{pmatrix}.$$

Fisher information matrix follows as $I(\theta) = -\mathsf{E}J(\theta|\mathbf{y})$. It holds that $\mathsf{E}\mathbf{y} = \mathbf{X}\beta$, and the Fisher information matrix is block diagonal. Further, we calculate

$$\mathsf{E}(\mathbf{y} - \mathbf{X}\beta)^T\Sigma^{-1}(\mathbf{y} - \mathbf{X}\beta) = \mathsf{E}\,\mathrm{tr}\left((\mathbf{y} - \mathbf{X}\beta)^T\Sigma^{-1}(\mathbf{y} - \mathbf{X}\beta)\right)$$

$$= \mathsf{E}\,\mathrm{tr}\left((\mathbf{y} - \mathbf{X}\beta)(\mathbf{y} - \mathbf{X}\beta)^T\Sigma^{-1}\right)$$

$$= \mathrm{tr}\left(\mathsf{E}(\mathbf{y} - \mathbf{X}\beta)(\mathbf{y} - \mathbf{X}\beta)^T\Sigma^{-1}\right)$$

$$= \mathrm{tr}(\mathrm{Cov}(\mathbf{y})\Sigma^{-1}) = \mathrm{tr}(\sigma^2\Sigma\Sigma^{-1}) = \sigma^2\mathrm{tr}(\mathbf{I}_n) = \sigma^2 n.$$

Thus we obtain the statement.

□

Corollary 6.3 now gives Jeffreys prior.

Corollary 6.3 *For* $\mathbf{y}|\theta \sim \mathsf{N}_n(\mathbf{X}\beta, \sigma^2\Sigma)$, *the Jeffreys prior of* $\theta = (\beta, \sigma^2)$ *is given as*

$$\pi(\theta) \propto \frac{1}{(\sigma^2)^{p/2+1}}. \tag{6.77}$$

PROOF: As $I(\theta)$ is a block diagonal matrix, we get see (see Seber, 2008)

$$\det(I(\theta)) = \det(\frac{1}{\sigma^2}\mathbf{X}^T\Sigma^{-1}\mathbf{X}) \cdot \frac{n}{2(\sigma^2)^2} \propto \frac{1}{(\sigma^2)^{p+2}}, \qquad (6.78)$$

so that the Jeffreys prior simplifies to

$$\pi(\theta) = \sqrt{\det(I(\theta))} \propto \sqrt{\frac{1}{(\sigma^2)^{p+2}}} \propto \frac{1}{(\sigma^2)^{p/2+1}}. \qquad (6.79)$$

$\square$

Following corollary provides Jeffreys prior assuming independence between location and scale parameter (see Example 3.24), i.e.,

$$\pi(\theta) = \pi(\beta)\pi(\sigma^2). \qquad (6.80)$$

Corollary 6.4 *For* $\mathbf{y}|\theta \sim N_n(\mathbf{X}\beta, \sigma^2\mathbf{I}_n)$*, the Jeffreys prior of* $\theta = (\beta, \sigma^2)$*, assuming (6.80), is*

$$\pi(\theta) \propto \frac{1}{\sigma^2}.$$

PROOF: Under independence, $\pi(\theta) = \pi(\beta, \sigma^2) = \pi(\beta)\pi(\sigma^2)$, where most of the computations are as in Theorem 6.6. Thus, we have for known σ^2

$$I(\beta) = \frac{1}{\sigma^2}\mathbf{X}^T\Sigma^{-1}\mathbf{X}$$

which is independent of β, so that $\pi(\beta) \propto 1$. Similarly for known β, it follows from

$$I(\sigma^2) = \frac{n}{2(\sigma^2)^2}$$

that

$$\pi(\sigma^2) = \sqrt{I(\sigma^2)} \propto \frac{1}{\sigma^2} \qquad (6.81)$$

so that $\pi(\theta) \propto 1/\sigma^2$, as needed to be proved.

$\square$

We observe that the Jeffreys prior under (6.80) is different from that in Corollary 6.3. In particular, it does not depend on p. In both cases, however, Jeffreys prior reduces to an improper prior for $\theta = (\beta, \sigma^2)$. But the posterior belongs to the conjugate family, as under the conjugate prior. This is stated in the following theorem. First recall that the maximum likelihood estimator in model

(6.74) is the generalized least-squares estimator

$$\widehat{\beta}_\Sigma = \arg\min_{\beta \in \mathbb{R}^p} (\mathbf{y} - \mathbf{X}\beta)^T \Sigma^{-1} (\mathbf{y} - \mathbf{X}\beta)$$
$$= (\mathbf{X}^T \Sigma^{-1} \mathbf{X})^{-1} \mathbf{X}^T \Sigma^{-1} \mathbf{y}. \tag{6.82}$$

Theorem 6.7 *Given* $\mathbf{y}|\theta \sim \mathsf{N}_n(\mathbf{X}\beta, \sigma^2\Sigma)$ *with* Σ *known. Then, under Jeffreys' priors*

$$\pi(\beta, \sigma^2) \propto (\sigma^2)^{-m}$$

with $2m = p + 2$ *in (6.77) and* $m = 1$ *in (6.81). It follows that*

$$(\beta, \sigma^2)|\mathbf{y} \sim \mathsf{NIG}(a_m, b, \gamma, \Gamma) \tag{6.83}$$

and

$$\beta|\mathbf{y} \sim \mathsf{t}_p(a_m, \widehat{\beta}_\Sigma, \frac{b}{a_m}(\mathbf{X}^T \Sigma^{-1} \mathbf{X})^{-1}) \tag{6.84}$$

with

$$a_m = 2m + n - p - 2$$
$$b = (\mathbf{y} - \mathbf{X}\widehat{\beta}_\Sigma)^T \Sigma^{-1} (\mathbf{y} - \mathbf{X}\widehat{\beta}_\Sigma)$$
$$\gamma = (\mathbf{X}^T \Sigma^{-1} \mathbf{X})^{-1} \mathbf{X}^T \Sigma^{-1} \mathbf{y} = \widehat{\beta}_\Sigma \tag{6.85}$$
$$\Gamma = (\mathbf{X}^T \Sigma^{-1} \mathbf{X})^{-1}.$$

PROOF: We calculate

$$\pi(\beta, \sigma^2|\mathbf{y}) \propto \pi(\beta, \sigma^2)\ell(\beta, \sigma^2|\mathbf{y})$$
$$\propto \frac{1}{(\sigma^2)^{n/2+m}} \exp\left(-\frac{1}{2\sigma^2}(\mathbf{y} - \mathbf{X}\beta)^T \Sigma^{-1}(\mathbf{y} - \mathbf{X}\beta)\right).$$

Since

$$\widehat{\beta}_\Sigma = (\mathbf{X}^T \Sigma^{-1} \mathbf{X})^{-1} \mathbf{X}^T \Sigma^{-1}(\mathbf{X}\beta + \epsilon)$$
$$= \beta + (\mathbf{X}^T \Sigma^{-1} \mathbf{X})^{-1} \mathbf{X}^T \Sigma^{-1}\epsilon,$$

we have

$$\mathbf{y} - \mathbf{X}\widehat{\beta}_\Sigma = \epsilon + \mathbf{X}(\beta - \widehat{\beta}_\Sigma)$$
$$= (\mathbf{I}_n - \mathbf{X}(\mathbf{X}^T \Sigma^{-1} \mathbf{X})^{-1} \mathbf{X}^T \Sigma^{-1})\epsilon.$$

The projection property of the generalized least-squares estimator implies that the mixed term is zero:

$$(\mathbf{y} - \mathbf{X}\widehat{\beta}_\Sigma)^T \Sigma^{-1} \mathbf{X}(\beta - \widehat{\beta}_\Sigma)$$
$$= \epsilon^T \left(\mathbf{I}_n - \Sigma^{-1}\mathbf{X}(\mathbf{X}^T\Sigma^{-1}\mathbf{X})^{-1}\mathbf{X}^T\right)\Sigma^{-1}\mathbf{X}(\mathbf{X}^T\Sigma^{-1}\mathbf{X})^{-1}\mathbf{X}^T\Sigma^{-1}\epsilon$$
$$= 0$$

Thus the following decomposition holds.

$$(\mathbf{y} - \mathbf{X}\beta)^T\Sigma^{-1}(\mathbf{y} - \mathbf{X}\beta)$$
$$= (\widehat{\beta}_\Sigma - \beta)^T\mathbf{X}^T\Sigma^{-1}\mathbf{X}(\widehat{\beta}_\Sigma - \beta) + (\mathbf{y} - \mathbf{X}\widehat{\beta}_\Sigma)^T\Sigma^{-1}(\mathbf{y} - \mathbf{X}\widehat{\beta}_\Sigma) \qquad (6.86)$$
$$= (\widehat{\beta}_\Sigma - \beta)^T\mathbf{X}^T\Sigma^{-1}\mathbf{X}(\widehat{\beta}_\Sigma - \beta) + b.$$

Using (6.86) we obtain

$$\pi(\beta, \sigma^2|\mathbf{y}) \quad \propto \quad (\sigma^2)^{-(m+n/2)}\exp\left(-\frac{1}{2\sigma^2}((\widehat{\beta}_\Sigma - \beta)^T\mathbf{X}^T\Sigma^{-1}\mathbf{X}(\widehat{\beta}_\Sigma - \beta) + b)\right),$$

which implies (6.83). Applying Lemma 6.5, we obtain (6.84).

$$\square$$

For $\Sigma = \mathbf{I}_n$, the posterior coincides with the least–squares solution, as stated in the following corollary.

Corollary 6.5 *Given* $\mathbf{y}|\theta \sim \mathsf{N}_n(\mathbf{X}\beta, \sigma^2\mathbf{I}_n)$. *Then, under Jeffreys prior*

$$\beta|\mathbf{y}, \sigma^2 \sim \mathsf{N}_p(\widehat{\beta}, \sigma^2(\mathbf{X}^T\mathbf{X})^{-1}),$$

with $\widehat{\beta} = (\mathbf{X}^T\mathbf{X})^{-1}\mathbf{X}^T\mathbf{y}$.

Example 6.11 (Life length vs. life line)
We continue with the centered model (6.34) in Examples 6.4 and 6.10, now using Jeffreys prior (6.81). Recall

$$\Sigma = \mathbf{I}_n, \quad \widehat{\beta} = \frac{s_{xy}}{s_{xx}}, \quad \mathbf{X}^T\Sigma^{-1}\mathbf{X} = s_{xx}, \quad s^2 = \frac{1}{n-1}\sum_{i=1}^{n}(y_i - \bar{y} - \widehat{\beta}(x_i - \bar{x}))^2.$$

We apply the posteriors from Theorem 6.7, i.e.,

$$\beta|\mathbf{y} \sim \mathsf{t}_1(n - 1, \widehat{\beta}, s^2/s_{xx}) \text{ and } \sigma^2|\mathbf{y} \sim \mathsf{InvGamma}((n - 1)/2, (n - 1)s^2/2).$$

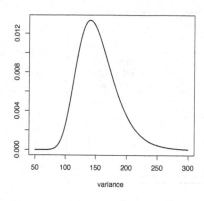 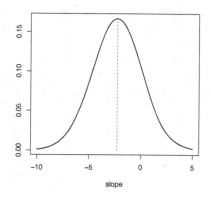

Figure 6.8: Example 6.11. Left: Posterior distribution $\mathsf{InvGamma}(23.5, 3456)$ of σ^2. Right: Posterior distribution $\mathsf{t}_1(47, -2.26, 2.38)$ of β with mean $\widehat{\beta}$ shown as vertical line.

In Example 6.3 we calculated $\widehat{\beta} = -2.26$, $s_{xx} = 61.65$ and $s^2 = 147,07$. Figure 6.8 depicts the posterior distributions.

$\square$

Example 6.12 (Corn plants)
Example 6.5 analyzes corn data (Example 6.1) using conjugate priors. Here, we re-analyze it for Jeffreys prior.
With $\mathbf{y} \sim \mathsf{N}(\mathbf{X}\beta, \sigma^2\mathbf{I})$, keeping $\sigma^2 = 300$ as before, the posterior distribution for $\beta = (\beta_0, \beta_1, \beta_2, \beta_3)^T$ under Jeffreys prior is computed as $\pi(\beta|\mathbf{y}, \sigma^2) \sim \mathsf{N}_4(\widehat{\beta}, \sigma^2(\mathbf{X}^T\mathbf{X})^{-1})$, where $\widehat{\beta}$ and $(\mathbf{X}^T\mathbf{X})^{-1}$ are given in Example 6.5. The posterior distributions for the three betas are shown in Figure 6.9, where vertical lines at the center of each curve are the mean values, $\widehat{\beta}_j$, $j = 1, 2, 3$. $\square$

6.3 Linear Mixed Models

In this section, we extend the general linear model, (6.2), to linear mixed model. For general theory of linear mixed models and their applications, see e.g., Demidenko (2013) and Searle et al. (2006).
Unlike model (6.2) a linear mixed model includes an additional random component besides the error variable. We consider the linear mixed model as defined in Demidenko (2013, Chapter 2):

$$\mathbf{y} = \mathbf{X}\beta + \mathbf{Z}\gamma + \epsilon, \quad \epsilon \sim \mathsf{N}_n(\mathbf{0}, \sigma^2\Sigma), \quad \gamma \sim \mathsf{N}_q(\mathbf{0}, \sigma^2\Delta), \qquad (6.87)$$

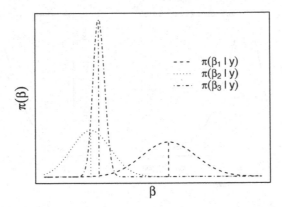

Figure 6.9: Example 6.12. Prior and posterior distributions of $\beta_1, \beta_2, \beta_3$ for corn data with Jeffreys prior.

where $\mathbf{y} \in \mathbb{R}^n$ is the observed vector of response variables, $\mathbf{X} \in \mathbb{R}^{n \times p}$ and $\mathbf{Z} \in \mathbb{R}^{n \times q}$ are the known design matrices for fixed and random parts of the model. The vector of unobserved random components $\gamma \in \mathbb{R}^q$ is a latent variable, ϵ is the vector of unobserved errors, where we assume that ϵ and γ are independent. The matrices $\Sigma \succ 0$ and $\Delta \succ 0$ are known. The unknown parameter $\theta = (\beta, \sigma^2)$ consists of the vector of regression parameters $\beta \in \mathbb{R}^p$ and the variance parameter $\sigma^2 > 0$. We assume that $\mathbf{X}$ and $\mathbf{Z}$ are full rank matrices, i.e., $r(\mathbf{X}) = p$ and $r(\mathbf{Z}) = q$, where $n > p$ and $n > q$.

Regarding notation, it may be emphasized that we use γ in (6.87) for random component, following the literature. In the literature, γ is often called random parameter and its distribution a prior. In our textbook we want to clarify that γ and its distribution are part of the statistical model and are not explained by a Bayes model. The Bayes model assumes priors on the parameter θ. Often, Model (6.87) is more convenient at observation level, i.e.,

$$y_i = \mathbf{x}_i^T \beta + \mathbf{z}_i^T \gamma + \epsilon_i, \qquad (6.88)$$

with $\mathbf{x}_i \in \mathbb{R}^p$ and $\mathbf{z}_i \in \mathbb{R}^q$, $i = 1, \ldots, n$. We illustrate it by the following example.

Example 6.13 (Hip operation)
We use a data set discussed in Crowder and Hand (1990, Example 5.4, p.78). The data consist of 4 repeated measurements taken on each of 30 hip operation patients, one before operation and three after it, with age recorded as a

covariate. The measurements are haematocrit (volume percentage of red blood cells measured in a blood test) as an indicator of patients' health. For male patients, values below 40.7 may indicate anemia, values above 50.3 are also a bad sign. The patients consist of two independent groups, 13 males and 17 females. For our purposes, we use male group data, removing three discordant observations and the corresponding columns. Hence, the analyzed data are of $n = 10$ patients at two time points. We consider age as fixed factor, and, for illustrative purposes, treat the two repeated observations as levels of random (time) factor ($q = 2$),

$$y_{ij} = \beta_0 + x_i \beta_1 + \gamma_j + \epsilon_{ij}, \quad i = 1, \ldots, 10, \quad j = 1, 2.$$

Collecting the observations in one column as

$$\mathbf{y} = (y_{11}, y_{12}, y_{21}, \ldots, y_{10,2})^T, \quad \mathbf{x} = (x_1, x_1, x_2, x_2, \ldots, x_{10})$$

and setting $\beta = (\beta_0, \beta_1)^T$,

$$\gamma = (\gamma_1, \gamma_2)^T, \quad \mathbf{Z} = (\mathbf{I}_2, \ldots, \mathbf{I}_2)^T$$

the model for this data can be stated as

$$\mathbf{y} = \mathbf{X}\beta + \mathbf{Z}\gamma + \epsilon, \quad \epsilon \sim \mathsf{N}_{20}(0, \sigma^2 \Sigma), \quad \gamma \sim \mathsf{N}_2(0, \sigma^2 \Delta)$$

where $\mathbf{X} \in \mathbb{R}^{20 \times 2}$ (with first column of 1s for intercept), $\mathbf{Z} \in \mathbb{R}^{20 \times 2}$ and $\epsilon \in \mathbb{R}^{20}$, hence, $\mathbf{y} \in \mathbb{R}^{20}$. The data are illustrated in Figure 6.10.   □

In model (6.87), we have two main tasks: estimation of θ and prediction of γ. Before we start with the Bayesian approach we present frequentist methods for estimation of β and prediction of γ.

Estimation of β

First we are only interested in estimating the regression parameter β. We eliminate the random component γ in the model. Setting

$$\xi = \mathbf{Z}\gamma + \epsilon, \quad \text{and} \quad \mathbf{V} = \mathbf{Z}\Delta\mathbf{Z}^T + \Sigma$$

we obtain a univariate linear model like (6.2)

$$\mathbf{y} = \mathbf{X}\beta + \xi, \quad \text{with} \quad \xi \sim \mathsf{N}_n(\mathbf{0}, \sigma^2 \mathbf{V}). \tag{6.89}$$

The model (6.89) is called *marginal model*. The maximum likelihood estimator of β follows as

$$\widehat{\beta}_V \;=\; (\mathbf{X}^T \mathbf{V}^{-1} \mathbf{X})^{-1} \mathbf{X}^T \mathbf{V}^{-1} \mathbf{y}. \tag{6.90}$$

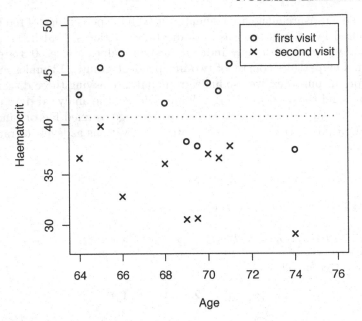

Figure 6.10: Illustration for the data set used in Example 6.13. Values under the broken line indicate anemia. But the blood values are improved for most of patients. Note that, the three patients with same age 70, are plotted as with age $69.5, 70, 70.5$.

Prediction of γ

First we re-formulate the model (6.87) in a hierarchical setting

$$
\begin{aligned}
\mathbf{y}|\gamma, \theta &\sim N_n(\mathbf{X}\beta + \mathbf{Z}\gamma, \sigma^2\Sigma) \\
\gamma|\theta &\sim N_q(\mathbf{0}, \sigma^2\Delta).
\end{aligned}
\tag{6.91}
$$

In the literature, this model formulation is called *two-level hierarchical* model. We consider the joint probability function as function of (β, γ):

$$
\begin{aligned}
p(\mathbf{y}, \gamma|\theta) &= p(\mathbf{y}|\gamma, \theta)p(\gamma|\theta) \\
&=: h(\beta, \gamma)
\end{aligned}
\tag{6.92}
$$

It holds that

$$
h(\beta, \gamma) \propto \exp\left(-\frac{1}{2\sigma^2}\left((\mathbf{y} - \mathbf{X}\beta - \mathbf{Z}\gamma)^T\Sigma^{-1}(\mathbf{y} - \mathbf{X}\beta - \mathbf{Z}\gamma) + \gamma^T\Delta^{-1}\gamma\right)\right).
$$

Maximizing $h(\beta, \gamma)$ with respect to (β, γ) delivers the least-squares minimization problem

$$
\min_{\beta, \gamma}\left((\mathbf{y} - \mathbf{X}\beta - \mathbf{Z}\gamma)^T\Sigma^{-1}(\mathbf{y} - \mathbf{X}\beta - \mathbf{Z}\gamma) + \gamma^T\Delta^{-1}\gamma\right).
$$

Putting the first derivatives to zero we obtain the normal equation system:

$$\mathbf{X}^T \Sigma^{-1} \mathbf{X} \beta + \mathbf{X}^T \Sigma^{-1} \mathbf{Z} \gamma = \mathbf{X}^T \Sigma^{-1} \mathbf{y} \qquad (6.93)$$

$$\mathbf{Z}^T \Sigma^{-1} \mathbf{X} \beta + (\mathbf{Z}^T \Sigma^{-1} \mathbf{Z} + \Delta^{-1}) \gamma = \mathbf{Z}^T \Sigma^{-1} \mathbf{y} \qquad (6.94)$$

The second equation (6.94) gives

$$\gamma = (\mathbf{Z}^T \Sigma^{-1} \mathbf{Z} + \Delta^{-1})^{-1} \mathbf{Z}^T \Sigma^{-1} (\mathbf{y} - \mathbf{X} \beta).$$

Setting this in (6.93) delivers

$$\mathbf{X}^T \mathbf{W} \mathbf{X}^T \beta = \mathbf{X}^T \mathbf{W} \mathbf{y}$$

with

$$\mathbf{W} = \Sigma^{-1} - \Sigma^{-1} \mathbf{Z}^T (\mathbf{Z}^T \Sigma^{-1} \mathbf{Z} + \Delta^{-1})^{-1} \mathbf{Z}^T \Sigma^{-1}.$$

Applying Lemma 6.3 we obtain

$$\mathbf{W} = (\mathbf{Z} \Delta \mathbf{Z}^T + \Sigma)^{-1} = \mathbf{V}^{-1}.$$

Thus the solution of the equation system for β coincides with $\widehat{\beta}_V$ in (6.90) and the prediction of γ is given by

$$\widehat{\gamma} = (\mathbf{Z}^T \Sigma^{-1} \mathbf{Z} + \Delta^{-1})^{-1} \mathbf{Z}^T \Sigma^{-1} (\mathbf{y} - \mathbf{X} \widehat{\beta}_V), \qquad (6.95)$$

which can also be expressed as

$$\widehat{\gamma} = \Delta \mathbf{Z}^T \mathbf{V}^{-1} (\mathbf{y} - \mathbf{X} \widehat{\beta}_V). \qquad (6.96)$$

It follows from

$$(\mathbf{Z}^T \Sigma^{-1} \mathbf{Z} + \Delta^{-1})^{-1} \mathbf{Z}^T \Sigma^{-1} = \Delta \mathbf{Z}^T \mathbf{V}^{-1},$$

with $\mathbf{Z} \Delta \mathbf{Z}^T + \Sigma = \mathbf{V}$, that

$$(\mathbf{Z}^T \Sigma^{-1} \mathbf{Z} + \Delta^{-1}) \Delta \mathbf{Z}^T \mathbf{V}^{-1} = \mathbf{Z}^T \Sigma^{-1} (\mathbf{Z} \Delta \mathbf{Z}^T + \Sigma) \mathbf{V}^{-1} = \mathbf{Z}^T \Sigma^{-1}.$$

Further the prediction (6.96) is a plug-in method. An application of Lemma 6.2 to (6.91) implies

$$\begin{pmatrix} \mathbf{y} \\ \gamma \end{pmatrix} | \theta \sim \mathsf{N}_{n+q} \left(\begin{pmatrix} \mathbf{X} \beta \\ \mathbf{0} \end{pmatrix}, \sigma^2 \begin{pmatrix} \mathbf{V} & \mathbf{Z} \Delta \\ \Delta \mathbf{Z}^T & \Delta \end{pmatrix} \right). \qquad (6.97)$$

Note that, the joint distribution (6.97) explains the name *marginal model* for (6.89). Lemma 6.1 implies

$$\mathsf{E}(\gamma | \mathbf{y}, \beta) = \Delta \mathbf{Z}^T \mathbf{V}^{-1} (\mathbf{y} - \mathbf{X} \beta).$$

Thus the prediction (6.96) coincides with the conditional expectation, with $\widehat{\beta}_V$ replacing β.

6.3.1 Bayes Linear Mixed Model, Marginal Model

In this section we consider the marginal model,

$$\mathcal{P} = \left\{ \mathsf{N}(\mathbf{X}\beta, \sigma^2 \mathbf{V}) : \ \mathbf{V} = \Sigma + \mathbf{Z}\Delta\mathbf{Z}^T, \ \theta = (\beta, \sigma^2) \in \mathbb{R}^p \times \mathbb{R}_+ \right\},$$

and derive the posterior distributions $\pi(\theta|\mathbf{y})$ under conjugate and non-informative priors. We start with the conjugate prior π_c, i.e.,

$$\theta \sim \mathsf{NIG}(a, b, \beta_0, \Gamma). \tag{6.98}$$

The Bayes model $\{\mathcal{P}, \pi_c\}$ is a univariate linear model with conjugate prior. We can apply Theorem 6.5 and obtain:

Theorem 6.8 *For the marginal model set up $\{\mathcal{P}, \pi_c\}$, it holds that*

$$\theta|\mathbf{y} \ \sim \mathsf{NIG}(a_1, b_1, \beta_1, \Gamma_1), \tag{6.99}$$

and

$$\beta|\mathbf{y} \ \sim \mathsf{t}_p(a_1, \beta_1, \frac{b_1}{a_1}\Gamma_1), \tag{6.100}$$

with

$$
\begin{aligned}
a_1 &= a + n \tag{6.101}\\
b_1 &= b + (\mathbf{y} - \mathbf{X}\beta_1)^T \mathbf{M}^{-1} (\mathbf{y} - \mathbf{X}\beta_1) \tag{6.102}\\
\beta_1 &= \beta_0 + \Gamma\mathbf{X}^T\mathbf{M}^{-1}(\mathbf{y} - \mathbf{X}\beta_0) \tag{6.103}\\
\Gamma_1 &= \Gamma - \Gamma\mathbf{X}^T\mathbf{M}^{-1}\mathbf{X}\Gamma \tag{6.104}\\
\mathbf{M} &= \mathbf{V} + \mathbf{X}\Gamma\mathbf{X}^T \tag{6.105}
\end{aligned}
$$

Now using the Jeffreys prior π_m

$$\pi_m(\theta) \propto (\sigma^2)^{-m}, \tag{6.106}$$

where $m = 1$ gives the prior, under the condition that location and scale parameter are independent. Otherwise, $2m = p + 2$ without this condition; see Section 6.2.3. The Bayes model $\{\mathcal{P}, \pi_m\}$ is a univariate linear model with non-informative prior. We apply Theorem 6.7 and obtain:

Theorem 6.9 *For the marginal model set up* $\{\mathcal{P}, \pi_m\}$ *it holds that*

$$\theta|\mathbf{y} \sim \mathsf{NIG}(a_m, b, \widehat{\beta}_V, \mathbf{X}^T\mathbf{V}^{-1}\mathbf{X}), \tag{6.107}$$

and

$$\beta|\mathbf{y} \sim \mathsf{t}_p(a_m, \widehat{\beta}_V, \frac{b}{a_m}\mathbf{X}^T\mathbf{V}^{-1}\mathbf{X}), \tag{6.108}$$

where

$$\widehat{\beta}_V = (\mathbf{X}^T\mathbf{V}^{-1}\mathbf{X})^{-1}\mathbf{V}^{-1}\mathbf{X}\mathbf{y}$$

$$
\begin{aligned}
a_m &= 2m+n-p-2 & \tag{6.109}\\
b &= (\mathbf{y}-\mathbf{X}\widehat{\beta}_V)^T\mathbf{V}^{-1}(\mathbf{y}-\mathbf{X}\widehat{\beta}_V). & \tag{6.110}
\end{aligned}
$$

6.3.2 Bayes Hierarchical Linear Mixed Model

Here, we consider linear mixed models (6.87) including the latent variable γ and its distribution. A Bayes model assumes a prior distribution on the parameter $\theta = (\beta, \sigma^2)$. Our aim is to derive posterior $\pi(\theta|\mathbf{y})$ for conjugate priors. Furthermore we are interested in the predictive distribution $p(\gamma|\mathbf{y})$.

We assume

$$\theta \sim \mathsf{NIG}(a, b, \beta_0, \Gamma). \tag{6.111}$$

The entire model set up can be formulated as

$$
\begin{aligned}
\mathbf{y}|\beta, \sigma^2, \gamma &\sim \mathsf{N}_n(\mathbf{X}\beta+\mathbf{Z}\gamma, \sigma^2\Sigma) & \tag{6.112}\\
\beta|\sigma^2 &\sim \mathsf{N}_p(\beta_0, \sigma^2\Gamma) & \tag{6.113}\\
\gamma|\sigma^2 &\sim \mathsf{N}_q(\mathbf{0}, \sigma^2\Delta) & \tag{6.114}\\
\sigma^2 &\sim \mathsf{InvGamma}(a/2, b/2) & \tag{6.115}
\end{aligned}
$$

where $\beta|\sigma^2$, and $\gamma|\sigma^2$ are assumed to be mutually independent. We see that under Bayes model set up the random component γ and the parameter θ have equivalent status. It also explains why often γ is called random parameter and its distribution prior. Set the joint parameter as $\alpha = (\beta, \gamma)$, $\alpha_0 = (\beta_0, \mathbf{0})$ and $\mathbf{U} = (\mathbf{X}\ \mathbf{Z})$, $\Psi = \mathsf{diag}(\Gamma\ \Delta)$. Then the model set up can also be written as

$$
\begin{aligned}
\mathbf{y}|\alpha, \sigma^2 &\sim \mathsf{N}_n(\mathbf{U}\alpha, \sigma^2\Sigma)\\
(\alpha, \sigma^2) &\sim \mathsf{NIG}(a, b, \alpha_0, \Psi).
\end{aligned}
\tag{6.116}
$$

Following theorem is a consequence of Theorem 6.5, as it sums up the posterior under (6.116).

Theorem 6.10 *For the model set up in (6.116), it holds that*

$$(\alpha, \sigma^2)|\mathbf{y} \sim \text{NIG}(a_1, b_1, \alpha_1, \Psi_1), \qquad (6.117)$$

and

$$\alpha|\mathbf{y} \sim t_{p+q}(a_1, \alpha_1, \frac{b_1}{a_1}\Psi_1), \qquad (6.118)$$

with

$$
\begin{aligned}
a_1 &= a + n & (6.119) \\
b_1 &= b + (\mathbf{y} - \mathbf{U}\alpha_1)^T \mathbf{M}^{-1}(\mathbf{y} - \mathbf{U}\alpha_1) & (6.120) \\
&= b + \mathbf{y}^T \Sigma^{-1}\mathbf{y} + \alpha_0^T \Psi^{-1}\alpha_0 - \alpha_1^T \Psi_1^{-1}\alpha_1 & (6.121) \\
\alpha_1 &= \alpha_0 + \Psi\mathbf{U}^T\mathbf{M}^{-1}(\mathbf{y} - \mathbf{U}\alpha_0) & (6.122) \\
&= \Psi_1(\mathbf{U}^T\Sigma^{-1}\mathbf{y} + \Psi^{-1}\alpha_0) & (6.123) \\
\Psi_1 &= \Psi - \Psi\mathbf{U}^T\mathbf{M}^{-1}\mathbf{U}\Psi & (6.124) \\
&= (\Psi^{-1} + \mathbf{U}^T\Sigma^{-1}\mathbf{U})^{-1} & (6.125) \\
\mathbf{M} &= \Sigma + \mathbf{U}\Psi\mathbf{U}^T & (6.126)
\end{aligned}
$$

where the component for $\beta|\mathbf{y}$ in $\alpha|\mathbf{y}$ is the posterior distribution of β and the component for $\gamma|\mathbf{y}$ is the predictive distribution for γ.

Note that the formulas for the posterior parameters can be presented in different ways as stated in Theorem 6.5. The presentation above is helpful for a separation between β and γ. Especially in (6.120) it holds for $\alpha_1^T = (\beta_1^T, \gamma_1^T)$ that

$$b_1 = b + (\mathbf{y} - \mathbf{X}\beta_1 - \mathbf{Z}\gamma_1)^T\mathbf{M}^{-1}(\mathbf{y} - \mathbf{X}\beta_1 - \mathbf{Z}\gamma_1)$$

and in (6.124)

$$\Psi_1 = \begin{pmatrix} \Gamma - \Gamma\mathbf{X}^T\mathbf{M}^{-1}\mathbf{X}\Gamma & -\Gamma\mathbf{X}^T\mathbf{M}^{-1}\mathbf{Z}\Delta \\ -\Delta\mathbf{Z}^T\mathbf{M}^{-1}\mathbf{X}\Gamma & \Delta - \Delta\mathbf{Z}^T\mathbf{M}^{-1}\mathbf{Z}\Delta \end{pmatrix}.$$

Using the presentation (6.125) we obtain

$$\Psi_1^{-1} = \begin{pmatrix} \Gamma + \mathbf{X}^T\Sigma^{-1}\mathbf{X} & \mathbf{X}^T\Sigma^{-1}\mathbf{Z} \\ \mathbf{Z}^T\Sigma^{-1}\mathbf{X} & \Delta + \mathbf{Z}^T\Sigma^{-1}\mathbf{Z} \end{pmatrix}. \qquad (6.127)$$

Further

$$\mathbf{M} = \Sigma + \mathbf{X}\Gamma\mathbf{X}^T + \mathbf{Z}\Delta\mathbf{Z}^T = \mathbf{V} + \mathbf{X}\Gamma\mathbf{X}^T$$

$$\beta_1 = \beta_0 + \Gamma \mathbf{X}^T \mathbf{M}^{-1}(\mathbf{y} - \mathbf{X}\beta_0)$$

and

$$\gamma_1 = \Delta \mathbf{Z}^T \mathbf{M}^{-1}\mathbf{y}.$$

We see that β_1 included in (6.122) coincides with β_1 in (6.103).

The practically most commonly used special case of Theorem 6.10 is when all covariance matrices are spherical and the design matrices are orthogonal. Furthermore we consider σ^2 as known. We state the result as a corollary to Theorem 6.10.

Corollary 6.6 *Given the model set up in (6.116), with* $\Gamma = r_\beta \mathbf{I}_p$, $\Delta = r_\gamma \mathbf{I}_q$ *and* $\Sigma = \mathbf{I}_n$, *where* $\mathbf{Z}^T \mathbf{X} = \mathbf{O}$, *then*

$$\beta | \mathbf{y}, \sigma^2 \sim N_p(\mu_{\beta|\mathbf{y}}, \sigma^2 \Sigma_{\beta|\mathbf{y}})$$
$$\gamma | \mathbf{y}, \sigma^2 \sim N_q(\mu_{\gamma|\mathbf{y}}, \sigma^2 \Sigma_{\gamma|\mathbf{y}})$$

(6.128)

Which are conditionally independent, with

$$\mu_{\beta|\mathbf{y}} = (r_\beta^{-1}\mathbf{I}_p + \mathbf{X}^T\mathbf{X})^{-1}(\mathbf{X}^T\mathbf{y} + r_\beta^{-1}\beta_0)$$
$$\Sigma_{\beta|\mathbf{y}} = (r_\beta^{-1}\mathbf{I}_p + \mathbf{X}^T\mathbf{X})^{-1}$$
$$\mu_{\gamma|\mathbf{y}} = (r_\gamma^{-1}\mathbf{I}_q + \mathbf{Z}^T\mathbf{Z})^{-1}\mathbf{Z}^T\mathbf{y}$$
$$\Sigma_{\gamma|\mathbf{y}} = (r_\gamma^{-1}\mathbf{I}_q + \mathbf{Z}^T\mathbf{Z})^{-1}.$$

(6.129)

PROOF: First, for the given set up, we have

$$\Psi^{-1} = \begin{pmatrix} r_\beta^{-1}\mathbf{I}_p & \mathbf{0} \\ \mathbf{0} & r_\gamma^{-1}\mathbf{I}_q \end{pmatrix}.$$

Using $\mathbf{U} = (\mathbf{X} \ \mathbf{Z})$ and the orthogonality of $\mathbf{X}$ and $\mathbf{Z}$

$$\mathbf{U}^T \Sigma^{-1} \mathbf{U} = \begin{pmatrix} \mathbf{X}^T\mathbf{X} & \mathbf{0} \\ \mathbf{0} & \mathbf{Z}^T\mathbf{Z} \end{pmatrix}.$$

Thus from (6.125) it follows that

$$\Psi_1 = \begin{pmatrix} (r_\beta^{-1}\mathbf{I}_p + \mathbf{X}^T\mathbf{X})^{-1} & \mathbf{O} \\ \mathbf{O} & (r_\gamma^{-1}\mathbf{I}_q + \mathbf{Z}^T\mathbf{Z})^{-1} \end{pmatrix}.$$

(6.130)

The block diagonal structure implies the independence of the posteriors given the parameter σ^2. Further we apply (6.123) and calculate

$$\alpha_1 = \begin{pmatrix} \beta_1 \\ \gamma_1 \end{pmatrix} = \Psi_1(\mathbf{U}^T\Sigma^{-1}\mathbf{y} + \Psi^{-1}\alpha_0)$$

with

$$\mathbf{U}^T\Sigma^{-1}\mathbf{y} + \Psi^{-1}\alpha_0 = \begin{pmatrix} \mathbf{X}^T\mathbf{y} + r_\beta^{-1}\beta_0 \\ \mathbf{Z}^T\mathbf{y} \end{pmatrix}$$

and obtain $\mu_{\beta|\mathbf{y}}$ and $\mu_{\gamma|\mathbf{y}}$.

$\square$

For a better understanding of the formulas above we discuss a simple special case.

Example 6.14 (Simple linear mixed model)
Suppose the following data generating equation:

$$y_i = x_i\beta + \gamma + \varepsilon_i, \quad \varepsilon_i \sim \mathsf{N}(0, \sigma^2), \quad \gamma \sim \mathsf{N}(0, \sigma^2\lambda), \quad i = 1, \ldots, n.$$

The parameter of interest is $\theta = (\beta, \sigma^2)$. The design points x_i are known and centered, $\bar{x} = 0$. The random variables γ and $\varepsilon_1, \ldots, \varepsilon_n$ are mutually independent and unobserved. The prior $\pi(\theta)$ is $\mathsf{NIG}(a, b, \beta_0, \tau)$. Because $\bar{x} = 0$ and $z = 1$ we have $\mathbf{X}^T\mathbf{Z} = 0$, and we apply Corollary 6.6 to obtain

$$\beta|\mathbf{y}, \sigma^2 \sim \mathsf{N}(\beta_1, \sigma^2\tau_1)$$

with

$$\beta_1 = \tau_1(s_{xy} + \tau^{-1}\beta_0)$$
$$\tau_1 = (s_{xx} + \tau^{-1})^{-1}$$

and

$$\gamma|\mathbf{y}, \sigma^2 \sim \mathsf{N}((n + \lambda^{-1})^{-1}n\bar{y}, \sigma^2(n + \lambda^{-1})^{-1}).$$

From Theorem 6.10 with (6.121), it follows that

$$\sigma^2|\mathbf{y} \sim \mathsf{InvGamma}((a + n)/2, b_1/2)$$

with

$$b_1 = b + n\overline{y^2} + \tau^{-1}\beta_0^2 - (s_{xx} + \tau^{-1})^{-1}(s_{xy} + \tau^{-1}\beta_0)^2 - (n + \lambda^{-1})^{-1}(n\bar{y})^2,$$

where $n\overline{y^2} = \sum_{i=1}^n y_i^2$. Summarizing it holds that

$$(\beta, \sigma^2)|\mathbf{y} \sim \mathsf{NIG}(a + n, b_1, \beta_1, \tau_1)$$

and Lemma 6.5 gives

$$\beta|\mathbf{y} \sim t_1(a + n, \beta_1, \frac{b_1}{a + n}\tau_1).$$

$\square$

Example 6.15 (Hip operation)

We continue with Example 6.13 and apply Theorem 6.10. Using conjugate prior $\mathsf{NIG}(a, b, \alpha_0, \mathbf{I}_4)$ with $a = 6, b = 40$ and $\alpha_0 = 0$, we obtain $a_1 = 26$, $b_1 = 451.6$ and

$$\alpha_1 = \begin{pmatrix} 2.62 \\ 0.50 \\ 4.86 \\ -2.23 \end{pmatrix} \quad \Psi_1 = \begin{pmatrix} 0.9668 & -0.0138 & -0.0166 & 0.0166 \\ -0.0138 & 0.0003 & -0.0069 & -0.0069 \\ -0.0166 & -0.0069 & 0.5371 & 0.4462 \\ -0.0166 & -0.0069 & 0.4462 & 0.5371 \end{pmatrix}.$$

The distributions of $\beta|\mathbf{y}$ and $\gamma|\mathbf{y}$ are

$$\beta|\mathbf{y} \sim t_2 \left(26, \begin{pmatrix} 2.62 \\ 0.50 \end{pmatrix}, \begin{pmatrix} 16.793 & -0.240 \\ -0.240 & 0.005 \end{pmatrix} \right)$$

and

$$\gamma|\mathbf{y} \sim t_2 \left(26, \begin{pmatrix} 4.86 \\ -2.23 \end{pmatrix}, \begin{pmatrix} 9.330 & 7.751 \\ 7.751 & 9.330 \end{pmatrix} \right).$$

$\square$

R Code 6.3.6. Hip data, Example 6.15.

```
# data
Age<-c(66,66,70,70,70,70,74,74,65,65,71,71,68,68,69,69,
64,64,70,70)
Haematocrit<-c(47.1,32.8,44.1,37,43.3,36.6,37.4,29.05,45.7,
39.8,46.05,37.8,42.1,36.05,38.25,30.5,43,36.65,37.8,30.6)
# joint design matrix
U<-matrix(rep(0,4*20),ncol=4)
U[,1]<-rep(1,20)
U[,2]<-Age
U[,3]<-c(1,0,1,0,1,0,1,0,1,0,1,0,1,0,1,0,1,0,1,0)
U[,4]<-c(0,1,0,1,0,1,0,1,0,1,0,1,0,1,0,1,0,1,0,1)
y<-Haematocrit
# prior: NIG(a0,b0,alpha0,Psi)
a0<-3*2
```

```
b0<-10*4
b0/(a0-2) # prior expectation variance
(b0/2)^2/((a0/2-1)^2*(a0/2-2))# prior variance of variance
Psi<-diag(4)
alpha0<-c(0,0,0,0)
# posterior: NIG(a1,b1,alpha1,Psi1)
UU<-t(U)%*%U
Psi1<-solve(solve(Psi)+UU)
alpha1<-Psi1%*%(t(U)%*%y+solve(Psi)%*%alpha0)
a1<-a0+length(Age)
b1<-b0+y%*%y+alpha0%*%solve(Psi)%*%alpha0
-alpha1%*%solve(Psi1)%*%alpha1
# t-distribution
s<-as.numeric(b1/a1)
s*Psi1
```

6.4 Multivariate Linear Models

Here, we extend the univariate general linear model in (6.2) to the multivariate case, i.e., when a vector of responses is observed on each unit. We give the results and the main arguments; for technical details we refer the reader to Zellner (1971, Chapter 8) and Box and Tiao (1973, Chapter 8).

We define the multivariate general linear model as

$$\mathbf{Y} = \mathbf{XB} + \mathbf{E}, \tag{6.131}$$

where $\mathbf{Y} = (y_{ij}) = (\mathbf{y}_1^T, \ldots, \mathbf{y}_n^T)^T \in \mathbb{R}^{n \times d}$ is the $n \times d$ matrix of responses, $\mathbf{E} = (\epsilon_{ij}) = (\epsilon_1^T, \ldots, \epsilon_n^T)^T$ is the $n \times d$ matrix of random errors, $\mathbf{B} = (\beta_1^T, \ldots, \beta_p^T)^T \in \mathbb{R}^{p \times d}$ is the $p \times d$ matrix of unknown parameters, and $\mathbf{X} = (\mathbf{x}_1^T, \ldots, \mathbf{x}_n^T)^T \in \mathbb{R}^{n \times p}$ is the $n \times p$ design matrix, assumed to be of full rank, i.e., $r(\mathbf{X}) = p$. The row vectors of the response matrix in (6.131),

$$\mathbf{y}_i = \mathbf{B}^T \mathbf{x}_i + \epsilon_i, \quad \epsilon_i \sim \mathsf{N}_d(0, \Sigma) \quad i = 1, \ldots, n \tag{6.132}$$

are i.i.d. This implies that the elements of the response matrix are normally distributed, where the elements in different columns are correlated, while elements in different rows are independent.

In order to describe this distribution in a closed form, we use the notion of a *matrix-variate normal distribution*, (Gupta and Nagar, 2000, Chapter 2).

Let $\mathbf{Z} = (z_{ij})$ be an $m \times k$ random matrix. Using the vec operator, $\text{vec}(\mathbf{Z}) = (z_{11}, \ldots, z_{m1}, z_{12}, \ldots, z_{mk})^T$, we assume

$$\text{vec}(\mathbf{Z}) \sim \mathsf{N}_{mk}(\text{vec}(\mathbf{M}), \mathbf{V} \otimes \mathbf{U})$$

where $\mathbf{M}$ ia an $m \times k$ location matrix, $\otimes$ denotes the Kronecker product and $\mathbf{V}$ is $k \times k$ positive definite scale matrix for the columns, $\mathbf{U}$ is $m \times m$ positive definite scale matrix for the rows. We write

$$\mathbf{Z} \sim \mathsf{MN}_{m,k}(\mathbf{M}, \mathbf{U}, \mathbf{V}). \tag{6.133}$$

For $\theta = (\mathbf{M}, \mathbf{U}, \mathbf{V})$, the density $f(\mathbf{Z}|\theta)$ is proportional to

$$|\mathbf{V}|^{-\frac{m}{2}} |\mathbf{U}|^{-\frac{k}{2}} \exp\left(-\frac{1}{2}\text{tr}\left(\mathbf{V}^{-1}(\mathbf{Z} - \mathbf{M})^T\mathbf{U}^{-1}(\mathbf{Z} - \mathbf{M})\right)\right). \tag{6.134}$$

This implies that, $\mathbf{E} \sim \mathsf{MN}_{n,d}(\mathbf{O}, \mathbf{I}_n, \Sigma)$, where $\mathbf{I}_n$ is the identity matrix. Correspondingly, then,

$$\mathbf{Y} \sim \mathsf{MN}_{n,d}(\mathbf{XB}, \mathbf{I}_n, \Sigma),$$

with density

$$f(\mathbf{Y}|\mathbf{B}, \Sigma) = (2\pi)^{-nd/2}|\Sigma|^{-n/2}\exp\left(-\frac{1}{2}\text{tr}\left(\Sigma^{-1}(\mathbf{Y} - \mathbf{XB})^T(\mathbf{Y} - \mathbf{XB})\right)\right). \tag{6.135}$$

The multivariate model (6.131) can be considered columnwise as a collection of d correlated univariate models. Let $\mathbf{y}_{(s)}$, $\beta_{(s)}$, and $\epsilon_{(s)}$ be the sth columns of the response matrix $\mathbf{Y}$, parameter matrix $\mathbf{B}$, and error matrix $\mathbf{E}$, respectively. Then

$$\mathbf{y}_{(s)} = \mathbf{X}\beta_{(s)} + \epsilon_{(s)}, \quad \epsilon_{(s)} \sim N_n(0, \sigma_s^2\mathbf{I}_n), \quad s = 1, \ldots, d \tag{6.136}$$

with $\mathsf{E}\mathbf{y}_{(s)} = \mathbf{X}\beta_{(s)}$, so that $\mathsf{E}\mathbf{Y} = \mathbf{XB} = (\mathbf{X}\beta_{(1)}, \ldots, \mathbf{X}\beta_{(d)})$. Note, however, that, $\epsilon_{(s)}$ are correlated, so that $\text{Cov}(\epsilon_{(s)}) = \sigma_s^2\mathbf{I}_n = \text{Cov}(\mathbf{y}_{(s)})$ and $\text{Cov}(\epsilon_{(s)}, \epsilon_{(l)}) = \sigma_{sl}\mathbf{I}_n$.

Under this setting, the model for $\mathbf{Y} \in \mathbb{R}^{n \times d}$ can be written as

$$\mathcal{P} = \left\{\mathsf{MN}_{n,d}(\mathbf{XB}, \mathbf{I}_n, \Sigma) : \mathbf{B} \in \mathbb{R}^{p \times d}, \Sigma \in \mathbb{R}^{d \times d}, \Sigma \succ 0\right\}. \tag{6.137}$$

For a better understanding of multivariate model, we go back to the corn data example.

Example 6.16 (Corn plants) We analyze complete corn data, using both dependent variables; see Example 6.1 on page 127. The three independent variables are the same as used in Examples 6.5 and 6.12, where the analysis with one dependent variable was discussed.

With $n = 17$ independent vectors, where $d = 2$ and $p = 4$, we have $\mathbf{Y}$ and $\mathbf{X}$ as 17×2 and 17×4 matrices, respectively. $\square$

We continue with analysis of the statistical model (6.137) with unknown parameter $\theta = (\mathbf{B}, \Sigma)$. Writing (6.135) as likelihood function the maximum likelihood estimation problem

$$\underset{\mathbf{B},\Sigma}{\arg\max} \left(-\frac{n}{2} \ln |\Sigma| - \frac{1}{2} \mathrm{tr}\left(\Sigma^{-1}(\mathbf{Y} - \mathbf{XB})^T(\mathbf{Y} - \mathbf{XB})\right) \right),$$

yields the MLEs of $\mathbf{B}$ and Σ, respectively, as

$$\widehat{\mathbf{B}} = (\mathbf{X}^T\mathbf{X})^{-1}\mathbf{X}^T\mathbf{Y} \tag{6.138}$$

$$\widehat{\Sigma} = \frac{1}{n}(\mathbf{Y} - \mathbf{X}\widehat{\mathbf{B}})^T(\mathbf{Y} - \mathbf{X}\widehat{\mathbf{B}}), \tag{6.139}$$

(see, e.g., Mardia et al., 1979, Chapter 6). The fact that each univariate model in (6.136) has the same design matrix $\mathbf{X}$ implies $\widehat{\mathbf{Y}} = \mathbf{X}\widehat{\mathbf{B}} = \mathbf{PY}$, with $\mathbf{P} = \mathbf{X}(\mathbf{X}^T\mathbf{X})^{-1}\mathbf{X}^T$ the same projection matrix as for the univariate case. Thus, we can write $\widehat{\mathbf{Y}} = (\mathbf{X}\widehat{\beta}_1, \dots, \mathbf{X}\widehat{\beta}_d)$ as a collection of d fitted univariate models, with the exception that they are possibly correlated. This also allows the similar orthogonal partition of the model matrices

$$(\mathbf{Y} - \mathbf{XB})^T(\mathbf{Y} - \mathbf{XB}) = \mathbf{S} + (\mathbf{B} - \widehat{\mathbf{B}})^T\mathbf{X}^T\mathbf{X}(\mathbf{B} - \widehat{\mathbf{B}}) \tag{6.140}$$

with $\mathbf{S} = (\mathbf{Y} - \mathbf{X}\widehat{\mathbf{B}})^T(\mathbf{Y} - \mathbf{X}\widehat{\mathbf{B}}) = n\widehat{\Sigma}$.

Finally, for $\widehat{\mathbf{B}}$ as linear function of $\mathbf{Y}$, the normality assumption implies

$$\widehat{\mathbf{B}} \sim \mathsf{MN}_{p,d}(\mathbf{B}, (\mathbf{X}^T\mathbf{X})^{-1}, \Sigma)$$

and for $n\widehat{\Sigma}$ we have the *Wishart distribution*,

$$n\widehat{\Sigma} \sim \mathsf{W}_d(n - p, \Sigma).$$

We hereby recall a general definition of the *Wishart distribution*.

The $k \times k$ positive definite random matrix, $\mathbf{W}$, follows the *Wishart distribution*

$$\mathbf{W} \sim \mathsf{W}_k(\nu, \mathbf{V}),$$

with $\nu > k - 1$ degrees of freedom and $k \times k$ positive definite scale matrix $\mathbf{V}$, iff its density is given by

$$f(\mathbf{W}|\mathbf{V}, \nu) = c_{\mathsf{W}}|\mathbf{V}|^{-\frac{\nu}{2}}|\mathbf{W}|^{\frac{\nu-k-1}{2}} \exp\left(-\frac{1}{2}\mathrm{tr}\left(\mathbf{W}\mathbf{V}^{-1}\right)\right), \quad (6.141)$$

where

$$c_{\mathsf{W}} = \left(2^{\frac{k\nu}{2}} \pi^{\frac{k(k-1)}{4}} \prod_{j=1}^{k} \Gamma(\frac{\nu+1-j}{2})\right)^{-1}. \quad (6.142)$$

6.5 Bayes Multivariate Linear Models

In the following, we discuss Bayes models $\{\mathcal{P}, \pi\}$ with $\mathcal{P}$ given in (6.137). We assume conjugate prior and Jeffreys prior to derive the posteriors over

$$\Theta = \left\{(\mathbf{B}, \Sigma) : \mathbf{B} \in \mathbb{R}^{p \times d}, \Sigma \in \mathbb{R}^{d \times d}, \Sigma \text{ positive definite}\right\}. \quad (6.143)$$

6.5.1 Conjugate Prior

To begin with conjugate prior, we introduce the conjugate family of *normal–inverse-Wishart distributions*. First we introduce the *inverse-Wishart distribution* $\mathsf{IW}_d(\nu, \mathbf{V})$ which is a multivariate extension of the inverse-gamma distribution.

The $k \times k$ positive definite random matrix, $\mathbf{W}$, has the *inverse-Wishart distribution*

$$\mathbf{W} \sim \mathsf{IW}_k(\nu, \mathbf{V}),$$

with degrees of freedom $\nu > k - 1$ and $k \times k$ positive definite scale matrix $\mathbf{V}$, iff

$$\mathbf{W}^{-1} \sim \mathsf{W}_k(\nu, \mathbf{V}^{-1}).$$

Its density is

$$f(\mathbf{W}|\mathbf{V}, \nu) = c_{\mathsf{IW}}|\mathbf{V}|^{\frac{\nu}{2}}|\mathbf{W}|^{-\frac{\nu+k+1}{2}} \exp\left(-\frac{1}{2}\mathrm{tr}\left(\mathbf{W}^{-1}\mathbf{V}\right)\right). \quad (6.144)$$

where $c_{\mathsf{IW}} = c_{\mathsf{W}}$ given in (6.142).

Set $k = 1$, $\mathbf{V} = \lambda$, then

$$\mathsf{IW}_1(\nu, \lambda) \equiv \mathsf{InvGamma}(\nu/2, \lambda/2).$$

The normal-inverse-Wishart distribution $\mathsf{NIW}(\nu, \mathbf{M}, \mathbf{U}, \mathbf{V})$ is then hierarchically defined as following.

Let $\mathbf{Z}$ be a $m \times k$ random matrix and let $\mathbf{W}$ be $k \times k$ positive definite random matrix. The random matrices $(\mathbf{Z}, \mathbf{W})$ have a joint distribution belonging to the family of *normal-inverse-Wishart distributions*

$$(\mathbf{Z}, \mathbf{W}) \sim \mathsf{NIW}(\nu, \mathbf{M}, \mathbf{U}, \mathbf{V})$$

iff

$$\mathbf{Z} | \mathbf{W} \sim \mathsf{MN}_{m,k}(\mathbf{M}, \mathbf{U}, \mathbf{W})$$
$$\mathbf{W} \sim \mathsf{IW}_k(\nu, \mathbf{V})$$

The joint density of $(\mathbf{Z}, \mathbf{W})$ is proportional to the kernel $k(\mathbf{Z}, \mathbf{W} | \theta)$, that is

$$|\mathbf{W}|^{-\frac{\nu+k+m+1}{2}} \exp\left(-\frac{1}{2}\mathrm{tr}\left(\mathbf{W}^{-1}\left(\mathbf{V} + (\mathbf{Z} - \mathbf{M})^T \mathbf{U}^{-1}(\mathbf{Z} - \mathbf{M})\right)\right)\right),$$
(6.145)

with the parameters, ν as degrees of freedom, $\mathbf{M}$ as $m \times k$ location matrix, $\mathbf{V}$ as $k \times k$ positive definite scale matrix, and $\mathbf{U}$ as $m \times m$ positive definite scale matrix. Thus the joint density is $f(\mathbf{Z}, \mathbf{W} | \theta) = c_{\mathsf{NIW}} \, k(\mathbf{Z}, \mathbf{W} | \theta)$ with

$$c_{\mathsf{NIW}} = \left(2^{mk} \pi^{\frac{k(k-1)+2mk}{4}} \prod_{j=1}^{k} \Gamma\left(\frac{\nu+1-j}{2}\right)\right)^{-1} |\mathbf{V}|^{\frac{\nu}{2}} |\mathbf{U}|^{-\frac{k}{2}}. \quad (6.146)$$

For $k = 1$, $\mathbf{M} = \mathbf{m}$ and $\mathbf{V} = \lambda$, it holds that

$$\mathsf{NIW}(\nu, \mathbf{m}, \mathbf{U}, \lambda) \equiv \mathsf{NIG}(\nu, \lambda, \mathbf{m}, \mathbf{U}).$$

There exists a general result on marginal distributions of NIW distributions. Before stating it we introduce the *matrix-variate* t-*distribution*. .

Let $\mathbf{Z}$ be a $m \times k$ random matrix. $\mathbf{Z}$ has a *matrix-variate* t-*distribution*

$$\mathbf{Z} \sim \mathsf{t}_{m,k}(\nu, \mathbf{M}, \mathbf{U}, \mathbf{V})$$

iff the density $f(\mathbf{Z}|\theta)$ is proportional to

$$|\mathbf{V}|^{-\frac{m}{2}}|\mathbf{U}|^{-\frac{k}{2}}|\mathbf{I}_k + \mathbf{V}^{-1}(\mathbf{Z} - \mathbf{M})^T\mathbf{U}^{-1}(\mathbf{Z} - \mathbf{M})|^{-\frac{\nu+m+k-1}{2}} \quad (6.147)$$

with the parameters, ν as degrees of freedom, $\mathbf{M}$ as $m \times k$ location matrix, $\mathbf{V}$ as $k \times k$ positive definite scale matrix, and $\mathbf{U}$ as $m \times m$ positive definite scale matrix.

Note that the parametrization is different from the multivariate t-distribution. For $k = 1$, $\mathbf{V} = \lambda$, and $\mathbf{M} = \mathbf{m}$, we have

$$\mathsf{t}_{m,1}(\nu, \mathbf{m}, \mathbf{U}, \lambda) \equiv \mathsf{t}_m(\nu, \mathbf{m}, \frac{\lambda}{\nu}\mathbf{U}). \quad (6.148)$$

We have

Lemma 6.6 *Assume*

$$(\mathbf{Z}, \mathbf{W}) \sim \mathsf{NIW}(\nu, \mathbf{M}, \mathbf{U}, \mathbf{V}).$$

Then

$$\mathbf{Z} \sim \mathsf{t}_{m,k}(\nu - k + 1, \mathbf{M}, \mathbf{U}, \mathbf{V}).$$

PROOF: The joint density of $(\mathbf{Z}, \mathbf{W})$ is proportional to

$$|\mathbf{U}|^{-\frac{k}{2}}|\mathbf{V}|^{\frac{\nu}{2}}|\mathbf{W}|^{-\frac{\nu+k+m+1}{2}} \exp\left(-\frac{1}{2}\mathrm{tr}\left(\mathbf{W}^{-1}\mathbf{A}\right)\right) \quad (6.149)$$

with

$$\mathbf{A} = \mathbf{V} + (\mathbf{Z} - \mathbf{M})^T \mathbf{U}^{-1} (\mathbf{Z} - \mathbf{M}).$$

Note that

$$K(\mathbf{W}) \propto |\mathbf{W}|^{-\frac{\nu+m+k+1}{2}}|\mathbf{A}|^{\frac{\nu+m}{2}} \exp\left(-\frac{1}{2}\mathrm{tr}\left(\mathbf{W}^{-1}\mathbf{A}\right)\right)$$

is kernel of $\mathsf{IW}(\nu + m, \mathbf{A})$. We have

$$f(\mathbf{Z}, \mathbf{W}) \propto |\mathbf{U}|^{-\frac{k}{2}}|\mathbf{V}|^{\frac{\nu}{2}}|\mathbf{A}|^{-\frac{\nu+m}{2}} K(\mathbf{W}).$$

Thus we can integrate out $\mathbf{W}$ and obtain

$$f(\mathbf{Z}) \propto |\mathbf{U}|^{-\frac{k}{2}}|\mathbf{V}|^{\frac{\nu}{2}}|\mathbf{A}|^{-\frac{\nu+m}{2}}$$
$$\propto |\mathbf{U}|^{-\frac{k}{2}}|\mathbf{V}|^{\frac{\nu}{2}}|\mathbf{V} + (\mathbf{Z} - \mathbf{M})^T \mathbf{U}^{-1} (\mathbf{Z} - \mathbf{M})|^{-\frac{\nu+m}{2}}$$
$$\propto |\mathbf{U}|^{-\frac{k}{2}}|\mathbf{V}|^{-\frac{m}{2}}|\mathbf{I}_k + \mathbf{V}^{-1} (\mathbf{Z} - \mathbf{M})^T \mathbf{U}^{-1} (\mathbf{Z} - \mathbf{M})|^{-\frac{\nu+m}{2}}.$$

This is the kernel of $\mathsf{t}_{m,k}(\nu - k + 1, \mathbf{U}, \mathbf{V})$.

□

We set the conjugate prior of $\theta = (\mathbf{B}, \Sigma)$ as

$$(\mathbf{B}, \Sigma) \sim \mathsf{NIW}(\nu_0, \mathbf{B}_0, \mathbf{C}_0, \Sigma_0), \tag{6.150}$$

where $\mathbf{B}_0$ is first guess for the matrix of regression parameters $\mathbf{B}$, $\mathbf{C}_0$ says something about the accuracy of the guess, Σ_0 includes subjective information on the correlation between different univariate models, and ν_0 measures how many observations the guesses are based on. The density of normal-inverse-Wishart (NIW) distribution, $\mathsf{NIW}(\nu_0, \mathbf{B}_0, \mathbf{C}_0, \Sigma_0)$, is:

$$\pi(\mathbf{B}, \Sigma) \propto |\Sigma_0|^{-\frac{\nu_0 + p + d + 1}{2}} \exp\left(-\frac{1}{2}\mathrm{tr}\left(\Sigma^{-1}(\Sigma_0 + (\mathbf{B} - \mathbf{B}_0)^{\mathbf{T}}\mathbf{C}_0^{-1}(\mathbf{B} - \mathbf{B}_0)))\right)\right) \tag{6.151}$$

The entire Bayes model setup can be stated as

$$\begin{aligned} \mathbf{Y}|\theta &\sim \mathsf{MN}_{n,d}(\mathbf{XB}, \mathbf{I}_n, \Sigma) \\ \mathbf{B}|\Sigma &\sim \mathsf{MN}_{p,d}(\mathbf{B}_0, \mathbf{C}_0, \Sigma) \\ \Sigma &\sim \mathsf{IW}_d(\nu_0, \Sigma_0). \end{aligned} \tag{6.152}$$

This leads to a normal-inverse-Wishart posterior, stated in the following theorem.

Theorem 6.11 *Given* $(\mathbf{B}, \Sigma) \sim \mathsf{NIW}(\nu_0, \mathbf{B}_0, \mathbf{C}_0, \Sigma_0)$, *then*

$$(\mathbf{B}, \Sigma)|\mathbf{Y} \sim \mathsf{NIW}(\nu_1, \mathbf{B}_1, \mathbf{C}_1, \Sigma_1) \ \ and \ \ \mathbf{B}|\mathbf{Y} \sim \mathsf{t}_{p,d}(\nu_1 - d + 1, \mathbf{B}_1, \mathbf{C}_1, \Sigma_1)$$

with $\nu_1 = \nu_0 + n$ *and*

$$\begin{aligned} \mathbf{B}_1 &= \mathbf{C}_1(\mathbf{X}^T\mathbf{Y} + \mathbf{C}_0^{-1}\mathbf{B}_0) \\ \mathbf{C}_1 &= \left(\mathbf{C}_0^{-1} + \mathbf{X}^T\mathbf{X}\right)^{-1} \\ \Sigma_1 &= \Sigma_0 + (\mathbf{Y} - \mathbf{XB}_1)^T(\mathbf{Y} - \mathbf{XB}_1) + (\mathbf{B}_0 - \mathbf{B}_1)^T\mathbf{C}_0^{-1}(\mathbf{B}_0 - \mathbf{B}_1). \end{aligned}$$

PROOF: Using $\pi(\theta|x) \propto \pi(\theta)\ell(\theta|x)$, where now the observations x are denoted by $\mathbf{Y}$ generated by the multivariate linear model in (6.131), we obtain

$$\pi(\theta|\mathbf{Y}) \propto |\Sigma|^{-\frac{\nu_0 + p + d + 1 + n}{2}} \exp\left(-\frac{1}{2}\mathrm{tr}\left(\Sigma^{-1}\mathrm{term}_1\right)\right)$$

with

$$\mathrm{term}_1 = \Sigma_0 + (\mathbf{Y} - \mathbf{XB})^T(\mathbf{Y} - \mathbf{XB}) + (\mathbf{B} - \mathbf{B}_0)^T\mathbf{C}_0^{-1}(\mathbf{B} - \mathbf{B}_0).$$

Completing the squares for $\mathbf{B}$ gives

$$\text{term}_1 = \Sigma_1 + (\mathbf{B} - \mathbf{B}_1)^T \mathbf{C}_1^{-1}(\mathbf{B} - \mathbf{B}_1),$$

where

$$\Sigma_1 = \Sigma_0 + (\mathbf{Y} - \mathbf{X}\mathbf{B}_1)^T(\mathbf{Y} - \mathbf{X}\mathbf{B}_1) + (\mathbf{B}_0 - \mathbf{B}_1)^T \mathbf{C}_0^{-1}(\mathbf{B}_0 - \mathbf{B}_1)$$

with

$$\mathbf{C}_1 = (\mathbf{C}_0^{-1} + \mathbf{X}^T\mathbf{X})^{-1} \quad \text{and} \quad \mathbf{B}_1 = \mathbf{C}_1(\mathbf{X}^T\mathbf{X}\widehat{\mathbf{B}} + \mathbf{C}_0^{-1}\mathbf{B}_0).$$

Summarizing, we obtain the posterior distribution proportional to

$$|\Sigma|^{-\frac{\nu_0 + n + d + p + 1}{2}} \exp\left(-\frac{1}{2}\text{tr}\left((\Sigma_1 + (\mathbf{B} - \mathbf{B}_1)^T \mathbf{C}_1^{-1}(\mathbf{B} - \mathbf{B}_1))\Sigma^{-1}\right)\right),$$

which is the kernel of the normal-inverse-Wishart distribution in the statement. The marginal posterior follows from Lemma 6.6.

$\square$

6.5.2 Jeffreys Prior

Assuming prior independence of $\mathbf{B}$ and Σ, the joint prior follows as

$$\pi_{\text{Jeff}}(\mathbf{B}, \Sigma) \propto \pi(\mathbf{B})\pi(\Sigma).$$

Consider first $\pi(\mathbf{B})$. The matrix $\mathbf{X}\mathbf{B}$ is the location parameter in (6.135). We apply the result on location models in Example 3.21 on $\text{vec}(\mathbf{Y}) \sim N_{nd}(\text{vec}(\mathbf{X}\mathbf{B}), \mathbf{I}_n \otimes \Sigma)$ and obtain

$$\pi(\mathbf{B}) \propto \text{const}.$$

The prior $\pi(\Sigma)$ is now related to model (6.132), where we set $\mathbf{B} = \mathbf{0}$. Thus we have to calculate the Fisher information of the model

$$\mathbf{y}_1, \ldots, \mathbf{y}_n \quad i.i.d. \quad N_d(0, \Sigma). \tag{6.153}$$

Recall the calculation of Fisher information in Section 6.2.3. It includes the calculation of the first derivative of the log-likelihood function with respect to the unknown parameters. Here in model (6.153) the unknown parameter is the symmetric matrix $\Sigma = (\sigma_{ij})$, with $\sigma_{ij} = \sigma_{ji}$, so that we have $m = d(d+1)/2$ different parameters. But the vectorization by vec operator includes all d^2 elements of the matrix,

$$\text{vec}(\Sigma) = (\sigma_{11}, \ldots \sigma_{d1}, \sigma_{12}, \ldots, \sigma_{d2}, \sigma_{13}, \ldots, \sigma_{dd})^T.$$

The way out is to use a vectorization of a symmetric matrix with the operator vech which stacks the columns by starting each column at its diagonal element, and includes m elements,

$$\text{vech}(\Sigma) = (\sigma_{11}, \ldots \sigma_{d1}, \sigma_{22}, \ldots, \sigma_{d2}, \sigma_{33}, \ldots, \sigma_{dd})^T.$$

It holds that

$$\text{vec}(\Sigma) = \mathbf{D}\,\text{vech}(\Sigma), \tag{6.154}$$

where $\mathbf{D}$ is the $d^2 \times m$ duplication matrix. Using this notation in Magnus and Neudecker (1999), where a calculus for matrix differentiation is developed, the Fisher information in model (6.153) is given by the $m \times m$ matrix (see also Magnus and Neudecker, 1980)

$$I(\Sigma) = \mathbf{D}^T(\Sigma^{-1} \otimes \Sigma^{-1})\mathbf{D}. \tag{6.155}$$

Further, the determinant of Fisher information is calculated as in (see Magnus and Neudecker, 1980)

$$|I(\Sigma)| = 2^m |\Sigma|^{-(d+1)}.$$

We obtain

$$\pi(\Sigma) \propto |I(\Sigma)|^{\frac{1}{2}} \propto |\Sigma|^{-\frac{d+1}{2}}.$$

Finally

$$\pi_{\text{Jeff}}(\mathbf{B}, \Sigma) \propto |\Sigma|^{-(d+1)/2}. \tag{6.156}$$

The following theorem gives the posterior and the marginal posterior.

Theorem 6.12 *For the Bayes model $\{\mathcal{P}, \pi_{\text{Jeff}}\}$, with $\mathcal{P}$ given in (6.137) and π_{Jeff} as defined in (6.156)*

$$(\mathbf{B}, \Sigma)|\mathbf{Y} \sim \text{NIW}(n, \widehat{\mathbf{B}}, (\mathbf{X}^T\mathbf{X})^{-1}, \mathbf{S})$$
$$\mathbf{B}|\mathbf{Y} \sim t_{p,d}(n - d + 1, \widehat{\mathbf{B}}, (\mathbf{X}^T\mathbf{X})^{-1}, \mathbf{S})$$

with

$$\widehat{\mathbf{B}} = (\mathbf{X}^T\mathbf{X})^{-1}\mathbf{X}^T\mathbf{Y}$$
$$\mathbf{S} = (\mathbf{Y} - \mathbf{X}\widehat{\mathbf{B}})^T(\mathbf{Y} - \mathbf{X}\widehat{\mathbf{B}}).$$

Recall that $\widehat{\mathbf{B}}$ is the least-squares estimator of $\mathbf{B}$ given in (6.138).
PROOF: Adding the normal likelihood for $\text{vec}(\mathbf{Y}) \sim \mathsf{N}_{nd}(\mathbf{X}\mathbf{B}, \mathbf{I}_n \otimes \Sigma)$ from (6.135), the joint posterior of $\theta = (\mathbf{B}, \Sigma)$ is

$$\pi(\theta|\mathbf{Y}) \propto |\Sigma|^{-(n+d+1)/2} \exp\left(-\frac{1}{2}\text{tr}\left(\Sigma^{-1}(\mathbf{Y} - \mathbf{X}\mathbf{B})^T(\mathbf{Y} - \mathbf{X}\mathbf{B})\right)\right). \tag{6.157}$$

Applying the orthogonal decomposition (6.140) we obtain

$$\pi(\theta|\mathbf{Y}) \propto |\Sigma|^{-(n+d+1)/2} \exp\left(-\frac{1}{2}\text{tr}\left(\Sigma^{-1}\Xi\right)\right)$$

with

$$\Xi = \mathbf{S} + (\mathbf{B} - \widehat{\mathbf{B}})^T\mathbf{X}^T\mathbf{X}(\mathbf{B} - \widehat{\mathbf{B}}).$$

This is the kernel of the NIW distribution above. The marginal posterior follows from Lemma 6.6.

$\square$

Example 6.17 (Corn plants)

We continue with Example 6.16, and use both non-informative and conjugate priors. First, using the least-squares estimation, we obtain:

$$\mathbf{X}^T\mathbf{X} = \begin{pmatrix} 17.0 & 188.2 & 700 & 2012 \\ 188.2 & 3602.8 & 8585 & 22231 \\ 700.0 & 8585.1 & 31712 & 84882 \\ 2012.0 & 22231.4 & 84882 & 267090 \end{pmatrix}, \quad \mathbf{X}^T\mathbf{Y} = \begin{pmatrix} 1295 & 1496 \\ 16204 & 17688 \\ 54081 & 65302 \\ 153606 & 182220 \end{pmatrix}.$$

Thus

$$\widehat{\mathbf{B}} = \begin{pmatrix} 64.440 & 26.823 \\ 1.301 & 0.090 \\ -0.130 & 1.189 \\ 0.023 & 0.095 \end{pmatrix}, \quad \mathbf{S} = \begin{pmatrix} 2087.17 & -349.79 \\ -349.79 & 3693.80 \end{pmatrix}.$$

We apply Theorem 6.12 with Jeffreys prior, $\pi(\Sigma) \propto |\Sigma|^{3/2}$, and assuming $\mathbf{Y} \sim \text{MN}_{17,2}(\mathbf{XB}, \mathbf{I}_{17}, \Sigma)$, the posterior of $\mathbf{B}$ is the multivariate t distribution, $t_{4,2}(16, \widehat{\mathbf{B}}, (\mathbf{X}^T\mathbf{X})^{-1}, \mathbf{S})$, the components of which are given above.

For the conjugate prior, we assume $\nu_0 = 5$, $\sigma_0^2 = 500$, with $\mathbf{B}|\Sigma \sim \text{MN}_{4,2}(\mathbf{0}, \mathbf{I}_4, \Sigma)$ and $\Sigma \sim \text{IW}(\nu_0, \sigma_0^2 \mathbf{I}_2)$, i.e., we set the prior $\text{NIW}(\nu_0, \mathbf{0}, \mathbf{I}_4, \sigma_0^2 \mathbf{I}_2)$. From Theorem 6.11, the joint posterior is $\text{NIW}(\nu_1, \mathbf{B}_1, \mathbf{C}_1, \Sigma_1)$, where $\nu_1 = 22$, and the moments are computed as

$$\mathbf{C}_1 = 10^{-5} \begin{pmatrix} 48367.52 & -49.42 & -602.68 & -168.71 \\ -49.42 & 79.06 & -24.73 & 1.65 \\ -602.68 & -24.73 & 36.75 & -5.08 \\ -168.71 & 1.65 & -5.08 & 3.12 \end{pmatrix}$$

$$\mathbf{B}_1 = \begin{pmatrix} 33.27 & 13.86 \\ 1.33 & 0.10 \\ 0.26 & 1.35 \\ 0.13 & 0.14 \end{pmatrix}, \quad \Sigma_1 = \begin{pmatrix} 4732.92 & 543.10 \\ 543.10 & 4567.11 \end{pmatrix}.$$

$\square$

6.6   List of Problems

1. Consider a multiple regression model

$$y_i = \beta_0 + x_i\beta_1 + z_i\beta_2 + \varepsilon_i, \quad i = 1, \ldots, n$$

with orthogonal design, i.e., $\sum_{i=1}^n x_i = 0$, $\sum_{i=1}^n z_i = 0$, $\sum_{i=1}^n x_i z_i = 0$, $\sum_{i=1}^n x_i^2 = n$ and $\sum_{i=1}^n z_i^2 = n$ where ε_i are i.i.d. normally distributed with expectation zero and variance $\sigma^2 = 1$. The unknown three dimensional parameter $\beta = (\beta_0, \beta_1, \beta_2)^T$ is normally distributed with mean $\mu = (1, 1, 1)^T$ and covariance matrix

$$\Sigma = \begin{bmatrix} 1 & 0.5 & 0 \\ 0.5 & 1 & 0 \\ 0 & 0 & 1 \end{bmatrix}.$$

(a) Write the model equation in matrix form.

(b) Determine the posterior distribution of β. Specify the expressions in (6.23) and (6.24).

2. Consider the simple linear regression model

$$y_i = \alpha + \beta x_i + \varepsilon_i, \quad i = 1, \ldots, n, \quad \varepsilon_i \sim N(0, \sigma^2) \; i.i.d. \tag{6.158}$$

The unknown parameter is $\theta = (\alpha, \beta, \sigma^2)$. We are mainly interested in the posterior of the slope β.

(a) Determine Jeffreys prior $\pi_{\text{Jeff}}(\theta)$ under prior independence of regression parameter and variance.

(b) Calculate the posterior distributions:

   i. $\pi((\alpha, \beta)|\mathbf{y}, \sigma^2)$.
   ii. $\pi(\beta|\mathbf{y}, \sigma^2)$
   iii. $\pi(\beta|\mathbf{y}, \alpha, \sigma^2)$.
   iv. $\pi(\sigma^2|\mathbf{y})$.
   v. $\pi(\beta|\mathbf{y})$.

3. Consider again the simple linear regression model (6.158) with unknown parameter $\theta = (\alpha, \beta, \sigma^2)$. We are mainly interested in the posterior of the variance σ^2. Let Jeffreys prior be $\pi_{\text{Jeff}}(\theta) \propto (\sigma^2)^{-1}$.

(a) Calculate the posterior distribution $\pi(\sigma^2|\mathbf{y}, \beta, \alpha)$.

(b) Compare it with $\pi(\sigma^2|\mathbf{y})$ in Problem 2 (b).

4. Assume again (6.158) and transform the model by centering the data, such that

$$y_i - \bar{y} = \beta(x_i - \bar{x}) + \xi_i, \quad i = 1, \ldots, n. \tag{6.159}$$

(a) Determine the relation between $\xi_1, \ldots, \xi_n$ and $\varepsilon_1, \ldots, \varepsilon_n$.

(b) Determine the covariance matrix of $\xi = (\xi_1, \ldots, \xi_n)$ and calculate its determinant.

(c) Set $\xi_{(-n)} = (\xi_1, \ldots, \xi_{n-1})^T$, i.e., drop the last observation. Calculate the determinant of the covariance matrix of $\xi_{(-n)}$.

(d) Assume model (6.159) with $i = 1, \ldots, n - 1$.

    i. Consider Jeffreys prior. Calculate $\pi(\beta|\mathbf{y}, \sigma^2)$ and $\pi(\sigma^2|\mathbf{y})$. Compare the results with those for model (6.158).

    ii. Let $\sigma^2 = 1$ and $\beta \sim N_1(\gamma_b, \lambda_2)$. Calculate $\pi(\beta|\mathbf{y})$. Using independent priors, $\beta \sim N_1(\gamma_b, \lambda_2)$ and $\alpha \sim N_1(\gamma_a, \lambda_1)$, compare the results with those for model (6.158).

5. Assume the univariate linear model (6.2) with $\epsilon \sim N_n(0, \sigma^2 \mathbf{I}_n)$. We are interested in the precision parameter σ^{-2}. Set $\theta = (\beta, \sigma^{-2})$. Determine the conjugate family.

6. Assume the univariate linear model (6.2) with $\epsilon \sim N_n(0, \sigma^2 \mathbf{I}_n)$. We are interested in $\eta = (\beta, \sigma)$, where σ is the scale parameter. Determine the Jeffreys priors with and without independence assumption ($\pi(\eta) = \pi(\beta)\pi(\sigma)$).

7. Consider three independent samples:

$$
\begin{aligned}
X &= (X_1, \ldots, X_m) \text{ i.i.d. sample from } N(\mu_1, \sigma^2 \lambda_1) \\
Y &= (Y_1, \ldots, Y_m) \text{ i.i.d. sample from } N(\mu_2, \sigma^2 \lambda_2) \\
Z &= (Z_1, \ldots, Z_m) \text{ i.i.d. sample from } N(\mu_3, \sigma^2 \lambda_3)
\end{aligned}
\tag{6.160}
$$

where $\lambda_1, \lambda_2, \lambda_3$ are known and $\mu_1 + \mu_2 + \mu_3 = 0$.

(a) Re-write the model (6.160) as univariate model. Determine the response vector $\mathbf{y}$, the design matrix $\mathbf{X}$ and the error covariance matrix $\sigma^2 \Sigma$.

(b) Suppose a conjugate prior for $\theta = (\mu_1, \mu_2, \sigma^2)$ gives the expressions for the posterior. Specify the posterior covariance of (μ_1, μ_2) given σ^2 under the prior covariance $\sigma^2 \mathbf{I}_2$ and $\lambda_1 = \lambda_2 = \lambda_3 = 1$. Set the prior expectation of (μ_1, μ_2) as zero.

(c) Consider only the first two samples X, Y. Compute the posterior, using the same prior as in (b).

(d) Compare the posterior covariance matrix of (μ_1, μ_2) given σ^2 based on two samples with the corresponding posterior covariance matrix based on three samples.

8. We are interested in two parallel regression lines:

$$
\begin{aligned}
y_i &= \alpha + \beta x_i + \varepsilon_i, \quad i = 1, \ldots, n, \quad \varepsilon_i \sim N(0, \sigma^2) \ i.i.d. \\
z_i &= \gamma + \beta x_i + \xi_i, \quad i = 1, \ldots, n, \quad \xi_i \sim N(0, \sigma^2) \ i.i.d.,
\end{aligned}
\tag{6.161}
$$

where ε_i and ξ_i are mutually independent and $\sum_{i=1}^{n} x_i = 0$. The unknown parameter is $\theta = (\alpha, \gamma, \beta, \sigma^2)$. Assume a conjugate prior $NIG(a, b, 0, \lambda^{-1}\mathbf{I}_3)$.

(a) Re-write the model (6.161) as univariate model. Determine the response vector $\mathbf{y}$, the design matrix $\mathbf{X}$ and the error covariance matrix $\sigma^2 \Sigma$.

(b) Calculate $\pi(\alpha, \gamma, \beta|\mathbf{y}, \sigma^2)$ and $\pi(\beta|\mathbf{y}, \sigma^2)$.

(c) Calculate $\pi(\sigma^2|\mathbf{y})$.

(d) Give the marginal posterior of the slope $\pi(\beta|\mathbf{y})$.

9. Consider the linear mixed model.

$$y_i = \beta x_i + \gamma z_i + \varepsilon_i, \quad i = 1, \ldots, n, \quad \varepsilon_i \sim \mathsf{N}(0,1), \ i.i.d., \ \gamma \sim \mathsf{N}(0,1),$$

where $\sum_{i=1}^n x_i^2 = n$ and $\sum_{i=1}^n z_i^2 = n$ but $\sum_{i=1}^n x_i z_i \neq 0$. Assume a conjugate prior for β and give an explicit formula for $\beta|\mathbf{y}$.

10. Assume two correlated regression lines

$$
\begin{aligned}
y_i &= \alpha_y + \beta_y x_i + \varepsilon_i, \quad i = 1, \ldots, n, \quad \varepsilon_i \sim \mathsf{N}(0, \sigma_1^2) \ i.i.d. \\
z_i &= \alpha_z + \beta_z x_i + \xi_i, \quad i = 1, \ldots, n, \quad \xi_i \sim \mathsf{N}(0, \sigma_2^2) \ i.i.d.,
\end{aligned}
\tag{6.162}
$$

where $\mathsf{Cov}(y, z) = \sigma_{1,2}$ and $\sum_{i=1}^n x_i = 0$, $\sum_{i=1}^n x_i^2 = n$. The unknown parameter is $\theta = (\alpha_y, \beta_y, \alpha_z, \beta_z, \sigma_1^2, \sigma_{12}, \sigma_2^2)$.

(a) Formulate the multivariate model.

(b) Assume a conjugate prior $\mathsf{NIW}(\nu, \mathbf{B}_0, \mathbf{C}_0, \Sigma_0)$, where $\mathbf{C}_0 = \mathrm{diag}(c_1, c_2)$ and $\mathbf{B}_0 = \mathbf{0}$, $\Sigma_0 = \mathbf{I}_2$. Give the posterior expectation of $\sigma_{1,2}$.

Chapter 7

Estimation

We consider the Bayes model $\{\mathcal{P}, \pi\}$, where $\mathcal{P} = \{\mathsf{P}_\theta : \theta \in \Theta\}$ and π is the known prior distribution of θ over Θ. The data generating distribution P_θ is the conditional distribution of $\mathbf{X}$ given θ and it is known up to θ.

Having observed $\mathbf{x}$ for $\mathbf{X} \sim \mathsf{P}_\theta$ we want to determine the underlying parameter θ – exploiting the model assumption $\{\mathcal{P}, \pi\}$.

Applying the Bayesian inference principle means that the posterior distribution takes over the role of the likelihood; see Chapter 2. All information we have about θ is included in the posterior. Depending on different estimation strategies we choose as estimator the

- mode,

- expectation, or

- median

of the posterior distribution. Before we discuss each method in more detail, we illustrate it by three examples. In the first example all estimators coincide.

Example 7.1 (Normal i.i.d. sample and normal prior)
In continuation of Example 2.12, we have an i.i.d. sample $\mathbf{X} = (X_1, \ldots, X_n)$ from $\mathsf{N}(\mu, \sigma^2)$ with known variance σ^2 and the prior of $\theta = \mu$ is $\mathsf{N}(\mu_0, \sigma_0^2)$. Then the posterior given in (2.8) is $\mathsf{N}(\mu_1, \sigma_1^2)$, with

$$\mu_1 = \frac{\bar{x} n \sigma_0^2 + \mu_0 \sigma^2}{n \sigma_0^2 + \sigma^2}$$

and

$$\sigma_1^2 = \frac{\sigma_0^2 \sigma^2}{n \sigma_0^2 + \sigma^2}.$$

The posterior is symmetric in μ_1; see also Figure 2.5. We have

$$\mathsf{Mode}(\theta|x) = \mathsf{E}(\theta|x) = \mathsf{Median}(\theta|x) = \frac{\bar{x} n \sigma_0^2 + \mu_0 \sigma^2}{n \sigma_0^2 + \sigma^2}.$$

$\square$

In the second example all estimators are different.

DOI: 10.1201/9781003221623-7

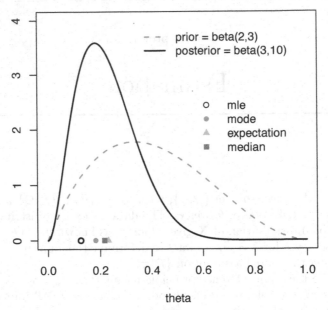

Figure 7.1: Example 7.2. All estimates are situated in the center of the posterior distribution: mle= 0.123, mode= 0.18, expectation= 0.23, median= 0.21.

Example 7.2 (Binomial distribution and beta prior)
In continuation of Example 2.11, we have $\mathbf{X}|\theta \sim \mathsf{Bin}(n,\theta)$ and $\theta \sim \mathsf{Beta}(\alpha_0, \beta_0)$, then

$$\theta|x \sim \mathsf{Beta}(\alpha_1, \beta_1), \quad \text{with } \alpha_1 = \alpha_0 + x, \quad \beta_1 = \beta_0 + n - x.$$

For $\alpha_1 \neq \beta_1$ the posterior density

$$B(\alpha_1, \beta_1)^{-1} \theta^{\alpha_1 - 1} (1 - \theta)^{\beta_1 - 1}$$

is not symmetric and mode, expectation and median differ. Particularly, we have

$$\mathsf{Mode}(\theta|x) = \frac{\alpha_1 - 1}{\alpha_1 + \beta_1 - 2}, \quad \mathsf{E}(\theta|x) = \frac{\alpha_1}{\alpha_1 + \beta_1}, \quad \mathsf{Median}(\theta|x) \approx \frac{\alpha_1 - \frac{1}{3}}{\alpha_1 + \beta_1 - \frac{2}{3}},$$

as illustrated in Figure 7.1, where for median and mode, $\alpha_1 > 1, \beta_1 > 1$. □

The Bayesian estimation approach has at least two advantages. One is the elegant handling of the nuisance parameters by exploring the marginal posterior of the parameter of interest, and the other is that the posterior distribution delivers information about the precision of the estimator. We demonstrate both properties in the following example.

Example 7.3 (Normal i.i.d. sample and conjugate prior)
We have an i.i.d. sample $\mathbf{X} = (X_1, \ldots, X_n)$ from $\mathsf{N}(\mu, \sigma^2)$ with unknown expectation μ and unknown variance σ^2, thus $\theta = (\mu, \sigma^2)$. The parameter of interest is μ; the variance is the nuisance parameter. In Chapter 6 the conjugate prior for the linear model is derived. This example is a special case with $p = 1$, where all design points equal 1. The conjugate family is the normal-inverse-gamma distribution, given in (6.41). We set the prior $\mathsf{NIG}(\alpha_0, \beta_0, \mu_0, \sigma_0^2)$. Recall, the prior of μ given σ^2 is $\mathsf{N}(\mu_0, \sigma^2 \sigma_0^2)$, the prior of σ^2 is the inverse-gamma distribution $\mathsf{InvGamma}(\frac{\alpha_0}{2}, \frac{\beta_0}{2})$. The prior parameters α_0, β_0, μ_0 and σ_0^2 are known. We have

$$\pi(\theta) \propto (\sigma^2)^{-\frac{\alpha_0+3}{2}} \exp\left(-\frac{1}{2\sigma^2}(\beta_0 + \frac{1}{\sigma_0^2}(\mu - \mu_0)^2) \right).$$

We give the main steps of deriving the posterior, as follows,

$$\pi(\theta|\mathbf{x}) \propto \pi(\theta)\ell(\theta|\mathbf{x})$$

$$\propto \pi(\theta)(\sigma^2)^{-\frac{n}{2}} \exp\left(-\frac{1}{2\sigma^2} \sum_{i=1}^{n}(x_i - \mu)^2 \right).$$

Let s^2 be the sample variance. Using

$$\sum_{i=1}^{n}(x_i - \mu)^2 = (n-1)s^2 + n(\mu - \bar{x})^2$$

and completing the squares such that

$$n(\mu - \bar{x})^2 + \frac{1}{\sigma_0^2}(\mu - \mu_0)^2 = \frac{1}{\sigma_1^2}(\mu - \mu_1)^2 + \frac{1}{\sigma_0^2}\mu_0^2 + n\bar{x}^2 - \frac{1}{\sigma_1^2}\mu_1^2,$$

with

$$\sigma_1^{-2} = \sigma_0^{-2} + n \quad \text{and} \quad \mu_1 = \sigma_1^2(\sigma_0^{-2}\mu_0 + n\bar{x});$$

we obtain the conditional posterior of μ given σ^2 by $\mathsf{N}(\mu_1, \sigma^2 \sigma_1^2)$. The marginal posterior of σ^2 is the inverse-gamma distribution with parameters $(\frac{\alpha_1}{2}, \frac{\beta_1}{2})$, where

$$\alpha_1 = \alpha_0 + n$$

$$\beta_1 = \beta_0 + (n-1)s^2 + \frac{1}{\sigma_0^2}\mu_0^2 + n\bar{x}^2 - \frac{1}{\sigma_1^2}\mu_1^2. \tag{7.1}$$

The parameter of interest is μ, therefore we are interested in the posterior

$$\pi(\mu|\mathbf{x}) = \int_0^\infty \pi(\mu|\sigma^2, x)\pi(\sigma^2|x)\, d\sigma^2.$$

From Lemma 6.5, the marginal distribution of a normal-inverse-gamma distribution $\mathsf{NIG}(\alpha_1, \beta_1, \mu_1, \sigma_1^2)$ is the scaled and shifted t-distribution with α_1

degrees of freedom, location parameter μ_1 and scale parameter $\sqrt{\frac{\beta_1}{\alpha_1}}\sigma_1$, whose density is

$$\pi(\mu|x) \propto \left(\frac{1}{\beta_1\sigma_1^2}\right)^{\frac{1}{2}} \left(1 + \frac{(\mu - \mu_1)^2}{\beta_1\sigma_1^2}\right)^{-\frac{a_1+1}{2}}. \tag{7.2}$$

The posterior is symmetric around μ_1, thus all estimates coincide with μ_1. As measure of precision we can use the posterior variance

$$\mathsf{Var}(\mu|\mathbf{x}) = \frac{\beta_1}{\alpha_1 - 2}\sigma_1^2.$$

For illustration, we simplify the example and set $\mu_0 = 0$ and $\sigma_0 = 1$. Then

$$\mu_1 = \frac{n}{n+1}\bar{x}$$

$$\sigma_1^2 = \frac{1}{n+1}$$

$$\alpha_1 = \alpha_0 + n$$

$$\beta_1 = \beta_0 + (n-1)s^2 + \frac{n}{(n+1)}\bar{x}^2$$

and

$$\mathsf{Var}(\mu|\mathbf{x}) = \frac{1}{\alpha_0 + n - 2}\left(\frac{\beta_0}{n+1} + \frac{n-1}{n+1}s^2 + \frac{n}{(n+1)^2}\bar{x}^2\right).$$

For $n \to \infty$ the leading term of the posterior variance is $\frac{1}{\alpha_0+n-2}s^2$. □

Now we are ready to discuss the above-mentioned three estimation methods in more detail. We begin with the mode.

7.1 Maximum a Posteriori (MAP) Estimator

In Chapter 2 the maximum likelihood principle is introduced, that the best explanation of data $\mathbf{x}$ is given by the **maximum likelihood estimator (MLE)**

$$\widehat{\theta}_{\mathrm{MLE}}(\mathbf{x}) \in \arg\max_{\theta \in \Theta} \ell(\theta|\mathbf{x}). \tag{7.3}$$

Combining the maximum likelihood principle and the principle of Bayesian inference, where the posterior is exploited instead of the likelihood, leads to following definition.

Definition 7.1 (Maximum a Posteriori Estimator) Assume   the Bayes model $\{\mathcal{P}, \pi\}$ with posterior $\pi(\theta|\mathbf{x})$. The **maximum a posteriori estimator (MAP)** $\widehat{\theta}_{\mathrm{MAP}}$ is defined as

$$\widehat{\theta}_{\mathrm{MAP}}(\mathbf{x}) \in \arg\max_{\theta \in \Theta} \pi(\theta|\mathbf{x}). \tag{7.4}$$

Recall Example 3.16 on lion's appetite. There the likelihood estimator and the MAP estimator give the same result: the lion was hungry before he had eaten three persons.

Note that, it is enough to know the kernel of $\pi(\theta|\mathbf{x}) \propto \pi(\theta)\ell(\theta|\mathbf{x})$. The integration step is not required. Otherwise, for unknown distributions we have to carry out a numerical maximization procedure. In case the posterior belongs to a known distribution family, we can apply mode's formula.

We continue now with two examples, the first of which uses mode's formula.

Example 7.4 (Normal i.i.d. sample and gamma prior)
In continuation of Example 2.13, we have an i.i.d. sample $\mathbf{X} = (X_1, \ldots, X_n)$ from $\mathsf{N}(0, \sigma^2)$. The parameter of interest is the precision parameter $\theta - \sigma^{-2}$ with gamma prior $\mathsf{Gamma}(\alpha_0, \beta_0)$. The posterior is $\mathsf{Gamma}(\alpha_0 + \frac{n}{2}, \beta_0 + \frac{1}{2}\sum_{i=1}^{n} x_i^2)$. Mode's formula of $\mathsf{Gamma}(\alpha, \beta)$ delivers

$$\text{Mode} = \frac{\alpha - 1}{\beta}, \quad \alpha \geq 1, \tag{7.5}$$

and we obtain, for $n > 1$,

$$\widehat{\theta}_{\text{MAP}}(\mathbf{x}) = \frac{2\alpha_0 + n - 2}{2\beta_0 + \sum_{i=1}^{n} x_i^2}.$$

For comparison, the MLE is

$$\widehat{\theta}_{\text{MLE}}(\mathbf{x}) = \arg\max_{\theta \in \Theta} \ell(\theta|\mathbf{x}) = \frac{n}{\sum_{i=1}^{n} x_i^2}.$$

□

In the next example we apply a numerical solution. Recall Example 3.7 on the weather experts.

Example 7.5 (Weather) Continuation of Example 3.15. Denote $\phi_{(m,\sigma^2)}(.)$ as density of $\mathsf{N}(m, \sigma^2)$. The conjugate prior is a normal mixture comprising of two different experts' reports

$$\pi(\theta) = \omega_1\, \phi_{(m_1, \tau_1^2)}(\theta) + \omega_2\, \phi_{(m_2, \tau_2^2)}(\theta);$$

see Figure 7.2. Then the posterior is

$$\pi(\theta|x) = \omega_1(x)\, \phi_{(m_1(x), \tau_{1,p}^2)}(\theta) + \omega_2(x)\, \phi_{(m_2(x), \tau_{2,p}^2)}(\theta)$$

with

$$m_i(x) = \rho_i(x\tau_i^2 + m_i\sigma^2), \quad \tau_{i,p}^2 = \rho_i\sigma\tau_i^2, \quad \rho_i = (\tau_i^2 + \sigma^2)^{-1}, \quad i = 1, 2,$$

Figure 7.2: Two weather experts.

and

$$\omega_i(x) \propto \omega_i \, \frac{\tau_i}{\tau_{i,p}} \exp\left(-\frac{1}{2\tau_i^2} m_i^2 + \frac{1}{2\tau_{i,p}^2} m_i(x)^2\right)$$

with $\omega_1(x) + \omega_2(x) = 1$. Figure 3.10 shows the prior, the posterior, and the observation. By searching for the maximum of the y-values we obtain $\widehat{\theta}_{\mathrm{MAP}}(4) = 5.33$.   □

The MAP estimators have a useful connection to regularized and restricted estimators in the linear model.

7.1.1 Regularized Estimators

Recall the normal linear model (6.2) in Chapter 6. Here we assume a Bayes linear model with general error distribution and general prior.

$$\mathbf{y} = \mathbf{X}\beta + \epsilon. \tag{7.6}$$

The parameter of interest is $\theta = \beta \in \mathbb{R}^p$. The unobserved error ϵ has expectation zero. We assume

$$p(\epsilon) \propto \exp(-L(\varepsilon)). \tag{7.7}$$

The function $L(\varepsilon)$ can be considered as a type of loss function; it is nonnegative, symmetric around zero and large for large errors. The prior has a similar structure and is given by

$$\pi(\beta) \propto \exp(-\text{pen}(\beta)). \tag{7.8}$$

The function $\text{pen}(\beta)$ can be considered as a type of penalty function, it is nonnegative, symmetric around zero and large for large parameters. Under this set up the likelihood is

$$\ell(\beta|\mathbf{y}) \propto \exp(-L(\mathbf{y} - \mathbf{X}\beta))$$

and the posterior

$$\pi(\beta|\mathbf{y}) \propto \exp\left(-L(\mathbf{y} - \mathbf{X}\beta) - \text{pen}(\beta)\right).$$

We obtain for

$$\widehat{\beta}_{\text{MAP}}(\mathbf{y}) \in \arg\max_{\beta \in \mathbb{R}^p} \pi(\beta|\mathbf{y}) \tag{7.9}$$

that it is equivalent to

$$\widehat{\beta}_{\text{MAP}}(\mathbf{y}) \in \arg\min_{\beta \in \mathbb{R}^p} \left(L(\mathbf{y} - \mathbf{X}\beta) + \text{pen}(\beta)\right). \tag{7.10}$$

It means, choosing the right Bayes model, the Bayes MAP estimator coincides with a **regularized estimator** in regression. It is a bridge between Bayes and frequentist approach. There is one more bridge. The optimization problem in (7.10) has an equivalent formulation that there exists a constant $k > 0$ such that

$$\widehat{\beta}_{\text{MAP}}(\mathbf{y}) \in \arg\min_{\{\beta : \text{pen}(\beta) \leq k\}} L(\mathbf{y} - \mathbf{X}\beta). \tag{7.11}$$

This means, choosing the right Bayes model, the Bayes MAP estimator coincides with a **restricted estimator** in regression. For short overview of regularized and restricted estimators, we recommend Zwanzig and Mahjani (2020, Section 6.4) and Hastie et al. (2015).

Now we consider popular cases for L and pen.

Ridge

We start with normal linear model with known covariance matrix Σ and conjugate prior:

$$\ell(\beta|\mathbf{y}) \propto \exp\left(-\frac{1}{2}(\mathbf{y} - \mathbf{X}\beta)^T \Sigma^{-1}(\mathbf{y} - \mathbf{X}\beta)\right)$$

$$\pi(\beta) \propto \exp\left(-\frac{1}{2}(\beta - \gamma)^T \Gamma^{-1}(\beta - \gamma)\right) \tag{7.12}$$

The posterior is $N(\mu_{\beta|y}, \Sigma_{\beta|y})$ given in Corollary 6.1 as

$$\pi(\beta|\mathbf{y}) \propto \exp\left(-\frac{1}{2}(\beta - \mu_{\beta|y})^T \Sigma_{\beta|y}^{-1}(\beta - \mu_{\beta|y})\right)$$

$$\mu_{\beta|y} = \Sigma_{\beta|y}(\mathbf{X}^T \Sigma^{-1}\mathbf{y} + \Gamma^{-1}\gamma) \tag{7.13}$$

$$\Sigma_{\beta|y} = (\mathbf{X}^T \Sigma^{-1}\mathbf{X} + \Gamma^{-1})^{-1}.$$

The posterior is symmetric around $\mu_{\beta|y}$, such that mode and expectation equal $\mu_{\beta|y}$. Applying (6.18) for $\mu_{\beta|y}$, we obtain

$$\widehat{\beta}_{\mathrm{MAP}}(\mathbf{y}) = \gamma + \Gamma \mathbf{X}^T (\mathbf{X}\Gamma\mathbf{X}^T + \Sigma)^{-1}(\mathbf{y} - \mathbf{X}\gamma).$$

This estimator is also known as **generalized ridge estimator**. Setting $\Sigma = \mathbf{I}$, $\gamma = 0$ and $\Gamma = \lambda^{-1}\mathbf{I}$, we obtain the classical ridge estimator

$$\widehat{\beta}_{\mathrm{ridge}} = (\mathbf{X}^T\mathbf{X} + \lambda\mathbf{I})^{-1}\mathbf{X}^T\mathbf{y}, \tag{7.14}$$

introduced by Hoerl and Kennard (1970). The idea that Hoerl and Kennard (1970) pursued was to augment $\mathbf{X}^T\mathbf{X}$ by $\lambda\mathbf{I}$, since $\mathbf{X}^T\mathbf{X} + \lambda\mathbf{I}$ is always invertible even when $\mathbf{X}^T\mathbf{X}$ is not. The main properties of the ridge estimator are presented in Zwanzig and Mahjani (2020, Section 6.4.1).

Lasso

Suppose a normal linear model with covariance matrix $\Sigma = \sigma^2\mathbf{I}$. The components of β are independent and Laplace distributed $\mathsf{La}(m, b)$ with location parameter $m = 0$ and scale parameter $b = \frac{2\sigma^2}{\lambda}$, so that

$$\ell(\beta|\mathbf{y}) \propto \exp\left(-\frac{1}{2\sigma^2}(\mathbf{y} - \mathbf{X}\beta)^T(\mathbf{y} - \mathbf{X}\beta)\right)$$

$$\pi(\beta) \propto \exp\left(-\frac{\lambda}{2\sigma^2}\sum_{j=1}^{p}|\beta_j|\right). \tag{7.15}$$

The posterior is

$$\pi(\beta|\mathbf{y}) \propto \exp\left(-\frac{1}{2\sigma^2}\left((\mathbf{y} - \mathbf{X}\beta)^T(\mathbf{y} - \mathbf{X}\beta) + \lambda\sum_{j=1}^{p}|\beta_j|\right)\right). \tag{7.16}$$

The $\widehat{\beta}_{\mathrm{MAP}}$ is equivalent to the **lasso estimator**, defined by Tibshirani (1996)

$$\widehat{\beta}_{\mathrm{lasso}}(\mathbf{y}) \in \arg\min_{\beta\in\mathbb{R}^p}(\mathbf{y} - \mathbf{X}\beta)^T(\mathbf{y} - \mathbf{X}\beta) + \lambda\sum_{j=1}^{p}|\beta_j|, \tag{7.17}$$

see also Zwanzig and Mahjani (2020, Section 6.4.2).

Zou and Hastie (2005) propose the following compromise between Lasso and ridge estimators.

Elastic Net

Consider a normal linear model with covariance matrix $\Sigma = \sigma^2 \mathbf{I}$. The components of β are i.i.d. from a prior distribution given by a compromise between $N(0, \frac{\sigma^2}{\lambda(1-\alpha)})$ and Laplace $La(0, \frac{\sigma^2}{\lambda\alpha^2})$ prior. For $\alpha = 1$, it is the Laplace prior $La(0, \frac{\sigma^2}{\lambda})$ and for $\alpha = 0$, it is the normal prior $N(0, \frac{\sigma^2}{\lambda})$; all other values $\alpha \in (0,1)$ lead to a compromise. It gives

$$\ell(\beta|\mathbf{y}) \propto \exp\left(-\frac{1}{2\sigma^2}(\mathbf{y} - \mathbf{X}\beta)^T(\mathbf{y} - \mathbf{X}\beta)\right)$$

$$\pi(\beta) \propto \exp\left(-\frac{\lambda}{2\sigma^2}\left((1-\alpha)\sum_{j=1}^{p}\beta_j^2 + \alpha\sum_{j=1}^{p}|\beta_j|\right)\right). \tag{7.18}$$

The $\widehat{\beta}_{\mathrm{MAP}}$ is equivalent to the **elastic net estimator**, defined by

$$\widehat{\beta}_{\mathrm{net}}(\mathbf{y}) \in \arg\min_{\beta \in \mathbb{R}^p}(\mathbf{y} - \mathbf{X}\beta)^T(\mathbf{y} - \mathbf{X}\beta) + \lambda\left((1-\alpha)\sum_{j=1}^{p}\beta_j^2 + \alpha\sum_{j=1}^{p}|\beta_j|\right)$$

$$\tag{7.19}$$

in Zou and Hastie (2005). The penalty term is a compromise between L_2-type ridge penalty and L_1-type lasso penalty.

Note that, the regularized estimators depend strongly on the weight λ assigned to the penalty term. There exists an extensive literature on adaptive methods for choosing the tuning parameter λ and respectively α, but this is beyond the scope of this book. In the Bayesian context the tuning parameters are hyperparameters of the prior. We refer to Chapter 3 where different proposals for the prior choice are presented.

We conclude the section with an illustrative example.

Example 7.6 (Regularized estimators)
We consider a polynomial regression model

$$y_i = \beta_0 + \beta_1 x_i + \beta_3 x_i^2 + \beta_4 x_i^4 + \varepsilon_i, \quad i = 1,\ldots,n.$$

A small data set is generated with 9 equidistant design points between 0 and 4; $\varepsilon_i \sim N(0, 0.5^2)$ *i.i.d.* and the true regression function

$$f(x) = 3 + x + 0.5x^2 - x^3 + 0.2x^4.$$

Using R the lasso estimate is calculated for $\lambda = 0.012$; the ridge estimator for $\lambda = 0.05$; and the elastic net estimator for $\alpha = 0.5$ and $\lambda = 0.02$.

The R-packages also include methods for an adaptive choice of the tuning parameters. Here for illustrative purposes the tuning parameters are chosen arbitrarily. Figure 7.3 shows different fitted polynomials.   □

R Code 7.1.7. Regularized estimators in Figure 7.3.

```
a0<-3; a1<-1; a2<-0.5; a3<--1; a4<-0.2 # true parameter
xx<-seq(0,4,0.01)
xx2<-xx*xx
xx3<-xx*xx*xx
xx4<-xx*xx*xx*xx
ff<-a0+a1*xx+a2*xx2+a3*xx3+a4*xx4
plot(xx,ff,"l",ylim=c(-2,5),xlab="",ylab="",lwd=2) # true
x<-seq(0,4,0.5) # design points
x2<-x*x
x3<-x*x*x
x4<-x*x*x*x
f<-a0+a1*x+a2*x2+a3*x3+a4*x4
y<-f+rnorm(9,0,0.5) # generated observations
points(x,y,col=1,lwd=3)
## lse
A<-coef(lm(y~x+x2+x3+x4))
flse<-A[1]+A[2]*xx+A[3]*xx2+A[4]*xx3+A[5]*xx4
lines(xx,flse,lwd=2,lty=1,col=gray(0.4)) # lse fit
## ridge
library(MASS)
Ridge<-lm.ridge(y~x+x2+x3+x4, lambda=0.05)
R<-coef(Ridge)
fridge<-R[1]+R[2]*xx+R[3]*xx2+R[4]*xx3+R[5]*xx4
lines(xx,fridge,col=gray(0.4),lwd=2,lty=2) # ridge fit
## lasso
library(lars)
X<-matrix(c(x,x2,x3,x4),ncol=4)
Lasso<-lars(X,y,type="lasso")
L<-Lasso$beta[9,]
flasso<-L[1]+L[2]*xx+L[3]*xx2+L[4]*xx3+L[5]*xx4
lines(xx,flasso,col=gray(0.4),lwd=2,lty=3) # lasso fit
## elastic net
library(glmnet)
MLSE<-glmnet(X,y,alpha=0.5,df=5)
N<-coef(MLSE,s=0.02)
fnet<-N[1]+N[2]*xx+N[3]*xx2+N[4]*xx3+N[5]*xx4
lines(xx,fnet,col=gray(0.4),lwd=2,lty=4) # elastic net fit
```

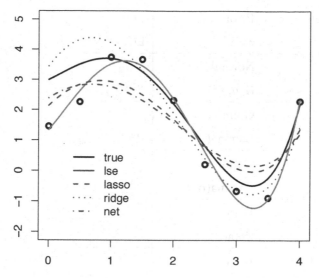

Figure 7.3: Example 7.6. Since the sample size $n = 9$ is small with respect to the number of parameters $p = 5$ the least-squares method overfits the data. Different regularized estimators correct the overfit.

7.2 Bayes Rules

In this section we discuss the methods based on expectation and median of the posterior. Both are Bayes rules given in Definition 4.2. Readers, interested in their optimality properties, are referred to Chapter 4. Here we present the application. Set the following notation

$$\widehat{\theta}_{L_2}(\mathbf{x}) = \mathsf{E}(\theta|\mathbf{x})$$
$$\widehat{\theta}_{L_1}(\mathbf{x}) = \mathsf{Median}(\theta|\mathbf{x}). \tag{7.20}$$

The subindex is related to the loss functions. The estimator $\widehat{\theta}_{L_2}$ obeys optimality properties with respect to L_2 loss and $\widehat{\theta}_{L_1}$ with respect to L_1 loss. We have already presented several examples related to these estimates. Besides the introductory Examples 7.1, 7.2 and 7.3, check Examples 4.6, 4.7 and 4.8 for $\widehat{\theta}_{L_2}$ and Example 4.9 for $\widehat{\theta}_{L_1}$.

In the following table priors and corresponding posterior expectations are collected related to some popular one parameter exponential families with parameter θ, assuming all other parameters known. The table is partly taken from Robert (2001), where posterior distributions are given for single x.

Distribution $p(x\|\theta)$	Prior $\pi(\theta)$	Posterior $\mathsf{E}(\theta\|x)$
Normal $\mathsf{N}(\theta,\sigma^2)$	Normal $\mathsf{N}(\mu,\tau^2)$	$\frac{\sigma^2\mu+\tau^2 x}{\sigma^2+\tau^2}$
Poisson $\mathsf{Poi}(\theta)$	Gamma $\mathsf{Gamma}(\alpha,\beta)$	$\frac{\alpha+x}{\beta+1}$
Gamma $\mathsf{Gamma}(\nu,\theta)$	Gamma $\mathsf{Gamma}(\alpha,\beta)$	$\frac{\alpha+\nu}{\beta+x}$
Binomial $\mathsf{Bin}(n,\theta)$	Beta $\mathsf{Beta}(\alpha,\beta)$	$\frac{\alpha+x}{\beta+n-x}$
Negative Binomial $\mathsf{NB}(m,\theta)$	Beta $\mathsf{Beta}(\alpha,\beta)$	$\frac{\alpha+m}{\alpha+\beta+x+m}$
Multinomial $\mathsf{Mult}_k(\theta_1,\ldots,\theta_k)$	Dirichlet $\mathsf{Dir}(\alpha_1,\ldots,\alpha_k)$	$\frac{\alpha_i+x_i}{\sum_j \alpha_j+n}$
Normal $\mathsf{N}\left(\mu,\frac{1}{\theta}\right)$	Gamma $\mathsf{Gamma}(\alpha,\beta)$	$\frac{2\alpha+1}{2\beta+(x-\mu)^2}$
Normal $\mathsf{N}\left(\mu,\theta\right)$	InvGamma $\mathsf{InvGamma}(\alpha,\beta)$	$\frac{2\beta+(\mu-x)^2}{2\alpha-1}$

Recall that for symmetric single mode posteriors both estimators in (7.20) coincide. In order to illustrate a posterior with two local maxima we come back to the classroom Example 3.7 now with two more controversial and more equally weighted weather experts.

Example 7.7 (Weather)

Continuation of Examples 3.7 and 3.15. Assume that both experts have very different subjective priors and they are closely weighted than in (3.7); we set the mixture prior

$$\pi(\theta) = 0.4\,\phi_{(-8,4)}(\theta) + 0.6\,\phi_{(8,10)}(\theta), \qquad (7.21)$$

where $\phi_{(\mu,\sigma^2)}$ denotes the density of $\mathsf{N}(\mu,\sigma^2)$. Applying (3.24) we obtain for $x=1$

$$\pi(\theta|x) = 0.34\,\phi_{(-2,3.33)}(\theta) + 0.66\,\phi_{(4.89,2.22)}(\theta).$$

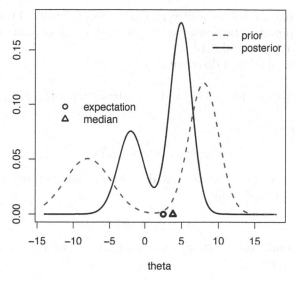

Figure 7.4: Example 7.7.

The posterior is more concentrated, but it is still a clear mixture of two components, as seen in Figure 7.4. The estimate

$$\widehat{\theta}_{L_2} = 0.34 \, (-2.3) + 0.66 \, (4.89) \approx 2.55$$

is the weighted average of the expectation of both components. The estimate

$$\widehat{\theta}_{L_1} \approx 3.81$$

is calculated by a simple searching algorithm. The neighbourhood around the posterior expectation has low posterior probability. Both estimates are illustrated in Figure 7.4.   □

We continue with more applications.

7.2.1   Estimation in Univariate Linear Models

We assume the linear model $\mathcal{P}$ given in (6.38):

$$\mathbf{y} = \mathbf{X}\beta + \epsilon, \quad \epsilon \sim \mathsf{N}_n(0, \sigma^2 \Sigma), \quad \theta = (\beta, \sigma^2),$$

under both Bayes models, $\{\mathcal{P}, \pi_c\}$, with conjugate prior, and $\{\mathcal{P}, \pi_{\text{Jeff}}\}$, with non-informative prior. The joint posterior distribution is a normal-inverse-gamma distribution,

$$\theta | \mathbf{y} \sim \mathsf{NIG}(a, b, \gamma, \Gamma),$$

see Theorems 6.5 and 6.7, respectively. Under the linear mixed model (6.87) we also obtain a posterior belonging to the normal-inverse-gamma distributions, see Theorems 6.8, 6.9 and 6.10.

This implies the marginal distributions

$$\beta|\mathbf{y} \sim t_p(a, \gamma, \frac{b}{a}\Gamma) \text{ and } \sigma^2 \sim \mathsf{InvGamma}(a/2, b/2).$$

In case of known σ^2 we use

$$\beta|(\mathbf{y}, \sigma^2) \sim \mathsf{N}_p(\gamma, \Gamma).$$

The t distribution and the normal distribution are symmetric around the location parameter. We have

$$\widehat{\beta}_{L_2}(\mathbf{y}) = \widehat{\beta}_{L_1}(\mathbf{y}) = \gamma.$$

The inverse-gamma distribution is not symmetric. We apply the formula for expectation and obtain

$$\widehat{\sigma}^2 = \widehat{\sigma}^2_{L_2}(\mathbf{y}) = \frac{b}{a-2}.$$

Unfortunately, the median is not given in a closed form. In case, the L_1-estimate of the variance is of main interest, numerical methods are needed. There is no difference in the estimate of β wether σ^2 is known or unknown. But its precision is different. It holds that

$$\mathsf{Cov}(\beta|\mathbf{y}, \sigma^2) = \sigma^2 \Gamma$$

and

$$\mathsf{Cov}(\beta|\mathbf{y}) = \frac{b}{a-2}\Gamma = \widehat{\sigma}^2 \Gamma.$$

Example 7.8 (Simple linear regression)
We assume

$$y_i = \beta x_i + \varepsilon_i, \quad i = 1, \dots, n, \quad \varepsilon_i \sim N(0, \sigma^2), \ i.i.d.$$

with $\theta = (\beta, \sigma^2)$. The prior $\mathsf{NIG}(a, b, \gamma, \Gamma)$ gives the posterior $\mathsf{NIG}(a_1, b_1, \gamma_1, \Gamma_1)$ given in Example 6.8 on page 151. The proposed Bayes estimates are

$$\widehat{\beta}_{L_2} = \gamma_1 \text{ and } \widehat{\sigma}^2 = \frac{b_1}{a+n-2}$$

with

$$\gamma_1 = \left(\sum_{i=1}^{n} x_i^2 + \lambda^{-1} \right)^{-1} \left(\sum_{i=1}^{n} x_i y_i + \lambda^{-1}\gamma \right)$$

$$b_1 = b + \sum_{i=1}^{n} y_i^2 + \lambda^{-1}\gamma^2 - \left(\sum_{i=1}^{n} x_i^2 + \lambda^{-1} \right)^{-1} \left(\sum_{i=1}^{n} x_i y_i + \lambda^{-1}\gamma \right)^2.$$

In case of Jeffreys prior $\pi_{\text{Jeff}}(\theta) \propto \sigma^{-2}$ we obtain

$$\widehat{\beta}_{\text{L}_2} = \frac{\sum_{i=1}^n x_i y_i}{\sum_{i=1}^n x_i^2} = \widehat{\beta} \quad \text{and} \quad \widehat{\sigma}_{\text{L}_2}^2 = \frac{1}{n-3} \sum_{i=1}^n (y_i - x_i \widehat{\beta})^2.$$

$\square$

We illustrates this simple model with the life-length data.

Example 7.9 (Life length vs. life line)

We continue with Example 6.4 based on data for Example 6.3. We study the centred model (6.34). Assuming $\sigma^2 = 160$ and $\beta \sim \text{N}(0, 100)$, we obtain $\widehat{\beta}_{\text{L}_2} = -2.257$. In Example 6.10 the variance is supposed to be unknown and a conjugate prior is assumed. Then the Bayes estimate is $\widehat{\beta}_{\text{L}_2} = -2.257$. The estimates are the same and give a negative trend, just opposite to the popular belief. The equality of estimates is not surprising because the expected value of the prior of the variance coincides with the supposed known value in the first case. We come back to this data and model later to assess the significance of this trend.

$\square$

Now let us compare the estimates under a simple linear model with intercept.

Example 7.10 (Life length vs. life line)

We continue with Example 6.4 based on data for Example 6.3 but now we assume model (6.33) with $\sigma^2 = 160$. Assuming the independent priors

$$\alpha \sim \text{N}(0, 100\,\sigma^2), \quad \beta \sim \text{N}(0, 100\,\sigma^2),$$

we obtain

$$\widehat{\alpha}_{\text{L}_2} = 86.69, \quad \widehat{\beta}_{\text{L}_2} = -2.13.$$

Alternatively, assuming

$$\alpha \sim \text{N}(100, 100\,\sigma^2), \quad \beta \sim \text{N}(0, 100\,\sigma^2),$$

again independent, we get

$$\widehat{\alpha}_{\text{L}_2} = 88.04, \quad \widehat{\beta}_{\text{L}_2} = -2.27.$$

The priors differ only in the expected value of the intercept. Also in this study we get a negative trend against the popular belief. We will revisit these results later.

$\square$

We consider an example of Bayes estimation in the linear mixed model.

Example 7.11 (Hip operation)
In Example 6.15 the posterior distributions are derived. The estimate for the slope β_1 is $\widehat{\beta}_1 = 0.5$ with $\text{Var}(\beta_1|\mathbf{y}) = 0.0054$. It implies that the covariate age has influence on the healing process. □

7.2.2　Estimation in Multivariate Linear Models

We consider the model $\mathcal{P}$ given in (6.131), i.e.,

$$\mathbf{Y} = \mathbf{XB} + \mathbf{E}, \quad \mathbf{E} \sim \text{MN}_{n,d}(\mathbf{O}, \mathbf{I}_n, \Sigma)$$

where $\mathbf{Y}$ is the $n \times d$ matrix of responses, $\mathbf{E}$ is the $n \times d$ matrix of random errors, $\mathbf{B}$ is the $p \times d$ matrix of unknown parameters and $\mathbf{X}$ is the $n \times p$ design matrix, assumed to be of full rank. The unknown parameter consists of $\mathbf{B}$ and Σ. Under both Bayes models $\{\mathcal{P}, \pi_c\}$, with conjugate prior, and $\{\mathcal{P}, \pi_{\text{Jeff}}\}$, with non-informative prior, the posterior distributions belong to the normal-inverse-Wishart distributions given in Theorems 6.11 and 6.12:

$$(\mathbf{B}, \Sigma)|\mathbf{Y} \sim \text{NIW}(\nu_1, \mathbf{B}_1, \mathbf{C}_1, \Sigma_1) \quad \text{and} \quad \mathbf{B}|\mathbf{Y} \sim t_{p,d}(\nu_1 - d + 1, \mathbf{B}_1, \mathbf{C}_1, \Sigma_1).$$

This delivers the estimates

$$\widehat{\mathbf{B}}_{L_2} = \mathbf{B}_1 \quad \text{and} \quad \widehat{\Sigma} = \frac{1}{\nu_1 - d - 1}\Sigma_1.$$

We illustrate the estimation of the covariance matrix Σ in the following example.

Example 7.12 (Corn plants)
Continuing with Example 6.16 on page 172, in model $\{\mathcal{P}, \pi_{\text{Jeff}}\}$ we obtain

$$\widehat{\Sigma} = \frac{1}{14}\mathbf{S} = \begin{pmatrix} 149.09 & -24.99 \\ -24.99 & 264.29 \end{pmatrix}.$$

For model $\{\mathcal{P}, \pi_c\}$ we get

$$\widehat{\Sigma} = \frac{1}{19}\Sigma_1 = \begin{pmatrix} 249.10 & 28.58 \\ 28.58 & 240.37 \end{pmatrix}.$$

□

7.3   Credible Sets

In this section we deal with Bayes confidence regions. We assume the Bayes model $\{\mathcal{P}, \pi\}$, where $\mathcal{P} = \{\mathsf{P}_\theta : \theta \in \Theta\}$ and π is the known prior distribution of θ over Θ. Having observed $\mathbf{x}$ for $\mathbf{X} \sim \mathsf{P}_\theta$ we want to determine a region $C_{\mathbf{x}} \in \Theta$ such that the underlying parameter $\theta \in C_{\mathbf{x}}$. According to the Bayesian inference principle that all information is included in the posterior $\pi(\theta|\mathbf{x})$, we define the set $C_{\mathbf{x}}$ as following, where **HPD** stands for **highest posterior density**.

Definition 7.2 (Credible Region) Assume the Bayes model $\{\mathcal{P}, \pi\}$ with posterior distribution $\mathsf{P}^\pi(.|\mathbf{x})$. A set $C_{\mathbf{x}}$ is a α-**credible region** iff

$$\mathsf{P}^\pi(C_{\mathbf{x}}|\mathbf{x}) \geq 1 - \alpha, \quad \alpha \in [0, 1]. \tag{7.22}$$

This region is called **HPD α-credible region** if it can be written as

$$\{\theta : \pi(\theta|\mathbf{x}) > k_\alpha\} \subseteq C_{\mathbf{x}} \subseteq \{\theta : \pi(\theta|\mathbf{x}) \geq k_\alpha\} \tag{7.23}$$

where k_α is the largest bound such that

$$\mathsf{P}^\pi(C_{\mathbf{x}}|\mathbf{x}) \geq 1 - \alpha.$$

In case the posterior is a continues distribution with density $\pi(\theta|\mathbf{x})$ the definition simplifies to

$$C_{\mathbf{x}} = \{\theta : \pi(\theta|\mathbf{x}) \geq k_\alpha\} \quad \text{with } \mathsf{P}^\pi(\pi(\theta|\mathbf{x}) \geq k_\alpha|\mathbf{x}) = 1 - \alpha. \tag{7.24}$$

We illustrate it by the following examples.

Example 7.13 (Normal i.i.d. sample and normal prior)
Consider an i.i.d. sample $\mathbf{X} = (X_1, \ldots, X_n)$ from $\mathsf{N}(\mu, \sigma^2)$ with known variance σ^2, thus $\theta = \mu$. We assume the prior $\theta \sim \mathsf{N}(\mu_0, \sigma_0^2)$. In Example 2.12 on page 16, the posterior is derived as

$$\mathsf{N}(\mu_1, \sigma_1^2), \text{ with } \mu_1 = \frac{\bar{x}n\sigma_0^2 + \mu_0\sigma^2}{n\sigma_0^2 + \sigma^2}, \text{ and } \sigma_1^2 = \frac{\sigma_0^2\sigma^2}{n\sigma_0^2 + \sigma^2}.$$

The posterior has a continuous density and is unimodal and symmetric around the Bayes estimate $\widehat{\theta} = \mu_1$. We calculate HPD α-credible interval by

$$\begin{aligned}
\{\theta : \pi(\theta|\mathbf{x}) \geq k_\alpha\} &= \{\theta : \frac{1}{\sqrt{2\pi}\sigma_1} \exp\left(-\frac{(\theta - \mu_1)^2}{\sigma_1^2}\right) \geq k_\alpha\} \\
&= \{\theta : \frac{(\theta - \mu_1)^2}{\sigma_1^2} \leq \ln(k_\alpha\sqrt{2\pi}\sigma_1)\}
\end{aligned} \tag{7.25}$$

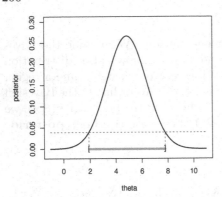
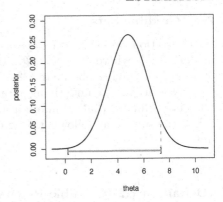

Figure 7.5: Example 7.25. Left: HPD interval for $\alpha = 0.5$ with interval length 5.88. Right: Credible interval for $\alpha = 0.5$, which is not HPD. The interval length is 7.12.

The bound k_α is calculated by

$$\ln(k_\alpha \sqrt{(2\pi)}\sigma_1) = (z_{1-\frac{\alpha}{2}})^2, \quad \text{where } P(Z < z_{1-\frac{\alpha}{2}}) = 1 - \frac{\alpha}{2}, \quad Z \sim N(0,1).$$

The HPD α-credible interval is given as

$$C_{\mathbf{x}} = \{\theta : \mu_1 - z_{1-\frac{\alpha}{2}}\sigma_1 \leq \theta \leq \mu_1 + z_{1-\frac{\alpha}{2}}\sigma_1\},$$

which has the same structure as the frequentist confidence interval, but now around the Bayes estimate.   □

Let us consider a case where the posterior density is skewed.

Example 7.14 (Normal i.i.d. sample and inverse-gamma prior)
Recall Example 2.14 on page 19. Consider an i.i.d. sample $\mathbf{X} = (X_1, \ldots, X_n)$ from $N(0, \sigma^2)$ with unknown variance σ^2. We set as prior an inverse-gamma distribution $\mathsf{InvGamma}(\alpha, \beta)$. Then the posterior is $\mathsf{InvGamma}(\alpha_1, \beta_1)$ with $\alpha_1 = \alpha + n/2$ and $\beta_1 = \beta + \sum_{i=1}^{n} x_i^2/2$ which has the density

$$f(\theta|\alpha_1, \beta_1) = \frac{\beta_1^{\alpha_1}}{\Gamma(\alpha_1)} \left(\frac{1}{\theta}\right)^{\alpha_1+1} \exp\left(-\frac{\beta_1}{\theta}\right). \tag{7.26}$$

The density is unimodal but skewed. The HPD α-credible interval is given by

$$\{\theta : \pi(\theta|\mathbf{x}) \geq k_\alpha\} = \{\theta : \left(\frac{1}{\theta}\right)^{\alpha_1+1} \exp\left(-\frac{\beta_1}{\theta}\right) \geq c_\alpha\} \tag{7.27}$$

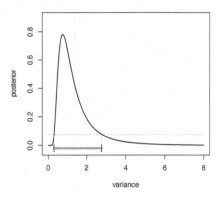 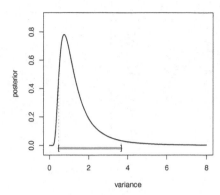

Figure 7.6: Example 7.14. Left: HPD interval for $\alpha = 0.1$ with interval length 2.46. The right tail probability is around 0.097. Right: Credible interval for $\alpha = 0.1$, which is equal–tailed. The interval length is 3.19.

where c_α is a constant calculated from k_α. The explicit interval requires a numerical solution; see R code and Figure 7.6. This interval is not symmetric around a Bayes estimator. It is not equal-tailed like the usual approximation in the frequentist approach; see Liero and Zwanzig (2011, p.171).   □

The following R code illustrates the construction of an HPD α-credible interval for a unimodal skewed density; see also Figure 7.7.

R Code 7.3.8. Example 7.14

```
library(invgamma)
a<-3; b<-3 #  posterior parameter
aa<-seq(0,0.1,0.000001) # confidence levels on the lower side
q1<-qinvgamma(aa,a,b) # quantile lower bound
q2<-qinvgamma(aa+0.9,a,b) # quantile upper bound
plot(aa,dinvgamma(q1,a,b),"l") # see Figure
lines(aa,dinvgamma(q2,a,b))
# Simple search algorithm gives the crossing point at 0.078
k<-0.078; lines(aa,rep(k,length(aa)))
# Simple searching algorithm gives approximative  bounds
L<-0.30155 ;  U<- 2.765,
# such that k=dinvgamma(L,a,b) and k=dinvgamma(U,a,b) .
```

Unfortunately in case of multi-modal densities, the interpretation of HPD credible intervals is not so convincing. We demonstrate it with the help of the classroom example (Example 3.7, page 37) about the two contradicting weather experts.

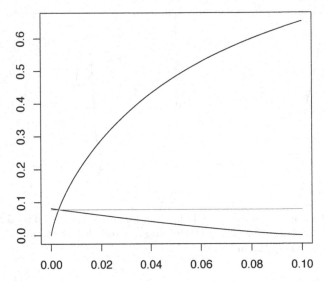

Figure 7.7: Illustration for R code related to Example 7.14.

Example 7.15 (Weather)

Recall Example 3.15 on page 48. Continuing with Example 7.7, the posterior is a normal mixture distribution shown in Figure 7.4. Depending on the confidence level α, it may happen that the α-credible region consists of two separate intervals as demonstrated in Figure 7.8.   □

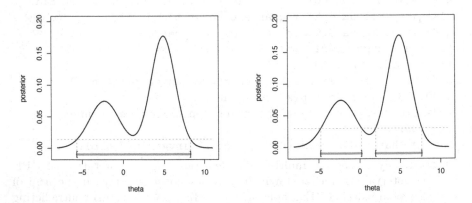

Figure 7.8: Example 7.15. Left: HPD interval for $\alpha = 0.05$. Right: HPD interval for $\alpha = 0.1$.

The interpretation of HPD credible regions can be complicated for discrete posterior distributions, i.e., for the parameter space $\Theta \subseteq \mathbb{Z}$. In this case the HPD α-credible regions do not have to be unique. We discuss it with the help of the lion example on page 7.

Example 7.16 (Lion's appetite)
Consider the model in Example 2.4 and the prior in Example 3.16, with $\pi(\theta_1) = \pi(\theta_2) = 0.1$. For $x = 1$ the likelihood is constant, such that prior and posterior coincide, i.e.,

	θ_1	θ_2	θ_3
$\pi(\theta\|x=1)$	0.1	0.1	0.8

Set $\alpha = 0.1$. We have two HPD α-credible regions: $C_1 = \{\theta_1, \theta_3\}$ and $C_2 = \{\theta_2, \theta_3\}$. The first region C_1 includes lion's hungry and lethargic modes; second region contains moderate and lethargic modes. $\quad\square$

7.3.1 Credible Sets in Linear Models

We assume the linear model

$$\mathbf{y} = \mathbf{X}\beta + \epsilon, \quad \epsilon \sim \mathsf{N}_n(0, \sigma^2 \mathbf{I}_n), \quad \theta = (\beta, \sigma^2).$$

Further we have that the posterior is $\theta|\mathbf{y} \sim \mathsf{NIG}(a_1, b_1, \gamma_1, \Gamma_1)$, i.e.,

$$\beta|\mathbf{y}, \sigma^2 \sim \mathsf{N}_p(\gamma_1, \sigma^2 \Gamma_1)$$

$$\beta|\mathbf{y} \sim \mathsf{t}(a_1, \gamma_1, \frac{b_1}{a_1}\Gamma_1) \qquad (7.28)$$

$$\sigma^2|\mathbf{y} \sim \mathsf{InvGamma}(a_1/2, b_1/2).$$

The construction of credible regions depends only on the posterior distribution. Therefore we can apply the calculations in Example 7.14 for credible intervals of σ^2. The posterior distribution of β in case of known σ^2 is normal; in case of unknown σ^2 it is a multivariate t-distribution. Both distributions are elliptical, i.e., the level sets are ellipsoids. Thus the HPD α-credible regions are ellipsoids, centered at the Bayes estimator γ_1. First we consider the case of known σ^2. It holds that there exists a constant c_α such that

$$\{\beta : \pi(\beta|\mathbf{y}, \sigma^2) \geq k_\alpha\} = \{\beta : \frac{1}{\sigma^2}(\beta - \gamma_1)^T \Gamma_1^{-1}(\beta - \gamma_1) \leq c_\alpha\}.$$

We have that the quadratic form

$$Q = \frac{1}{\sigma^2}(\beta - \gamma_1)^T \Gamma_1^{-1}(\beta - \gamma_1) \sim \chi_p^2,$$

where χ_p^2 denotes the χ^2–distribution with p degrees of freedom. Thus the HPD α-credible region is the ellipsoid

$$C_{\mathbf{y}} = \{\beta : \frac{1}{\sigma^2}(\beta - \gamma_1)^T \Gamma_1^{-1}(\beta - \gamma_1) \leq \chi_p^2(1 - \alpha)\}$$

where $\chi_p^2(1 - \alpha)$ is the $(1 - \alpha)$-quantile of χ^2–distribution with p degrees of freedom. In case of unknown σ^2 we have

$$\{\beta : \pi(\beta|\mathbf{y}) \geq k_\alpha\} = \{\beta : f(\beta) \geq k_\alpha\} \qquad (7.29)$$

where f is the density of $t(a_1, \gamma_1, \frac{b_1}{a_1}\Gamma_1)$ given in (6.51). Thus there exists a constant c_α such that

$$C_{\mathbf{y}} = \{\beta : \frac{1}{b_1}(\beta - \gamma_1)^T \Gamma_1^{-1}(\beta - \gamma_1) \leq c_\alpha\}. \qquad (7.30)$$

We apply the following relation between t-distribution and F-distribution; see Lin (1972). If $X \sim t_q(\nu, \mu, \Sigma)$ then

$$\frac{1}{q}(X - \mu)^T \Sigma^{-1}(X - \mu) \sim F_{q,\nu}, \qquad (7.31)$$

where $F_{q,\nu}$ denotes the F-distribution with q and ν degrees of freedom. Let $F_{p,a_1}(1 - \alpha)$ be the $1 - \alpha$ quantile of F-distribution with p and a_1 degrees of freedom. The HPD α-credible set is given by

$$C_{\mathbf{y}} = \{\beta : \frac{a_1}{p\,b_1}(\beta - \gamma_1)^T \Gamma_1^{-1}(\beta - \gamma_1) \leq F_{p,a_1}(1 - \alpha)\}. \qquad (7.32)$$

Recall that the Bayes estimate of σ^2 is $b_1/(a_1 - 2)$.

Example 7.17 (Life length vs. life line)

We analyze the data for Example 6.3 on page 130 under the simple linear regression model with unknown variance; $\theta = (\alpha, \beta, \sigma^2)$. In Example 6.4 on page 139, the posterior distribution is derived for $\sigma^2 = 160$. Here we additionally set a prior on σ^2, so that

$$\begin{pmatrix} \alpha \\ \beta \end{pmatrix} \sim N_2 \left(\begin{pmatrix} 100 \\ 0 \end{pmatrix}, \sigma^2 \begin{pmatrix} 100 & 0 \\ 0 & 100 \end{pmatrix} \right) \quad \text{and} \quad \sigma^2 \sim \text{InvGamma}(5, 640).$$

with $E(\sigma^2) = 160$. Applying Theorem 6.5 we obtain

$$\theta|\mathbf{y} \sim \text{NIG}(a_1, b_1, \gamma_1, \Gamma_1)$$

with $a_1 = a + n = 58$, $b_1 = 7554$ and

$$\Gamma_1 = \begin{pmatrix} 1.346 & -0.146 \\ -0.146 & 0.016 \end{pmatrix}, \gamma_1 = \begin{pmatrix} 88.04 \\ -2.27 \end{pmatrix}.$$

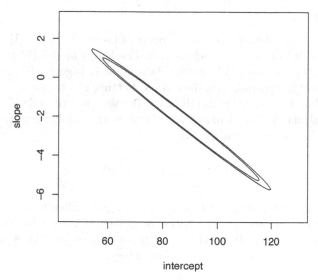

Figure 7.9: Example 7.17. The larger ellipse is the HPD credible region for $\alpha = 0.05$; the smaller is for $\alpha = 0.1$.

The credible region for (α, β) is an ellipse given in (7.32) where the F–quantiles are $F_{2,a_1}(0.9) = 2.39$ and $F_{2,a_1}(0.95) = 3.15$, illustrated in Figure 7.9. $\square$

R Code 7.3.9. Credible Region, Figure 7.9.

```
qf1<-qf(0.95,2,a1) # quantile F distribution
qf2<-qf(0.9,2,a1)
b<-gamma1 # Bayes estimate
S<-Gamma1
Q<-function(x,y){a1/(2*b1)*t(c(x,y)-b)%*%solve(S)%*%(c(x,y)-b)}
#Quadratic form
x1<-seq(45,130,1) # intercept
x2<-seq(-7,3,0.1) # slope
Z<-as.matrix(outer(xx1,xx2))
for (i in 1:length(xx1))
{ for (j in 1: length(xx2))
{ Z[i,j]<-Q(x1[i],x2[j])
}}
contour(x1,x2,Z,xlab="intercept",ylab="slope",level=c(qf1,qf2),
drawlabels=FALSE)
```

7.4 Prediction

We consider the problem to predict a future observation $x_f \in \mathcal{X}_f$. We refer to Section 4.3.3 for theoretical background. The Bayes model $\{\mathcal{P}, \pi\}$ with the posterior $\pi(\theta|\mathbf{x})$ is assumed. The future data point x_f is generated by a distribution Q_θ, possibly depending on the data $\mathbf{x}$. Setting $z = x_f$, we have $q(z|\theta, \mathbf{x})$. The main tool is the predictive distribution. It takes over the role of the posterior. The **predictive distribution** is defined as the conditional distribution of the future data point given the data $\mathbf{x}$;

$$\pi(z|\mathbf{x}) = \int_\Theta q(z|\theta, \mathbf{x})\pi(\theta|\mathbf{x})\, d\theta. \tag{7.33}$$

Analogous to the estimation problem, depending on different strategies, the Bayes predictor can be mode, expectation or median of the predictive distribution. In Section 4.3.3 it is shown that the expectation is the **Bayes predictor** which is optimal with respect to a quadratic loss,

$$\widehat{x}_f(\mathbf{x}) = \int_{\mathcal{X}_f} z\, \pi(z|\mathbf{x})\, dz. \tag{7.34}$$

The **prediction regions** $\mathcal{X}_{\text{pred}} \subset \mathcal{X}_f$ of level $1 - \alpha$ are defined by

$$P(\mathcal{X}_{\text{pred}}|\mathbf{x}) = \int_{\mathcal{X}_{\text{pred}}} \pi(z|\mathbf{x})\, dz = 1 - \alpha. \tag{7.35}$$

The best choice with minimal volume are the sets with highest predictive probability;

$$\int_{\mathcal{X}_{\text{pred}}} \pi(z|\mathbf{x})\, dz = 1 - \alpha \text{ and } \pi(z|\mathbf{x}) \geq \pi(z'|\mathbf{x}) \text{ for } z \in \mathcal{X}_{\text{pred}}, z' \notin \mathcal{X}_{\text{pred}}. \tag{7.36}$$

The main problem is now to derive the predictive distribution. In this section we present several cases, where it is possible to derive explicit formulas. Otherwise computer-intensive methods, explained in Chapter 9, can be used. In the following example we have an i.i.d. sample and want to predict the next observation.

Example 7.18 (Poisson distribution and gamma prior)
Consider an i.i.d. sample $\mathbf{X} = (X_1, \ldots, X_n)$ from $\text{Poi}(\lambda)$ such that

$$p(\mathbf{x}|\lambda) = \prod_{i=1}^n \left(\frac{\lambda^{x_i}}{x_i!} \exp(-\lambda)\right), \tag{7.37}$$

and for $\theta = \lambda$

$$\ell(\theta|\mathbf{x}) \propto \theta^{\sum_{i=1}^n x_i} \exp(-n\theta). \tag{7.38}$$

Assume the conjugate prior $\mathsf{Gamma}(\alpha, \beta)$

$$\pi(\theta) = \frac{\beta^\alpha}{\Gamma(\alpha)} \theta^{\alpha-1} \exp(-\beta\theta), \tag{7.39}$$

where α and β are known; α/β stands for a guess of θ and β for the sample size this guess is based on. We obtain

$$\begin{aligned}
\pi(\theta|\mathbf{x}) &\propto \pi(\theta)\ell(\theta|\mathbf{x}) \\
&\propto \theta^{\alpha-1} \exp(-\beta\theta)\, \theta^{\sum_{i=1}^n x_i} \exp(-n\theta) \\
&\propto \theta^{\alpha + \sum_{i=1}^n x_i - 1} \exp(-(\beta+n)\theta),
\end{aligned} \tag{7.40}$$

hence the posterior is $\mathsf{Gamma}(\alpha + \sum_{i=1}^n x_i, \beta + n)$. Summarizing for an i.i.d. Poisson sample we have

$$\theta \sim \mathsf{Gamma}(\alpha, \beta) \quad \text{and} \quad \theta|\mathbf{x} \sim \mathsf{Gamma}(\alpha + \sum_{i=1}^n x_i, \beta + n); \tag{7.41}$$

see Figure 7.10. The Bayes estimator of θ is the expectation of the posterior,

$$\widehat{\theta}_{\mathrm{L}_2} = \frac{\alpha + \sum_{i=1}^n x_i}{\beta + n}.$$

For the predictive distribution, set $s_n = \sum_{i=1}^n x_i$. We have $X_{\mathrm{f}} \sim \mathsf{Poi}(\theta)$, independent of $\mathbf{x}$, and

$$\begin{aligned}
P(X_{\mathrm{f}} = k|\mathbf{x}) &= \int_\Theta q(k|\theta, \mathbf{x})\pi(\theta|\mathbf{x})\, d\theta \\
&= \int_0^\infty \frac{\theta^k}{k!} \exp(-\theta) \frac{(\beta+n)^{\alpha+s_n}}{\Gamma(\alpha+s_n)} \theta^{\alpha+s_n-1} \exp(-(\beta+n)\theta)\, d\theta \\
&= \frac{(\beta+n)^{\alpha+s_n}}{\Gamma(\alpha+s_n)k!} \int_0^\infty \theta^{\alpha+s_n+k-1} \exp(-(\beta+n+1)\theta)\, d\theta.
\end{aligned} \tag{7.42}$$

Applying the integral

$$\int_0^\infty x^{a-1} \exp(-b\,x)\, dx = \Gamma(a)b^{-a} \tag{7.43}$$

for $a = \alpha + s_n + k$ and $b = \beta + n + 1$, we obtain the predictive distribution

$$P(X_{\mathrm{f}} = k|\mathbf{x}) = \frac{\Gamma(\alpha + s_n + k)}{\Gamma(\alpha + s_n)k!} \left(\frac{\beta+n}{\beta+n+1}\right)^{\alpha+s_n} \left(\frac{1}{\beta+n+1}\right)^k. \tag{7.44}$$

Set $r = \alpha + s_n$ and $1 - p = (\beta + n + 1)^{-1}$. The predictive distribution is the

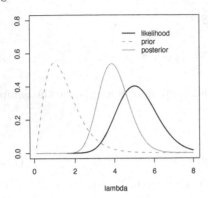
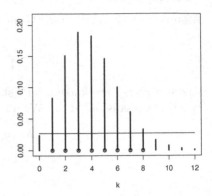

Figure 7.10: Example 7.18. Left: Likelihood function of an i.i.d Poisson sample with $n = 5$, prior $\mathsf{Gamma}(3,2)$ and the posterior $\mathsf{Gamma}(28,7)$. Right: The predictive distribution $\mathsf{NB}(28, 0.875)$. The mode is 3. The predictive region $\{1,2,3,4,5,6,7,8\}$ has the level 0.946.

generalized negative binomial distribution $\mathsf{NB}(r,p)$, where r is positive and real, and the expected value is $\frac{(1-p)r}{p}$. We obtain the Bayes predictor

$$\widehat{x}_{\mathsf{f}}(\mathbf{x}) = \frac{1}{\beta + n}(\alpha + \sum_{i=1}^{n} x_i), \tag{7.45}$$

so that Bayes predictor and Bayes estimator coincide. This is no surprise, since the parameter $\theta = \lambda$ is the expectation of $\mathsf{Poi}(\lambda)$. In general, the expectation is not an integer. We recommend instead the mode of the predictive distribution. The mode of the generalized negative binomial distribution $\mathsf{NB}(r,p)$ is the largest integer equal or less than $\frac{p(r-1)}{1-p}, r > 1$. It is illustrated in Figure 7.10.   □

The next example is related to a dynamic model where the future observation depends on the past.

Example 7.19 (Autoregressive model AR(1))
Set x_0 as given. We observe $\mathbf{x}_T = (x_1, \ldots, x_T)$ from $\mathbf{X}_T = (X_1, \ldots, X_T)$ generated by the first-order autoregressive model, given as

$$X_t = \rho x_{t-1} + \varepsilon_t, \quad \varepsilon_t \sim \mathsf{N}(0, \sigma^2) \text{ i.i.d.}, \quad t = 1, \ldots, T. \tag{7.46}$$

The unknown parameter is $\theta = (\rho, \sigma^2)$. It holds that

$$\mathsf{Var}(X_t|\theta) = \rho^2 \mathsf{Var}(X_{t-1}|\theta) + \sigma^2 = \frac{1 - \rho^{2t}}{1 - \rho^2}\sigma^2, \quad |\rho| < 1. \tag{7.47}$$

We assume the conjugate prior $\mathsf{NIG}(a_0, b_0, \gamma_0, \kappa_0)$, so that

$$\rho|\sigma^2 \sim \mathsf{N}(\gamma_0, \sigma^2 \kappa_0), \quad \text{and} \quad \sigma^2 \sim \mathsf{InvGamma}(\frac{a_0}{2}, \frac{b_0}{2}). \qquad (7.48)$$

We derive an iterative formula for the posterior by setting the prior equal to the previous posterior: see Section 2.3.1. Set $T = 1$. Applying Theorem 6.5 we obtain the posterior $\mathsf{NIG}(a_1, b_1, \gamma_1, \kappa_1)$, i.e.,

$$\rho|\sigma^2, x_1 \sim \mathsf{N}(\gamma_1, \sigma^2 \kappa_1), \quad \text{and} \quad \sigma^2|x_1 \sim \mathsf{InvGamma}(\frac{a_1}{2}, \frac{b_1}{2})$$

with

$$a_1 = a_0 + 1$$

$$b_1 = b_0 + (x_1 - \gamma_0 x_0)^2 (1 + \frac{x_0^2}{\kappa_0})^{-1}$$

$$\gamma_1 = \kappa_1 (x_0 x_1 + \frac{\gamma_0}{\kappa_0}) \qquad (7.49)$$

$$\kappa_1 = (x_0^2 + \frac{1}{\kappa_0})^{-1}.$$

Note that, for $x_0 = 0$ and $\gamma_0 = 0$ we have

$$a_1 = a_0 + 1, \quad b_1 = b_0 + x_1^2, \quad \gamma_1 = 0, \quad \text{and} \quad \kappa_1 = \kappa_0. \qquad (7.50)$$

Set $T = 2$. Applying Theorem 6.5 we obtain the posterior

$$\theta|(x_0, x_1, x_2) \sim \mathsf{NIG}(a_2, b_2, \gamma_2, \kappa_2)$$

with

$$a_2 = a_1 + 1$$

$$b_2 = b_1 + (x_2 - \gamma_1 x_1)^2 (1 + \frac{x_1^2}{\kappa_1})^{-1}$$

$$\gamma_2 = \kappa_2 (x_1 x_2 + \frac{\gamma_1}{\kappa_1}) \qquad (7.51)$$

$$\kappa_2 = (x_1^2 + \frac{1}{\kappa_1})^{-1}.$$

Note that, also for $\gamma_1 = 0$ we have $\gamma_2 \neq 0$ a.s. In general we have the iterative formula for the posterior

$$\theta|(x_0, \ldots, x_T) \sim \mathsf{NIG}(a_T, b_T, \gamma_T, \kappa_T) \qquad (7.52)$$

with

$$a_T = a_{T-1} + 1$$

$$b_T = b_{T-1} + (x_T - \gamma_{T-1} x_{T-1})^2 (1 + \frac{x_{T-1}^2}{\kappa_{T-1}})^{-1}$$

$$\gamma_T = \kappa_T (x_{T-1} x_T + \frac{\gamma_{T-1}}{\kappa_{T-1}}) \qquad (7.53)$$

$$\kappa_T = (x_{T-1}^2 + \frac{1}{\kappa_{T-1}})^{-1}.$$

We are interested in the distribution of X_{T+1} given $\mathbf{x}_T = (x_0, \ldots, x_T)$. The posterior $\theta | \mathbf{x}_T \sim \mathsf{NIG}(a_T, b_T, \gamma_T, \kappa_T)$ and (7.46) imply

$$\rho | \sigma^2, \mathbf{x}_T \sim \mathsf{N}(\gamma_T, \sigma^2 \kappa_T)$$
$$X_{T+1} | (\rho, \sigma^2, \mathbf{x}_T) \sim \mathsf{N}(\rho x_T, \sigma^2). \tag{7.54}$$

From Theorem 6.1, we get

$$X_{T+1} | (\sigma^2, \mathbf{x}_T) \sim \mathsf{N}(\gamma_T x_T, \sigma^2(\kappa_T x_T^2 + 1)). \tag{7.55}$$

Further, we have

$$\sigma^2 | \mathbf{x}_T \sim \mathsf{InvGamma}(\frac{a_T}{2}, \frac{b_T}{2}).$$

Applying Lemma 6.5, we obtain the predictive distribution

$$X_{T+1} | \mathbf{x}_T \sim \mathsf{t}_1(a_T, \gamma_T x_T, \frac{b_T}{a_T}(\kappa_T x_T^2 + 1)). \tag{7.56}$$

Prediction regions are of the form

$$\{x : (x - \gamma_T x_T)^2 \frac{a_T}{b_T}(\kappa_T x_T^2 + 1)^{-1} \le \text{const}\}. \tag{7.57}$$

Recall that,

$$\text{for } X \sim \mathsf{t}_1(\nu, \mu, \sigma^2), \text{ it holds } \frac{(X - \mu)}{\sigma} \sim \mathsf{t}_\nu, \tag{7.58}$$

where t_ν is students t-distribution with ν degrees of freedom. Setting $\mathsf{t}_{\nu, 1 - \frac{\alpha}{2}}$ for the $(1 - \frac{\alpha}{2})$–quantile, we obtain the prediction interval

$$\left[\gamma_T x_T - \mathsf{t}_{a_T, 1 - \frac{\alpha}{2}} \sqrt{\frac{b_T}{a_T}(\kappa_T x_T^2 + 1)}, \gamma_T x_T + \mathsf{t}_{a_T, 1 - \frac{\alpha}{2}} \sqrt{\frac{b_T}{a_T}(\kappa_T x_T^2 + 1)} \right].$$
$$\tag{7.59}$$

$\square$

7.4.1 Prediction in Linear Models

We assume the linear model

$$\mathbf{y} = \mathbf{X}\beta + \epsilon, \quad \epsilon \sim \mathsf{N}_n(\mathbf{0}, \sigma^2 \mathbf{I}_n), \quad \theta = (\beta, \sigma^2),$$

with the posterior $\theta | \mathbf{y} \sim \mathsf{NIG}(a_1, b_1, \gamma_1, \Gamma_1)$, i.e.,

$$\beta | \mathbf{y}, \sigma^2 \sim \mathsf{N}_p(\gamma_1, \sigma^2 \Gamma_1)$$
$$\sigma^2 | \mathbf{y} \sim \mathsf{InvGamma}(a_1/2, b_1/2). \tag{7.60}$$

We are interested in the prediction of q future observations generated by

$$\mathbf{y}_f = \mathbf{Z}\beta + \varepsilon, \quad \varepsilon \sim N_q(0, \sigma^2 \Sigma_f), \qquad (7.61)$$

where the $q \times p$ matrix $\mathbf{Z}$ is known, the parameter $\theta = (\beta, \sigma^2)$ is the same as above. The error term ε is independent of ϵ. We have

$$\begin{aligned}
\mathbf{y}_f | (\beta, \sigma^2, \mathbf{y}) &\sim N_q(\mathbf{Z}\beta, \sigma^2 \Sigma_f) \\
\beta | \mathbf{y}, \sigma^2 &\sim N_p(\gamma_1, \sigma^2 \Gamma_1).
\end{aligned} \qquad (7.62)$$

From Theorem 6.1, we obtain

$$\mathbf{y}_f | \sigma^2, \mathbf{y} \sim N_q \left(\mathbf{Z}\gamma_1, \sigma^2 (\mathbf{Z}\Gamma_1 \mathbf{Z}^T + \Sigma_f) \right). \qquad (7.63)$$

Further, we have

$$\sigma^2 | \mathbf{y} \sim \mathsf{InvGamma}(a_1/2, b_1/2).$$

Applying Lemma 6.5, we obtain the predictive distribution

$$\mathbf{y}_f | \mathbf{y} \sim t_q(a_1, \mathbf{Z}\gamma_1, \frac{b_1}{a_1}(\mathbf{Z}\Gamma_1 \mathbf{Z}^T + \Sigma_f)). \qquad (7.64)$$

The multivariate t-distribution $t_q(\nu, \mu, \Sigma)$ belongs to the elliptical class. The level sets are ellipsoids around the location μ; the matrix Σ gives the shape of the ellipsoid. Using (7.31) the prediction region is

$$\{\mathbf{z} : \frac{a_1}{qb_1}(\mathbf{z} - \mathbf{Z}\gamma_1)^T (\mathbf{Z}\Gamma_1 \mathbf{Z}^T + \Sigma_f)^{-1}(\mathbf{z} - \mathbf{Z}\gamma_1) \le F_{q,a_1}(1 - \alpha)\} \qquad (7.65)$$

where $F_{q,a_1}(1 - \alpha)$ is the $1 - \alpha$ quantile of F_{q,a_1}. For $q = 1$ we have

$$\mathbf{y}_f | \mathbf{y} \sim t_1(a_1, \widehat{y}_f, \sigma_f^2), \text{, with } \widehat{y}_f = \mathbf{Z}\gamma_1, \ \sigma_f^2 = \frac{b_1}{a_1}(\mathbf{Z}\Gamma_1 \mathbf{Z}^T + \Sigma_f) \qquad (7.66)$$

and

$$\frac{\mathbf{y}_f - \widehat{y}_f}{\sigma_f} \sim t_{a_1} \qquad (7.67)$$

where t_{a_1} is the t-distribution with a_1 degrees of freedom. Let $t_{a_1}(1 - \alpha)$ be the quantile of t_{a_1}, then we obtain the prediction interval:

$$[\widehat{y}_f - t_{a_1}(1 - \frac{\alpha}{2})\sigma_f, \ \widehat{y}_f + t_{a_1}(1 - \frac{\alpha}{2})\sigma_f] \qquad (7.68)$$

Example 7.20 (Prediction in quadratic regression)
We consider a polynomial regression relation

$$y_i = x_i + x_i^2 + \varepsilon_i, \quad i = 1, \ldots, 21. \qquad (7.69)$$

A data set, using (7.69), is generated with 21 equidistant design points between -1 and 1; $\varepsilon_i \sim N(0, 0.5^2)$ i.i.d. The generated data are

x	-1	-0.9	-0.8	-0.7	-0.6	-0.5	-0.4
y	0.470	0.098	-0.680	0.144	-0.486	-1.407	0.439
x	-0.3	-0.2	-0.1	0	0.1	0.2	0.3
y	-0.632	-0.641	-0.644	-0.688	0.265	0.670	0.373
x	0.4	0.5	0.6	0.7	0.8	0.9	1
y	-0.157	0.703	0.546	1.632	0.790	1.968	1.887

We consider the quadratic regression model with no intercept

$$y_i = \beta_1 x_i + \beta_2 x_i^2 + \varepsilon_i, \quad i = 1, \dots, 21, \varepsilon_i \sim N(0, 0.5^2), \quad i.i.d.$$

and set the prior $\mathsf{NIG}(a_0, b_0, \gamma_0, \Gamma_0)$ with $a_0 = 10$, $b_0 = 1$, $\gamma_0 = (0, 0)^T$, and $\Gamma_0 = 2\mathbf{I}_2$. The posterior is $\mathsf{NIG}(a_1, b_1, \gamma_1, \Gamma_1)$ with $a_1 = 31$, $b_1 = 7.12$, $\gamma_1 = (0.903, 0.869)^T$, and

$$\Gamma_1 = \begin{pmatrix} 0.122 & \approx 0 \\ \approx 0 & 0.180 \end{pmatrix}.$$

We set $\alpha = 0.1$, then the quantile is $t_{a_1} \approx 1.70$. Applying (7.68) the prediction interval is calculated at $x_f = 1.1$ as $[1.08, 3.01]$ and at $x_f = 1.4$ as $[1.84, 4.10]$. Figure 7.11 illustrates the example. $\square$

7.5   List of Problems

1. Consider an i.i.d. sample $\mathbf{X} = (X_1, \dots, X_n)$ from $\mathsf{Gamma}(\nu, \theta)$. The parameter of interest is the rate parameter θ. Assume $\theta \sim \mathsf{Gamma}(\alpha, \beta)$.

   (a) Derive the posterior.

   (b) Determine the L_2 estimator for θ.

   (c) Determine the MAP estimator for θ.

   (d) Give a procedure for determining the HPD α-credible interval.

2. Consider an i.i.d. sample $\mathbf{X} = (X_1, \dots, X_n)$ from $N(\mu, \sigma^2)$. The parameter of interest is $\theta = (\mu, \sigma^2)$. Assume a conjugate prior.

   (a) Determine the maximum likelihood estimators for θ.

   (b) Are the maximum likelihood estimators of μ and σ^2 correlated? Calculate $\mathsf{Cov}(\widehat{\mu}_{\mathrm{MLE}}, \widehat{\sigma}^2_{\mathrm{MLE}} | \theta)$.

   (c) Determine the MAP estimator for θ.

   (d) Are the MAP-estimators of μ and σ^2 correlated? Calculate $\mathsf{Cov}(\widehat{\mu}_{\mathrm{MAP}}, \widehat{\sigma}^2_{\mathrm{MAP}} | \theta)$ for $\theta = (0, 1)$.

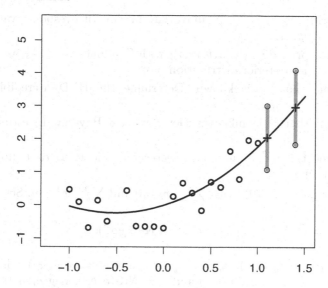

Figure 7.11: Example 7.20.

3. Consider the simple linear regression model with $\sum_{i=1}^{n} x_i = 0$

$$y_i = \alpha + \beta x_i + \varepsilon_i, \quad i = 1, \dots, n, \quad \varepsilon_i \sim \mathsf{N}(0, \sigma^2) \ i.i.d.$$

The unknown parameter is $\theta = (\alpha, \beta) \sim \mathsf{N}_2(\gamma, \sigma^2 \mathrm{diag}(\lambda_1, \lambda_2))$, the error variance is known.

(a) Derive the expression for an α_0-credible interval $C(z)$ for $\mu(z) = \alpha + \beta z$ such that

$$\mathsf{P}(\mu(z) \in C(z)|\mathbf{y}) \geq 1 - \alpha_0$$

and the width of the band is as small as possible.

(b) Compare $C(z)$ with the respective prediction interval at z.

4. Let $X|\theta \sim \mathsf{Bin}(n, \theta)$ be observed and $\theta \sim \mathsf{Beta}(\alpha_0, \beta_0)$. The future data point has the distribution $X_f|\theta \sim \mathsf{Bin}(n_f, \theta)$.

(a) Determine the predictive distribution $\pi(x_f|x)$.

(b) Set $\alpha_0 = \beta_0 = 1$, $n = 5$, $x = 3$ and $n_f = 1$. Calculate the predictive distribution. (Use R.)

5. We are interested in two regression lines

$$\begin{aligned} y_i &= \alpha_y + \beta_y x_i + \varepsilon_i, \quad i = 1, \dots, n, \quad \varepsilon_i \sim \mathsf{N}(0, \sigma^2) \ i.i.d. \\ z_i &= \alpha_z + \beta_z x_i + \xi_i, \quad i = 1, \dots, n, \quad \xi_i \sim \mathsf{N}(0, \sigma^2) \ i.i.d., \end{aligned} \quad (7.70)$$

where ε_i and ξ_i are mutually independent and $\sum_{i=1}^{n} x_i = 0$, $\sum_{i=1}^{n} x_i^2 = n$. The unknown parameter is $\theta = (\alpha_y, \beta_y, \alpha_z, \beta_z, \sigma^2)$. We are mainly interested in $\eta = \frac{1}{2}(\beta_y + \beta_z)$.

(a) Re–write model (7.70) as univariate model with response variable $\mathbf{y} = (y_1, \ldots, y_n, z_1, \ldots, z_n)^T$.

(b) Assume prior $\theta \sim \text{NIG}(a, b, m, \Gamma)$ with $\Gamma = \text{diag}(\lambda_1, \lambda_2, \lambda_3, \lambda_4)$. Derive the posterior distribution of η.

(c) Assume that σ^2 is known. Determine the HPD α-credible interval $C_1(\mathbf{y}, \sigma^2)$ for η.

(d) Assume that σ^2 is unknown. Determine the Bayesian L_2 estimate $\tilde{\sigma}_1^2$ for σ^2.

(e) Assume that σ^2 is unknown. Determine the HPD α-credible interval $C_1(\mathbf{y})$ for η.

6. Consider model (7.70), with $\sum_{i=1}^n x_i = 0$ and $\sum_{i=1}^n x_i^2 = n$. Set

$$u_i = \frac{1}{2}(y_i + z_i) = \alpha_u + \beta_u x_i + \epsilon_i, \ \epsilon_i \sim \text{N}(0, \frac{1}{2}\sigma^2), \ i.i.d., \ i = 1, \ldots, n, \ (7.71)$$

where $\alpha_u = \frac{1}{2}(\alpha_y + \alpha_z)$, $\beta_u = \frac{1}{2}(\beta_y + \beta_z)$ and $\epsilon_i = \frac{1}{2}(\varepsilon_i + \xi_i)$. The unknown parameter is $\theta_u = (\alpha_u, \beta_u, \sigma^2)$ and $\theta_u \sim \text{NIG}(a, b, 0, \text{diag}(c_1, c_2))$.

(a) Derive the posterior distribution of β_u.

(b) Assume that σ^2 is known. Determine the HPD α-credible interval $C_2(\mathbf{u}, \sigma^2)$ for β_u.

(c) Assume that σ^2 is unknown. Determine the Bayesian L_2 estimate $\tilde{\sigma}_2^2$ for σ^2.

(d) Assume that σ^2 is unknown. Determine the HPD α-credible interval $C_2(\mathbf{u})$ for β_u.

7. Consider the two models (7.70) and (7.71). The parameter of interest is $\eta = \frac{1}{2}(\beta_y + \beta_z) = \beta_u$. Compare the results in Problems 5 and 6, particularly:

(a) Specify prior distributions such that (η, σ^2) have the same prior in both models.

(b) Assume that σ^2 is known. Compare the HPD α-credible intervals for η.

(c) Assume that σ^2 is unknown. Compare the Bayesian L_2 estimates for σ^2.

(d) Assume that σ^2 is unknown. Compare the HPD α-credible intervals for η.

8. Related to Problem 10 in Chapter 6. Assume two correlated regression lines

$$\begin{aligned} y_i &= \alpha_y + \beta_y x_i + \varepsilon_i, \ i = 1, \ldots, n, \ \varepsilon_i \sim \text{N}(0, \sigma_1^2) \ i.i.d. \\ z_i &= \alpha_z + \beta_z x_i + \xi_i, \ i = 1, \ldots, n, \ \xi_i \sim \text{N}(0, \sigma_2^2) \ i.i.d., \end{aligned} \quad (7.72)$$

where $\text{Cov}(y, z) = \sigma_{12}$ with $\sum_{i=1}^n x_i = 0$, $\sum_{i=1}^n x_i^2 = n$. The unknown parameter is $\theta = (\alpha_y, \beta_y, \alpha_z, \beta_z, \sigma_1^2, \sigma_{12}, \sigma_2^2)$. Assume the same conjugate prior for θ as in Problem 10 of Chapter 6, $\theta \sim \text{NIW}(\nu, \mathbf{0}, \text{diag}(\lambda_1, \lambda_2), \mathbf{I}_2)$. We are mainly interested in $\eta = \frac{1}{2}(\beta_y + \beta_z)$.

(a) Determine the estimators of β_y, β_z and η.

(b) Derive the posterior distribution of η given Σ. Give the HPD α-credible interval $C_3(\mathbf{Y}, \Sigma)$ for η for known Σ. Setting $\sigma_1 = \sigma_2 = \sigma$ known and $\sigma_{12} = 0$, compare this HPD α-credible interval $C_3(\mathbf{Y}, \Sigma)$ with the interval $C_1(\mathbf{y}, \sigma)$ in model (7.70). Hint:

$$\text{If } \mathbf{Z} \sim \mathsf{MN}(\mathbf{M}, \mathbf{U}, \mathbf{W}) \text{ then } \mathbf{AZB} \sim \mathsf{MN}(\mathbf{AMB}, \mathbf{AUA}^T, \mathbf{BWB}^T).$$
$$(7.73)$$

(c) Derive the posterior distribution of η. Hint: Use the result Gupta and Nagar (2000, Theorem 4.38), which says that if

$$\mathbf{T} \sim \mathsf{t}_{m,k}(\nu, \mathbf{M}, \mathbf{U}, \mathbf{V}) \text{ then } \mathbf{ATC} \sim \mathsf{t}_{s,l}(\nu, \mathbf{AMC}, \mathbf{AUA}^T, \mathbf{BVB}^T).$$
$$(7.74)$$

(d) Determine the HPD α-credible interval for η.

## Chapter 8

Testing and Model Comparison

This chapter deals with the Bayesian approach for hypotheses testing. The hypotheses are two alternative Bayes models. The goal is to figure out which of the models can be the right one for the observed data. We set

$$\mathsf{H}_0 : \mathcal{M}_0 = \{\mathcal{P}_0, \pi_0\} \quad \text{versus} \quad \mathsf{H}_1 : \mathcal{M}_1 = \{\mathcal{P}_1, \pi_1\} \tag{8.1}$$

where

$$\mathcal{P}_j = \{\mathsf{P}_{j,\theta_j} : \theta_j \in \Theta_j\}, \quad \theta_j \sim \pi_j, \quad j = 0, 1. \tag{8.2}$$

It is possible that the statistical models are different including the parameter spaces. Note that, θ_j is not a component of the parameter θ, rather another parameter. We give an example for two different Bayes models.

Example 8.1 (Two alternative models) In a medical study, 1000 patients are randomly chosen and their blood samples are tested for multiple drug resistance. 15 persons got a positive test result. From earlier studies it is known that the risk of infection is around 2%. Let X be the number of infected patients. Two different Bayes models are proposed:

$$\mathcal{M}_0 : \quad X \sim \mathsf{Bin}(1000, p), \quad p \sim \mathsf{Beta}(2, 100)$$
$$\mathcal{M}_1 : \quad X \sim \mathsf{Poi}(\lambda), \quad \lambda \sim \mathsf{Gamma}(20, 1)$$

$\square$

In this chapter we take up three different approaches:

- The goal is to decide after experiment which model is the right one. Here we suppose that both models, $\mathcal{M}_0$ and $\mathcal{M}_1$, are generated by splitting of the common model $\mathcal{M} = \{\mathcal{P}, \pi\}$ with $\mathcal{P} = \{\mathsf{P}_\theta, \theta \in \Theta\}$. We apply decision theoretic results derived in Chapter 4.

- The goal is to obtain evidence against the null hypothesis. We introduce a model indicator $k \in \{0, 1\}$ and embed both models $\mathcal{M}_0$ and $\mathcal{M}_1$ in a common model. The Bayesian principle recommends the model with a higher posterior probability of its model indicator. The main tool is the Bayes factor.

DOI: 10.1201/9781003221623-8

- The goal is to compare models by empirical methods which includes model fit and model complexity. We present the Bayesian information criterion (BIC) and the deviance information criterion (DIC).

8.1 Bayes Rule

In this section we treat the decision theoretic approach. We consider a more specific test problem. The Bayes model $\mathcal{M} = \{\mathcal{P}, \pi\}$ is split as follows. Set $\mathcal{P} = \{\mathsf{P}_\theta : \theta \in \Theta\}$ and $\Theta = \Theta_0 \cup \Theta_1$ with $\Theta_0 \cap \Theta_1 = \varnothing$, and define

$$\mathcal{P}_j = \{\mathsf{P}_\theta : \theta \in \Theta_j\}, \quad j = 0, 1.$$

We assume that the prior probability, $\mathsf{P}^\pi(\Theta_0) > 0$. The prior π on Θ is decomposed as

$$\pi(\theta) = \mathsf{P}^\pi(\Theta_0)\pi_0(\theta) + \mathsf{P}^\pi(\Theta_1)\pi_1(\theta)$$

with $\mathsf{P}^\pi(\Theta_0) + \mathsf{P}^\pi(\Theta_1) = 1$, and

$$\pi_0(\theta) = \begin{cases} \frac{\pi(\theta)}{\mathsf{P}^\pi(\Theta_0)} & \text{for} \quad \theta \in \Theta_0 \\ 0 & \text{else} \end{cases} \qquad \pi_1(\theta) = \begin{cases} \frac{\pi(\theta)}{\mathsf{P}^\pi(\Theta_1)} & \text{for} \quad \theta \in \Theta_1 \\ 0 & \text{else} \end{cases}.$$

The test problem (8.1) is:

$$\mathsf{H}_0 : \mathcal{M}_0 = \{\mathcal{P}_0, \pi_0\} \quad \text{versus} \quad \mathsf{H}_1 : \mathcal{M}_1 = \{\mathcal{P}_1, \pi_1\}$$

The model $\mathcal{M}_j$ is true, if the data generating function is element of $\mathcal{P}_j$ for $j = 0, 1$. Simplified, the test problem usually is written as

$$\mathsf{H}_0 : \Theta_0 \quad \text{versus} \quad \mathsf{H}_1 : \Theta_1. \tag{8.3}$$

We introduce a test as the decision rule depending on $\mathbf{x} \in \mathcal{X}$.

Definition 8.1 (Test) A test φ is a statistic from the sample space $\mathcal{X}$ to $\{0, 1\}$:

$$\varphi(\mathbf{x}) = \begin{cases} 1 & \text{if} \quad \mathbf{x} \in C_1 \quad (\text{reject } \mathsf{H}_0) \\ 0 & \text{if} \quad \mathbf{x} \in C_0 \quad (\text{do not reject } \mathsf{H}_0) \end{cases}$$

where $\mathcal{X} = C_1 \cup C_0$, with $C_1 \cap C_0 = \varnothing$.

In Chapter 4 on decision theory we already introduced the test (4.29) as Bayes rule. We can make two different types of error: the error of type I is that $\mathcal{M}_0$ is true but we decide for $\mathcal{M}_1$. This error gets the loss value a_0. The opposite error is the error of type II, that we wrongly decide for $\mathcal{M}_0$, gets the loss value a_1. The Bayes rule related to this asymmetric loss function can be derived analogous to Theorem 4.5. Here we give the optimal Bayes test.

Set $P^\pi(.|\mathbf{x})$ for the posterior distribution in the common model $\{\{P_\theta : \theta \in \Theta\}, \pi\}$. Then Bayes test is

$$\varphi(\mathbf{x}) = \begin{cases} 1 & \text{if} \quad P^\pi(\Theta_0|\mathbf{x}) < \frac{a_1}{a_0+a_1} \\ 0 & \text{if} \quad P^\pi(\Theta_0|\mathbf{x}) \geq \frac{a_1}{a_0+a_1} \end{cases}. \qquad (8.4)$$

Note that $P^\pi(\Theta_0|\mathbf{x}) + P^\pi(\Theta_1|\mathbf{x}) = 1$. For $a_0 = a_1$ the model, whose parameter space has higher posterior probability, is accepted. This procedure has a nice heuristic background. Following the Bayesian inference principle, we apply the ratio of posterior probabilities instead of likelihood ratio. But note that, it does not give the same answer as in the Neyman–Pearson theory, where first the probability of type I error is bounded and then the probability of type II error is minimized. The test in (8.4) treats both errors simultaneously, only corrected by different weights.

Example 8.2 (Binomial distribution and beta prior)
We consider the Bayes model in Example 2.11, on page 16, given by $X|\theta \sim$ $\text{Bin}(n, \theta)$ and $\theta \sim \text{Beta}(\alpha_0, \beta_0)$ with $\alpha_0 > 1$ and $\beta_0 > 1$. Then $\theta|x \sim \text{Beta}(\alpha_0 + x, \beta_0 + n - x)$. We are interested in the testing problem:

$$H_0 : \theta \geq \frac{1}{2} \quad \text{versus} \quad H_1 : \theta < \frac{1}{2}.$$

Both types of errors should have the same weight; we set $a_0 = a_1$. Recall the properties of $\text{Beta}(\alpha, \beta)$. For $\alpha = \beta$ the distribution is symmetric around 0.5, for $1 < \alpha < \beta$ the distribution is unimodal and negatively skewed. This implies, for $\alpha_0 + x < \beta_0 + n - x$, we reject the null hypothesis. We obtain the test

$$\varphi(x) = \begin{cases} 1 & \text{if } x < \frac{1}{2}(\beta_0 - \alpha_0 + n) \\ 0 & \text{otherwise} \end{cases} \qquad (8.5)$$

$\square$

We continue with binomial data.

Example 8.3 (Sex ratio at birth)
The sex ratio at birth (SRB) is defined as male births per female births. The WHO (World Health Organization) determines the expected SRB by 106 boys per 100 girls. The number of boys has $\text{Bin}(n, p)$ distribution with success probability $p = 106/206 = 0.5145631$. The Official Statistics of Sweden counted in 2021 $x = 58485$ male births and $n - x = 55778$ female births, which gives an actual sex rate at birth as 1.04855. For uniform prior, $\alpha_0 = 1$ and $\beta_0 = 1$, we obtain the posterior $\text{Beta}(58486, 55779)$. We consider the testing problem

$$H_0 : p \geq 0.5145631 \quad \text{versus} \quad H_1 : p < 0.5145631.$$

Figure 8.1: Example 8.3. The white storks decide the SRB.

The posterior probability of the null hypothesis is $P^\pi(\Theta_0|\mathbf{x}) = 1 - 0.9669733 = 0.03302665$. We conclude that the Swedish sex ratio at birth in 2021 was significant less than the value of WHO. This clear conclusion is possible because, due to high number of births, the posterior variance is very small; see Figure 8.4.   □

Example 8.4 (Normal distribution and normal prior)
Consider $X \sim N(\mu, \sigma^2)$ with known variance σ^2; $\theta = \mu$ with normal prior $\theta \sim N(\mu_0, \sigma_0^2)$. In Example 2.12, on page 16 with $n = 1$ we derived the posterior distribution

$$N(\mu_1, \sigma_1^2), \text{ with } \mu_1 = \frac{x\sigma_0^2 + \mu_0\sigma^2}{\sigma_0^2 + \sigma^2} \text{ and } \sigma_1^2 = \frac{\sigma_0^2\sigma^2}{\sigma_0^2 + \sigma^2}. \tag{8.6}$$

We are interested in the testing problem:

$$H_0 : \theta \leq 0 \text{ versus } H_1 : \theta > 0.$$

Calculate

$$P^\pi(\theta \leq 0|x) = P^\pi\left(\frac{\theta - \mu_1}{\sigma_1} \leq \frac{-\mu_1}{\sigma_1}|x\right) = \Phi\left(\frac{-\mu_1}{\sigma_1}\right)$$

where Φ is the distribution function of $N(0,1)$. Setting z_a for the quantile

$$\Phi(z_a) = \frac{a_1}{a_0 + a_1},$$

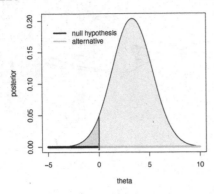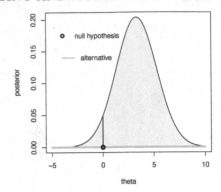

Figure 8.2: Left: Example 8.4. The null hypothesis is rejected. Right: Illustration for point-null hypothesis. The null hypothesis is always rejected for any experiment's result.

we obtain the test

$$\varphi(x) = \begin{cases} 1 & \text{if } x\sigma_0^2 + \mu_0\sigma^2 > -z_a\,\sigma\sigma_0\sqrt{\sigma_0^2 + \sigma^2} \\ 0 & \text{otherwise.} \end{cases} \tag{8.7}$$

For $a_0 = a_1$ the test is illustrated in Figure 8.2.   □

We continue with analysis of the life length data set.

Example 8.5 (Life length vs. life line)
We analyse the data presented in Example 6.3, on page 130. In particular, we want to know if the length of life line is an indicator of life length. This is formulated with respect to the slope β of the linear regression line as

$$H_0 : \beta \geq 0 \quad \text{versus} \quad H_1 : \beta < 0.$$

In Example 7.17 we obtained that the posterior of intercept α and slope β given the variance σ^2 is the two dimensional normal distribution, $N_2(\gamma_1, \sigma^2\Gamma_1)$, so that the marginal posterior of β given σ^2 is $N(-2.27, \sigma^2\,0.016)$. Further, $\sigma^2|\mathbf{y} \sim \mathsf{InvGamma}(a_1/2, b_1/2)$ with $a_1 = 58$ and $b_1 = 7554$. Applying Lemma 6.5 we obtain $\beta|\mathbf{y} \sim t_1(58, -2.27, b_1/a_1\,0.016)$. Thus, in this model, $P^\pi(\Theta_0|\mathbf{y}) = 0.06$ and we reject the popular belief. See Figure 8.3.   □

8.2   Bayes Factor

In this section we consider the approach which formulates evidence against the null hypothesis. Assume the general test problem with two alternative

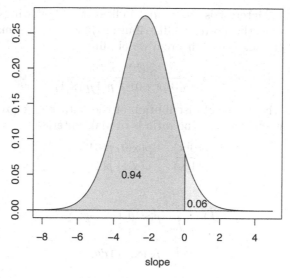

Figure 8.3: Example 8.5. Bayes model with $\theta = (\alpha, \beta, \sigma^2)$. The null hypothesis is clearly rejected.

Bayes models,

$$\mathsf{H}_0 : \mathcal{M}_0 = \{\mathcal{P}_0, \pi_0\} \quad \text{versus} \quad \mathsf{H}_1 : \mathcal{M}_1 = \{\mathcal{P}_1, \pi_1\} \tag{8.8}$$

with $\mathcal{P}_j = \{\mathsf{P}_{j,\theta_j} : \theta_j \in \Theta_j\}$.

We start with the construction of a common Bayes model $\mathcal{M}_{\mathrm{com}}$ which embed both models. We introduce an additional parameter, the *model indicator* and a prior on it:

$$k \in \{0, 1\}, \quad \pi(k) = p_k, \quad p_0 + p_1 = 1, \quad p_0 > 0. \tag{8.9}$$

The parameter space of the common model $\mathcal{P}_{\mathrm{com}}$ is

$$\Theta_{\mathrm{com}} = \{0\} \times \Theta_0 \cup \{1\} \times \Theta_1$$

with parameter $\theta = (k, \theta_k) \in \Theta_{\mathrm{com}}$, such that

$$\mathcal{P}_{\mathrm{com}} = \{\mathcal{P}_0 \cup \mathcal{P}_1 : \theta \in \Theta_{\mathrm{com}}\}.$$

The common prior is the mixture

$$\begin{aligned}
\pi(\theta) &= \pi((k, \theta_k)) \\
&= \pi((k, \theta_k)|k = 0)\pi(k = 0) + \pi((k, \theta_k)|k = 1)\pi(k = 1) \\
&= \pi_0(\theta_0)p_0 + \pi_1(\theta_1)p_1.
\end{aligned} \tag{8.10}$$

We get the common Bayes model

$$\mathcal{M}_{\mathrm{com}} = \{\mathcal{P}_{\mathrm{com}}, \pi\}. \tag{8.11}$$

The parameter of interest is the model indicator k. Following the Bayesian inference principle, the posterior distribution $p(k|\mathbf{x})$ is the main source of information. Applying Bayes theorem we obtain

$$p(k|\mathbf{x}) = \frac{p(k)p(\mathbf{x}|k)}{p(0)p(\mathbf{x}|0) + p(1)p(\mathbf{x}|1)}, \qquad (8.12)$$

where $p(\mathbf{x}|k)$ is the marginal distribution of the data $\mathbf{x} \in \mathcal{X}$ given the model $\mathcal{M}_k$. To compare the models, the ratio is of main interest is

$$\frac{p(0|\mathbf{x})}{p(1|\mathbf{x})} = \frac{p(\mathbf{x}|0)}{p(\mathbf{x}|1)} \frac{p(0)}{p(1)}. \qquad (8.13)$$

We have

$$p(\mathbf{x}|k) = \int_{\Theta_k} p(\mathbf{x}, \theta_k|k)\, d\theta_k$$

$$= \int_{\Theta_k} p_k(\mathbf{x}, \theta_k)\, d\theta_k \qquad (8.14)$$

$$= \int_{\Theta_k} \ell_k(\theta_k|\mathbf{x})\pi_k(\theta_k)\, d\theta_k$$

We get

$$\frac{p(0|\mathbf{x})}{p(1|\mathbf{x})} = \frac{\int_{\Theta_0} \ell_0(\theta_0|\mathbf{x})\pi_0(\theta_0)\, d\theta_0}{\int_{\Theta_1} \ell_1(\theta_1|\mathbf{x})\pi_1(\theta_1)\, d\theta_1} \frac{p(0)}{p(1)}. \qquad (8.15)$$

The information from the data is included in the Bayes factor, defined as follows.

Definition 8.2 (Bayes factor) Assume the test problem (8.8). The Bayes factor is
$$B_{01} = B_{01}^{\pi} = \frac{\int_{\Theta_0} \ell_0(\theta_0|\mathbf{x})\pi_0(\theta_0)\, d\theta_0}{\int_{\Theta_1} \ell_1(\theta_1|\mathbf{x})\pi_1(\theta_1)\, d\theta_1}.$$

Applying (8.15) we get the relation between the *prior odds ratio*,

$$\frac{p(0)}{p(1)} = \frac{p(0)}{1 - p(0)}$$

and the *posterior odds ratio*,

$$\frac{p(0|\mathbf{x})}{p(1|\mathbf{x})} = \frac{p(0|\mathbf{x})}{1 - p(0|\mathbf{x})}$$

with the help of the Bayes factor:

$$\frac{p(0|x)}{1 - p(0|x)} = B_{01} \frac{p(0)}{1 - p(0)}$$

i.e.,

$$\boxed{\text{posterior odds} = \text{Bayes factor} \times \text{prior odds}}$$

The Bayes factor measures the change in the odds. Small Bayes factor B_{01} means evidence against the null hypothesis. There are different empirical scales for judging the size of Bayes factor. We quote here a table from Kass and Raftery (1997). Note that, the table is given for $B_{10} = (B_{01})^{-1}$.

$2\ln(B_{10})$	B_{10}	Evidence against H_0
0 to 2	1 to 3	Not worth more than a bare mention
2 to 6	3 to 20	Positive
6 to 10	20 to 150	Strong
> 10	> 150	Very strong

Example 8.6 (Two alternative models)
We continue with Example 8.1 and calculate the Bayes factor $B_{01} = \frac{A}{B}$. The prior in model $\mathcal{M}_0$ is $\text{Beta}(\alpha_0, \beta_0)$. It holds that

$$
\begin{aligned}
A :=& \int_{\Theta_0} \ell_0(\theta_0|\mathbf{x})\pi_0(\theta_0)\, d\theta_0 \\
=& \int_0^1 \binom{n}{x}\theta^x(1-\theta)^{n-x}\frac{1}{B(\alpha_0,\beta_0)}\theta^{\alpha_0-1}(1-\theta)^{\beta_0-1}\, d\theta \\
=& \binom{n}{x}\frac{1}{B(\alpha_0,\beta_0)}\int_0^1 \theta^{\alpha_0+x-1}(1-\theta)^{\beta_0+n-x-1}\, d\theta \\
=& \binom{n}{x}\frac{B(\alpha_0+x,\beta_0+n-x)}{B(\alpha_0,\beta_0)}.
\end{aligned}
$$

Using

$$
\binom{n}{x} = \frac{1}{n+1}\frac{1}{B(x+1,n-x+1)},
$$

we obtain

$$
A = \frac{1}{n+1}\frac{B(\alpha_0+x,\beta_0+n-x)}{B(x+1,n-x+1)B(\alpha_0,\beta_0)}.
$$

The prior in model $\mathcal{M}_1$ is $\mathsf{Gamma}(\alpha_1, \beta_1)$. It holds that

$$B := \int_{\Theta_1} \ell_1(\theta_1|\mathbf{x})\pi_1(\theta_1)\, d\theta_1$$

$$= \int_0^\infty \frac{\lambda^x}{x!} \exp(-\lambda)\frac{\beta_1^{\alpha_1}}{\Gamma(\alpha_1)}\lambda^{\alpha_1-1}\exp(-\beta_1\lambda)\, d\lambda$$

$$= \frac{\beta_1^{\alpha_1}}{x!\Gamma(\alpha_1)}\int_0^\infty \lambda^{x+\alpha_1-1}\exp(-\lambda(\beta_1+1))\, d\lambda.$$

Using

$$\int_0^\infty \lambda^{x+\alpha_1-1}\exp(-\lambda(\beta_1+1))\, d\lambda = \frac{\Gamma(x+\alpha_1)}{(\beta_1+1)^{\alpha_1+x}},$$

we obtain

$$B = \frac{\beta_1^{\alpha_1}}{x!\Gamma(\alpha_1)}\frac{\Gamma(x+\alpha_1)}{(\beta_1+1)^{\alpha_1+x}}.$$

In general the Bayes factor A/B has a difficult expression which divides large values. In case $\alpha_0 = \beta_0 = 1$, we can use $B(1,1) = 1$ and can simplify A to $1/(n+1)$. Setting also $\alpha_1 = \beta_1 = 1$ in B, we obtain

$$\mathsf{B}_{01} = \frac{A}{B} = \frac{1}{n+1}2^{x+1}.$$

$\square$

In the special case where we formulate the alternatives by splitting the common model $\{\mathcal{P}, \pi\}$, as in Section 8.1 we can re-formulate the Bayes factor as follows. The prior probability p_k of the submodel $\mathcal{M}_k$ coincides with $\mathsf{P}^\pi(\Theta_k)$, $k = 0, 1$ and

$$\int_{\Theta_k} \ell_k(\theta_k|\mathbf{x})\pi_k(\theta_k)\, d\theta_k = \frac{1}{\mathsf{P}^\pi(\Theta_k)}\int_{\Theta_k}\ell(\theta|\mathbf{x})\pi(\theta)\, d\theta$$

$$= \frac{m(\mathbf{x})}{\mathsf{P}^\pi(\Theta_k)}\int_{\Theta_k}\pi(\theta|\mathbf{x})\, d\theta \qquad (8.16)$$

$$= \frac{m(\mathbf{x})}{\mathsf{P}^\pi(\Theta_k)}\mathsf{P}^\pi(\Theta_k|\mathbf{x})$$

where $m(\mathbf{x}) = \int_\Theta \ell(\theta|\mathbf{x})\pi(\theta)\, d\theta$. We obtain, for the test problem (8.3), the Bayes factor as

$$\mathsf{B}_{01}^\pi = \frac{\mathsf{P}^\pi(\Theta_0|\mathbf{x})}{\mathsf{P}^\pi(\Theta_1|\mathbf{x})}\frac{\mathsf{P}^\pi(\Theta_1)}{\mathsf{P}^\pi(\Theta_0)}.$$

Note that, P^π and $\mathsf{P}^\pi(.|\mathbf{x})$ are the prior and posterior probabilities related to the common model $\{\mathcal{P}, \pi\}$. In this case (8.4) is the optimal decision rule. Both approaches, decision making or evidence finding, can be transformed to each other. We have the relation:

$$\text{Reject } \mathsf{H}_0 \Leftrightarrow \mathsf{P}^\pi(\Theta_0|\mathbf{x}) < \frac{a_1}{a_0 + a_1}$$

$$\Leftrightarrow \mathsf{B}_{01}^\pi < \frac{a_1}{a_0}\frac{p_1}{p_0}.$$

8.2.1 Point Null Hypothesis

Given $\{\mathcal{P}, \pi\}$ with $\mathcal{P} = \{P_\theta : \theta \in \Theta\}$, a point-null hypothesis is defined by splitting $\Theta = \Theta_1 \cup \{\theta_0\}$, where $\Theta_1 = \Theta \setminus \{\theta_0\}$, written as

$$\mathsf{H}_0 : \theta = \theta_0 \text{ versus } \mathsf{H}_1 : \theta \neq \theta_0.$$

Under H_0 we have $\mathcal{P}_0 = \{P_{\theta_0}\}$ and under the alternative $\mathcal{P}_1 = \{P_\theta : \theta \in \Theta_1\}$. The Bayes test problem of a point-null hypothesis corresponds to the comparison of the Bayes models

$$\mathsf{H}_0 : \mathcal{M}_0 = \{\mathcal{P}_0, \pi_0\} \text{ versus } \mathsf{H}_1 : \mathcal{M}_1 = \{\mathcal{P}_1, \pi_1\}$$

with two different priors. The prior of the point-null model is the Dirac measure on θ_0,

$$\pi_0(\theta) = \begin{cases} 1 & \text{if } \theta = \theta_0 \\ 0 & \text{otherwise.} \end{cases}$$

The prior π_1 of the alternative is defined on Θ_1; for continuous priors it is defined on Θ as well by $\pi_1(\theta) = \pi(\theta)$. Note that, the priors π_0 and π_1 are not generated by a split of a common prior π as in (8.10). A formal application of the test (8.4) gives a useless result, because for continuous posteriors, $\mathsf{P}^\pi(\Theta_0|\mathbf{x}) = 0$ and the null hypothesis is never accepted; see Figure 8.2.
The way out is the concept of Bayes factor and using the model indicator $k = 0, 1$. We set the prior probability p_0 on $k = 0$. Then the common prior, mixing π_0 and π_1, is

$$\pi_{\text{com}}(\theta) = p_0 \pi_0(\theta) + (1 - p_0) \pi_1(\theta). \tag{8.17}$$

This prior distribution is called zero-inflated. It has a distribution function with a jump of height p_0 at θ_0. For continuous prior π_1 we have

$$\int_{\Theta_1} p(\mathbf{x}|\theta) \pi_1(\theta) \, d\theta = \int_\Theta p(\mathbf{x}|\theta) \pi(\theta) \, d\theta = m(\mathbf{x}).$$

Further

$$\int_{\Theta_0} p(\mathbf{x}|\theta) \pi_0(\theta) \, d\theta = p(\mathbf{x}|\theta_0),$$

hence

$$B_{01} = \frac{p(\mathbf{x}|\theta_0)}{m(\mathbf{x})}. \tag{8.18}$$

Example 8.7 (Binomial distribution and beta prior)
We continue with Example 8.2. Now we are interested in the two-sided testing problem:

$$\mathsf{H}_0 : \theta = \frac{1}{2} \text{ versus } \mathsf{H}_1 : \theta \neq \frac{1}{2}.$$

The prior on H_0 is given by $\rho_0 = \frac{1}{2}$. The prior under the alternative is $U(0,1)$. We have

$$p(x|\theta_0) = \binom{n}{x}\left(\frac{1}{2}\right)^n$$

and

$$m(x) = \int_0^1 \binom{n}{x}\theta^x(1-\theta)^{n-x}\,d\theta = \binom{n}{x}B(x+1, n-x+1).$$

Using

$$\binom{n}{k} = \frac{1}{(n+1)B(k+1, n-k+1)},$$

we obtain $m(x) = \frac{1}{n+1}$ and

$$B_{01} = (n+1)\binom{n}{x}\left(\frac{1}{2}\right)^n.$$

$\square$

In Example 8.3 we studied the sex ratio at birth in Sweden and concluded that the rate is less than the expected rate given by WHO. Now we want to check the related one–point hypothesis.

Example 8.8 (Sex ratio at birth)
We continue with Example 8.3 and set $\theta_0 = 0.5145631$. We are interested in the testing problem

$$H_0 : \theta_0 \quad \text{versus} \quad H_1 : \theta \neq \theta_0.$$

We take a uniform prior with $\alpha_0 = 1$ and $\beta_0 = 1$ and set $p_0 = \frac{1}{2}$. It holds that

$$p(x|\theta_0) = \binom{n}{x}\theta_0^x(1-\theta_0)^{n-x} \approx 0.000436.$$

Thus the Bayes factor

$$B_{01} = p(x|\theta_0)(n+1) = 114264 \times 0.000436 \approx 49.82, \quad B_{10} = 0.02$$

Applying the table above we have no evidence against H_0. $\square$

In likelihood testing theory we can test a one–point hypothesis with the help of a confidence region. The same method is possible in Bayes inference applying HPD credible regions. Assume that

$$\theta_0 \in C_{\mathbf{x}} = \{\theta : \pi(\theta|\mathbf{x}) \geq k\}, \quad P^\pi(\pi(\theta|\mathbf{x}) \geq k|\mathbf{x}) = 1 - \alpha(k),$$

then

$$\pi(\theta_0|\mathbf{x}) = \frac{p(\mathbf{x}|\theta_0)\pi(\theta_0)}{m(\mathbf{x})} > k$$

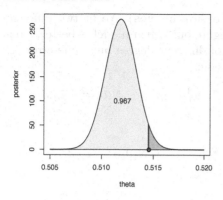 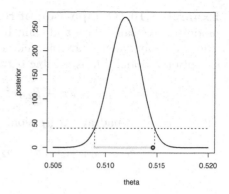

Figure 8.4: Sex ratio at birth (SRB) data. Left: Example 8.3: The one-sided test rejects the null hypothesis. The SRB in Sweden is significantly less than the value of WHO. Right: Example 8.9: The WHO value lies inside of credible region. There is no evidence against the one–point null hypothesis.

and

$$\mathsf{B}_{01} = \frac{p(\mathbf{x}|\theta_0)}{m(\mathbf{x})} > \frac{k}{\pi(\theta_0)}.$$

Small B_{01} delivers evidence against H_0, but here we have a lower bound. Roughly speaking, if $\theta_0 \in C_{\mathbf{x}}$ we will find no evidence against H_0.
We illustrate it with the Swedish SRB data.

Example 8.9 (Sex ratio at birth)
We continue with the example above. The posterior distribution is concentrated on a small interval between 0.505 and 0.520. The HPD α-credible interval for $\alpha = 0.05$ is $[0.5089, 0.5147]$. It includes the theoretical value $\theta_0 = 0.5145$. We get no evidence against H_0; see Figure 8.4. □

8.2.2 Bayes Factor in Linear Model

We consider two alternative linear models, which differ in the regression part, while the error distributions are the same:

$$\mathsf{H}_0 : \mathcal{M}_0 = \{\mathcal{P}_0, \pi_0\} \ \text{versus} \ \mathsf{H}_1 : \mathcal{M}_1 = \{\mathcal{P}_1, \pi_1\} \tag{8.19}$$

where, for $j = 0, 1$,

$$\mathcal{P}_j = \left\{ \mathsf{N}_n(\mathbf{X}_{(j)}\beta_j, \sigma^2\Sigma) : \ \theta_j = (\beta_j, \sigma^2), \ \beta_j \in \mathbb{R}^{p_j}, \ \sigma^2 \in \mathbb{R}_+ \right\}$$

and

$$\pi_j(\theta_j) = \pi_j(\beta_j|\sigma^2)\pi(\sigma^2).$$

This set up can be illustrated as follows.

Example 8.10 (Comparison of regression curves) We fit two different models to the same data and want to figure out which model is better. Consider two polynomial regression curves of different degree but same error assumptions. Model $\mathcal{M}_0$ is a cubic regression:

$$\mathcal{P}_0 : y_i = \beta_{0,0} + \beta_{1,0} x_i + \beta_{2,0} x_i^2 + \beta_{3,0} x_i^3 + \varepsilon_i, \ \varepsilon_i \sim N(0, \sigma^2), \ i.i.d.$$

Model $\mathcal{M}_1$ is a quadratic regression:

$$\mathcal{P}_1 : y_i = \beta_{0,1} + \beta_{1,1} x_i + \beta_{2,1} x_i^2 + \varepsilon_i, \ \varepsilon_i \sim N(0, \sigma^2), \ i.i.d.$$

$\square$

To simplify the calculations let us consider an arbitrary linear model

$$\left\{ \left\{ N_n(\mathbf{X}\beta, \sigma^2 \Sigma) : \ \theta = (\beta, \sigma^2), \ \beta \in \mathbb{R}^p, \ \sigma^2 \in \mathbb{R}_+, \right\}, \pi \right\}$$

with posterior $NIG(a, b, \gamma, \Gamma)$. Using the known posterior we calculate the integral

$$m(\mathbf{y}) = \int_\Theta \ell(\theta|\mathbf{y}) \pi(\theta) d\theta.$$

Set

$$\ell(\theta|\mathbf{y}) \pi(\theta) = c_0 c_\pi \mathrm{ker}(\theta|\mathbf{y})$$

where c_π is the constant from the prior and

$$c_0 = \left(\frac{1}{\sqrt{2\pi}} \right)^n \frac{1}{|\Sigma|^{1/2}}.$$

Further, for

$$\pi(\theta|\mathbf{y}) = c_1 \mathrm{ker}(\theta|\mathbf{y})$$

with the same kernel function $\mathrm{ker}(\theta|\mathbf{y})$ and

$$c_1 = \left(\frac{1}{\sqrt{2\pi}} \right)^p \frac{1}{|\Gamma|^{1/2}} \left(\frac{b}{2} \right)^{\frac{a}{2}} \frac{1}{\Gamma(\frac{a}{2})}$$

we obtain

$$m(\mathbf{y}) = \frac{c_0 \, c_\pi}{c_1}.$$

Note that, the constant c_0 is the same in both models. Summarizing, we obtain the Bayes factor

$$B_{01} = \frac{m^{(0)}(\mathbf{y})}{m^{(1)}(\mathbf{y})} = \frac{c_\pi^{(0)}}{c_\pi^{(1)}} \frac{c_1^{(1)}}{c_1^{(0)}},$$

where the superscript indicates the corresponding model. We consider the conjugate prior and Jeffreys prior. For conjugate priors, we assume

$$\beta_j | \sigma^2 \sim N_{p_j}(\gamma_0^{(j)}, \sigma^2 \Gamma_0^{(j)}), \quad \sigma^2 \sim \text{InvGamma}(a_0/2, b_0/2).$$

The constants of variance prior are the same in both models. We get

$$\frac{c_\pi^{(0)}}{c_\pi^{(1)}} = (\sqrt{2\pi})^{p_1 - p_0} \left(\frac{|\Gamma_0^{(1)}|}{|\Gamma_0^{(0)}|} \right)^{\frac{1}{2}}.$$

Now we calculate the ratio of the constants of $\text{NIG}(a^{(0)}, b^{(0)}, \gamma^{(0)}, \Gamma^{(0)})$ and $\text{NIG}(a^{(1)}, b^{(1)}, \gamma^{(1)}, \Gamma^{(1)})$. Note that, $a^{(0)} = a^{(1)} = a_0 + n$, where all other parameters differ. We get

$$\frac{c_1^{(1)}}{c_1^{(0)}} = (\sqrt{2\pi})^{p_0 - p_1} \left(\frac{|\Gamma^{(0)}|}{|\Gamma^{(1)}|} \right)^{\frac{1}{2}} \left(\frac{b^{(1)}}{b^{(0)}} \right)^{\frac{a_0 + n}{2}}.$$

Summarizing, we have

$$B_{01} = \left(\frac{|\Gamma^{(0)}|}{|\Gamma_0^{(0)}|} \right)^{\frac{1}{2}} \left(\frac{|\Gamma_0^{(1)}|}{|\Gamma^{(1)}|} \right)^{\frac{1}{2}} \left(\frac{b^{(1)}}{b^{(0)}} \right)^{\frac{a_0 + n}{2}}. \tag{8.20}$$

Applying (6.67) we have

$$b^{(j)} = b_0 + \text{Ridge}^{(j)}$$

with

$$\text{Ridge}^{(j)} = (\mathbf{y} - \hat{\mathbf{y}}^{(j)})^T \Sigma^{-1} (\mathbf{y} - \hat{\mathbf{y}}^{(j)}) + \text{pen}^{(j)}$$

where

$$\hat{\mathbf{y}}^{(j)} = \mathbf{X}_{(j)} \gamma^{(j)}, \quad \text{pen}^{(j)} = (\gamma_0^{(j)} - \gamma^{(j)})^T (\Gamma_0^{(j)})^{-1} (\gamma_0^{(j)} - \gamma^{(j)}).$$

We assume, there exist positive matrices $\mathbf{C}^{(j)}$, $j = 0, 1$ such that

$$\frac{1}{n} \mathbf{X}_{(j)}^T \Sigma^{-1} \mathbf{X}_{(j)} \to \mathbf{C}^{(j)}, \quad j = 0, 1. \tag{8.21}$$

From (6.64) we get

$$\frac{|\Gamma^{(0)}|}{|\Gamma^{(1)}|} = \frac{|(\Gamma_0^{(1)})^{-1} + \mathbf{X}_{(1)}^T \Sigma^{-1} \mathbf{X}_{(1)}|}{|(\Gamma_0^{(0)})^{-1} + \mathbf{X}_{(0)}^T \Sigma^{-1} \mathbf{X}_{(0)}|} = n^{(p_1 - p_0)} \left(\frac{|\mathbf{C}^{(1)}|}{|\mathbf{C}^{(0)}|} + o(1) \right).$$

We transform the Bayes factor and take only the leading terms for $n \to \infty$ into account, such that

$$2 \ln(B_{01}) \approx (a_0 + n) \left(\ln(\text{Ridge}^{(1)} + b_0) - \ln(\text{Ridge}^{(0)} + b_0) \right) + (p_1 - p_0) \ln(n).$$

Hence, we prefer the model with the smaller Ridge criterion, penalized by the number of parameters.

Now, for Jeffreys prior we assume the same prior in both models:

$$\pi(\theta_j) \propto \frac{1}{\sigma^2}.$$

The Bayes factor is calculated by the ratio of the posterior constants. Under Jeffreys prior all posterior parameters are different. We obtain

$$B_{01} = \frac{c_1^{(1)}}{c_1^{(0)}} = \sqrt{\pi}^{-(p_0 - p_1)} \left(\frac{|\Gamma^{(0)}|}{|\Gamma^{(1)}|} \right)^{\frac{1}{2}} \frac{(\frac{b^{(1)}}{2})^{\frac{n-p_1}{2}}}{(\frac{b^{(0)}}{2})^{\frac{n-p_0}{2}}} \frac{\Gamma(\frac{n-p_0}{2})}{\Gamma(\frac{n-p_1}{2})}.$$

Applying (6.85) and setting $(\mathbf{y} - \hat{\mathbf{y}}^{(j)})^T \Sigma^{-1} (\mathbf{y} - \hat{\mathbf{y}}^{(j)}) = \text{RSS}^{(j)}$ we have

$$(b^{(j)})^{\frac{n-p_j}{2}} = n^{\frac{n-p_j}{2}} \left(\frac{1}{n} \text{RSS}^{(j)} \right)^{\frac{n-p_j}{2}}.$$

Using the approximation,

$$\Gamma(x + \alpha) \approx \Gamma(x) x^{\alpha}, \text{ for } x \to \infty,$$

we get

$$\frac{\Gamma(\frac{n-p_0}{2})}{\Gamma(\frac{n-p_1}{2})} \approx \left(\frac{n}{2} \right)^{\frac{p_1 - p_0}{2}}.$$

Under (8.21) we have

$$\frac{|\Gamma^{(0)}|}{|\Gamma^{(1)}|} = \frac{|\mathbf{X}_{(1)}^T \Sigma^{-1} \mathbf{X}_{(1)}|}{|\mathbf{X}_{(0)}^T \Sigma^{-1} \mathbf{X}_{(0)}|} = n^{(p_1 - p_0)} \left(\frac{|\mathbf{C}^{(1)}|}{|\mathbf{C}^{(0)}|} + o(1) \right).$$

We transform the Bayes factor and take only the leading terms for $n \to \infty$ into account, so that

$$2 \ln(B_{01}) \approx n(\ln(\text{RSS}^{(1)}/n) - \ln(\text{RSS}^{(0)}/n)) + (p_1 - p_0) \ln(n).$$

Hence we prefer the model with the smaller BIC (Bayes information criterion, or Schwarz criterion; Schwarz (1978)). In the linear normal model, BIC is given as (see Subsection 8.3.1)

$$\text{BIC} = n \ln(\frac{\text{RSS}}{n}) + p \ln(n).$$

8.2.3 *Improper Prior*

We can also apply improper priors, given that posterior is proper. A surprising fact, however is that the test based on an improper prior cannot be approximated by tests based on priors with increasing variances. This is known as

Jeffreys–Lindley paradox. We illustrate one of the complications by the following example.

Example 8.11 (Improper prior)

Let a single observation x from $X \sim \mathsf{N}(\theta, 1)$ and consider Jeffreys prior $\pi(\theta) \propto 1$. The test problem is

$$\mathsf{H}_0 : \theta = 0 \text{ versus } \mathsf{H}_1 : \theta \neq 0.$$

The Bayes factor for point-null hypothesis given in (8.18) is

$$\mathsf{B}_{01} = \frac{p(x|\theta = 0)}{m(x)}$$

with

$$m(x) = \int_{-\infty}^{\infty} p(x|\theta)\, d\theta = \int_{-\infty}^{\infty} \frac{1}{\sqrt{2\pi}} \exp(-\frac{(x - \theta)^2}{2})\, d\theta = 1$$

and

$$p(x|\theta = 0) = \frac{1}{\sqrt{2\pi}} \exp(-\frac{x^2}{2}) \leq \frac{1}{\sqrt{2\pi}}.$$

Hence

$$\mathsf{B}_{01} \leq \frac{1}{\sqrt{2\pi}} \approx 0.4.$$

This procedure favours rejecting H_0.   □

The *pseudo Bayes factor* delivers a possible way out from the problems regarding improper priors. The idea is to explore the iterative structure of Bayes methods. The sample is split into a training sample, sufficiently large to derive a proper posterior, and this posterior is set as new proper prior. The training sample should be chosen as small as possible. Following example demonstrates the applicability of this idea.

Example 8.12 (Training sample)

Consider two observations (x_1, x_2), i.i.d, from $X \sim \mathsf{N}(\mu, \sigma^2)$. Set $\theta = (\mu, \sigma^2)$ and assume Jeffreys prior $\pi(\theta) \propto \frac{1}{\sigma^2}$. The posterior is proper, if the integral $m(\mathbf{x})$ exists. Using $(x_1 - \mu)^2 + (x_2 - \mu)^2 = s^2 + 2(\bar{x} - \mu)^2$, we have

$$m(\mathbf{x}) = \int \ell(\theta|\mathbf{x})\pi(\theta)\, d\theta$$

$$\propto \int \frac{1}{\sigma^2} \exp\left(-\frac{(x_1 - \mu)^2 + (x_2 - \mu)^2}{2\sigma^2}\right) \frac{1}{\sigma^2}\, d\theta$$

$$\propto \int_0^{\infty} \frac{1}{\sigma^3} \exp\left(-\frac{s^2}{2\sigma^2}\right) \int_{-\infty}^{\infty} \exp\left(-\frac{(\bar{x} - \mu)^2}{\sigma^2}\right) \frac{1}{\sigma}\, d\theta.$$

Applying

$$\int_{-\infty}^{\infty} \exp\left(-\frac{(\bar{\mathbf{x}} - \mu)^2}{\sigma^2}\right) \frac{1}{\sigma} d\mu = \sqrt{2\pi},$$

and

$$\int_{0}^{\infty} \frac{1}{\sigma^3} \exp\left(-\frac{s^2}{2\sigma^2}\right) d\sigma^2 = \left(\frac{2}{s^2}\right)^{\frac{1}{2}} \Gamma(\frac{1}{2}),$$

we obtain $m(\mathbf{x}) < \infty$. The sample size $n = 2$ implies a proper posterior.   □

8.3   Bayes Information

The goal of information criteria is to summarize the quality of a model in one number. When comparing models, the model with the smaller number is preferred. The common structure of information criteria is that the quality of the model fit is penalized by the complexity of the model. The main idea is to support less complex models with a sufficient good model fit.

8.3.1   Bayesian Information Criterion (BIC)

Assume a Bayes model $\{\mathcal{P}, \pi\}$ with $\mathcal{P} = \{P_\theta : \theta \in \Theta \in \mathbb{R}^p\}$ and non-informative prior π. Given the data $\mathbf{x}$, the model fit is measured using the maximum of the log-likelihood function and the model complexity is the weighted dimension of the parameter space. Schwarz (1978) introduced the Bayes information criterion, also known as *Schwarz information criterion*, as follows.

Definition 8.3 (Bayes information criterion) The Bayes information criterion $\mathsf{BIC} = \mathsf{BIC}(\mathbf{x})$ is defined as

$$\mathsf{BIC} = -2 \max_{\theta \in \Theta} \ln(\ell(\theta|\mathbf{x})) + p\ln(n) + \text{const.}$$

The term const is the same for two competing models, such that it does not change the result. We calculate BIC for the normal linear model with Jeffreys prior

$$\left\{\left\{\mathsf{N}_n(\mathbf{X}\beta, \sigma^2\Sigma) : \theta = (\beta, \sigma^2),\ \beta \in \mathbb{R}^p,\ \sigma^2 \in \mathbb{R}_+\right\}, \pi_{\text{Jeff}}\right\};\quad \pi_{\text{Jeff}} \propto \frac{1}{\sigma^2}.$$

Recall that, the data are now $\mathbf{y}$ and $\mathbf{X}$ denotes the design matrix. The log-likelihood $l(\theta|\mathbf{y}) = \ln \ell(\theta|\mathbf{y})$ is

$$l(\theta|\mathbf{y}) = -\frac{n}{2}\ln(2\pi\sigma^2) - \frac{1}{2}\ln(|\Sigma|) - \frac{1}{2\sigma^2}(\mathbf{y} - \mathbf{X}\beta)^T\Sigma^{-1}(\mathbf{y} - \mathbf{X}\beta). \quad (8.22)$$

Its maximum is attained at

$$\widehat{\beta}_\Sigma = (\mathbf{X}^T\Sigma^{-1}\mathbf{X})^{-1}\mathbf{X}^T\Sigma^{-1}\mathbf{y},$$

and

$$\widehat{\sigma}^2 = \frac{1}{n}\mathsf{RSS}, \quad \mathsf{RSS} = (\mathbf{y} - \mathbf{X}\widehat{\beta}_\Sigma)^T\Sigma^{-1}(\mathbf{y} - \mathbf{X}\widehat{\beta}_\Sigma).$$

Thus

$$-2\max_{\theta\in\Theta} l(\theta|\mathbf{x}) = n\ln(\frac{1}{n}\mathsf{RSS}) + n\ln(2\pi) + \ln(|\Sigma|) + n$$

and with $\mathrm{const} = -n - \ln(|\Sigma| - n\ln(2\pi))$ we obtain

$$\mathsf{BIC} = n\ln(\frac{1}{n}\mathsf{RSS}) + p\ln(n). \quad (8.23)$$

8.3.2 Deviance Information Criterion (DIC)

This popular criterion was introduced by Spiegelhalter et al. (2002). Assume a Bayes model $\{\mathcal{P}, \pi\}$ with $\mathcal{P} = \{\mathsf{P}_\theta : \theta \in \Theta \in \mathbb{R}^p\}$. The deviance is defined by the log-likelihood $l(\theta|\mathbf{x}) = \ln\ell(\theta|\mathbf{x})$ as

$$D(\theta) = -2l(\theta|\mathbf{x}) + \mathrm{const} \quad (8.24)$$

where the constant term is independent of the model and has no influence to model choice. The model fit is measured by the posterior expectation of the deviance. The complexity of the model is given by

$$p_D = \mathsf{E}(D(\theta)|\mathbf{x}) - D(\mathsf{E}(\theta|\mathbf{x})). \quad (8.25)$$

Definition 8.4 (Deviance information criterion) The Deviance information criterion $\mathsf{DIC} = \mathsf{DIC}(\mathbf{x})$ is defined as

$$\mathsf{DIC} = \mathsf{E}(D(\theta)|\mathbf{x}) + p_D.$$

We calculate DIC for the normal linear model

$$\mathbf{y} = \mathbf{X}\beta + \varepsilon, \quad \varepsilon \sim \mathsf{N}_n(0, \sigma^2\Sigma).$$

Theorem 8.1

Assume the Bayes model

$$\left\{ \left\{ \mathsf{N}_n(\mathbf{X}\beta, \sigma^2\Sigma) : \theta = (\beta, \sigma^2), \beta \in \mathbb{R}^p, \sigma^2 \in \mathbb{R}_+ \right\}, \pi \right\}$$

with a prior π such that

$$\theta|\mathbf{y} \sim \mathsf{NIG}(a, b, \gamma, \Gamma).$$

Then

$$\mathsf{DIC} = n\ln(\frac{b}{a-2}) - 2n\Psi(\frac{a}{2}) + \frac{a+2}{b}\mathsf{FIT} + 2\,\mathrm{tr}\left(\Gamma\mathbf{X}^T\Sigma^{-1}\mathbf{X}\right) + \mathrm{const}, \quad (8.26)$$

where $\Psi(.)$ is the digamma function and

$$\mathsf{FIT} = (\mathbf{y} - \mathbf{X}\gamma)^T\Sigma^{-1}(\mathbf{y} - \mathbf{X}\gamma). \tag{8.27}$$

PROOF: The log-likelihood is given in (8.22), so that

$$D(\theta) = n\ln(\sigma^2) + \frac{1}{\sigma^2}(\mathbf{y} - \mathbf{X}\beta)^T\Sigma^{-1}(\mathbf{y} - \mathbf{X}\beta) + \mathrm{const}$$

with $\mathrm{const} = nln(2\pi) + \ln(|\Sigma|)$. In the following we apply the posterior moments

$$\mathsf{E}(\beta|\mathbf{y}) = \gamma, \quad \mathsf{E}(\sigma^2|\mathbf{y}) = \frac{b}{a-2}, \quad \mathrm{Cov}(\beta|\mathbf{y}, \sigma^2) = \sigma^2\Gamma$$

and

$$\mathsf{E}(\sigma^{-2}|\mathbf{y}) = \frac{a}{b}, \quad \mathsf{E}(\ln(\sigma^2)|\mathbf{y}) = \ln(\frac{b}{2}) - \Psi(\frac{a}{2}). \tag{8.28}$$

We decompose the quadratic form

$$\begin{aligned}
(\mathbf{y} - \mathbf{X}\beta)^T\Sigma^{-1}(\mathbf{y} - \mathbf{X}\beta) =&(\mathbf{y} - \mathbf{X}\gamma)^T\Sigma^{-1}(\mathbf{y} - \mathbf{X}\gamma) \\
&+ 2(\mathbf{y} - \mathbf{X}\gamma)^T\Sigma^{-1}\mathbf{X}(\gamma - \beta) \\
&+ (\gamma - \beta)^T\mathbf{X}^T\Sigma^{-1}\mathbf{X}(\gamma - \beta)
\end{aligned}$$

and obtain

$$\mathsf{E}((\mathbf{y} - \mathbf{X}\beta)^T\Sigma^{-1}(\mathbf{y} - \mathbf{X}\beta)|\mathbf{y}, \sigma^2) = \mathsf{FIT} + \mathsf{E}((\gamma - \beta)^T\mathbf{X}^T\Sigma^{-1}\mathbf{X}(\gamma - \beta)|\mathbf{y}, \sigma^2).$$

Further

$$\begin{aligned}
\mathsf{E}((\gamma - \beta)^T\mathbf{X}^T\Sigma^{-1}\mathbf{X}(\gamma - \beta)|\mathbf{y}, \sigma^2) =&\mathsf{E}(\mathrm{tr}\left((\gamma - \beta)(\gamma - \beta)^T\mathbf{X}^T\Sigma^{-1}\mathbf{X}|\mathbf{y}, \sigma^2\right)) \\
=&\mathrm{tr}\left(\mathrm{Cov}(\beta|\mathbf{y}, \sigma^2)\mathbf{X}^T\Sigma^{-1}\mathbf{X}\right) \\
=&\sigma^2\mathrm{tr}\left(\Gamma\mathbf{X}^T\Sigma^{-1}\mathbf{X}\right).
\end{aligned}$$

It follows that

$$E(D(\theta)|\mathbf{y}) = nE(\ln(\sigma^2)|\mathbf{y}) + E(\frac{1}{\sigma^2}(\mathbf{y} - \mathbf{X}\beta)^T \Sigma^{-1}(\mathbf{y} - \mathbf{X}\beta)|\mathbf{y}) + \text{const}$$

$$= n(\ln(\frac{b}{2}) - \Psi(\frac{a}{2})) + \frac{a}{b}\text{FIT} + \text{tr}\left(\Gamma \mathbf{X}^T \Sigma^{-1} \mathbf{X}\right) + \text{const}.$$

$$(8.29)$$

Further

$$p_D = n(\ln(\frac{2}{a-2}) - \Psi(\frac{a}{2})) + \frac{2}{b}\text{FIT} + \text{tr}\left(\Gamma \mathbf{X}^T \Sigma^{-1} \mathbf{X}\right). \qquad (8.30)$$

Summarizing, we obtain the statement.

□

In the following we calculate the difference of DIC between two Bayes linear models

$$\Delta\text{DIC} = \text{DIC}^{(0)} - \text{DIC}^{(1)}.$$

We assume the same set up as in Subsection 8.2.2, for comparing models with different linear regression functions but the same error distribution.

$$\mathcal{M}_0 : \mathbf{y} = \mathbf{X}_{(0)}\beta_0 + \varepsilon, \quad \mathcal{M}_1 : \mathbf{y} = \mathbf{X}_{(1)}\beta_1 + \varepsilon, \qquad (8.31)$$

with $\varepsilon \sim N_n(0, \sigma^2\Sigma)$ and $\beta_0 \in \mathbb{R}^{p_0}$, $\beta_1 \in \mathbb{R}^{p_1}$. We consider conjugate priors and Jeffreys prior. First we assume conjugate priors which fulfill $\pi_j(\theta_j) = \pi_j(\beta_j|\sigma^2)\pi(\sigma^2)$, $j = 0, 1$. Thus

$$\theta_j \sim \text{NIG}(a_0, b_0, \gamma_0^{(j)}, \Gamma_0^{(j)}), \quad \theta_j|\mathbf{y} \sim \text{NIG}(a_1, b_1^{(j)}, \gamma_1^{(j)}, \Gamma_1^{(j)}), \quad j = 0, 1.$$

The expressions for the posteriors are given in Theorem 6.5. Note that the parameters a_0 and a_1 are the same in both models. Applying Theorem 8.1 for $j = 0, 1$ we calculate

$$\Delta\text{DIC} = D^{(0)} - D^{(1)}$$

where

$$D^{(j)} = n\ln(b_1^{(j)}) + \frac{a_1 + 2}{b_1^{(j)}}\text{FIT}^{(j)} + 2\text{tr}\left(\Gamma_1^{(j)} \mathbf{X}_{(j)}^T \Sigma^{-1} \mathbf{X}_{(j)}\right). \qquad (8.32)$$

Especially

$$b_1^{(j)} = b_0 + \text{FIT}^{(j)} + (\gamma_0^{(j)} - \gamma_1^{(j)})^T (\Gamma_0^{(j)})^{-1}(\gamma_0^{(j)} - \gamma_1^{(j)}), \quad a_1 = a_0 + n$$

and

$$\Gamma_1^{(j)} = ((\Gamma_0^{(j)})^{-1} + \mathbf{X}_{(j)}^T \Sigma^{-1} \mathbf{X}_{(j)})^{-1}$$

with

$$\gamma_1^{(j)} = \Gamma_1^{(j)} \left(\mathbf{X}_{(j)}^T \Sigma^{-1}\mathbf{y} + (\Gamma_0^{(j)})^{-1}\gamma_0^{(j)}\right).$$

Under (8.21) we have

$$n\Gamma_1^{(j)} = \left(\frac{1}{n}(\Gamma_0^{(j)})^{-1} + \frac{1}{n}\mathbf{X}_{(j)}^T \Sigma^{-1} \mathbf{X}_{(j)} \right)^{-1} = \mathbf{C}_{(j)}^{-1} + o(1)$$

and

$$\text{tr}\left(\Gamma_1^{(j)} \mathbf{X}_{(j)}^T \Sigma^{-1} \mathbf{X}_{(j)} \right) = \text{tr}\left(n\Gamma_1^{(j)} \frac{1}{n}\mathbf{X}_{(j)}^T \Sigma^{-1} \mathbf{X}_{(j)} \right) = p_j + o(1).$$

Secondly we assume Jeffreys prior

$$\pi_j(\theta_j) \propto \left(\frac{1}{\sigma^2} \right)^{\frac{p_j+2}{2}}.$$

The posterior is derived in Theorem 6.7, where

$$b_1^{(j)} = \text{RSS}^{(j)}, \quad a_1^{(j)} = n, \quad (\Gamma_1^{(j)})^{-1} = \mathbf{X}_{(j)}^T \Sigma^{-1} \mathbf{X}_{(j)}$$

Applying Theorem 8.1 we get

$$\Delta\text{DIC} = D^{(0)} - D^{(1)}$$

where, for $j = 0, 1$,

$$D^{(j)} = n \ln \left(\frac{\text{RSS}^{(j)}}{n} \right) + 2\,p_j. \tag{8.33}$$

Note that, in this case, DIC coincides with *Akaike information criterion* (AIC).

8.4 List of Problems

1. Consider an i.i.d. sample $X_1, \ldots, X_n$ from $\text{Poi}(\theta)$ and a conjugate prior.
   (a) Calculate the posterior distribution.
   (b) For a symmetric loss function, derive the Bayes rule for testing

$$H_0 : \theta \geq 1 \quad \text{versus} \quad H_1 : \theta < 1. \tag{8.34}$$

   (c) Set the prior as $\text{Gamma}(2, 1)$. Given $n = 20$, $\sum_{i=1}^n x_i = 15$, carry out the test.
2. Consider an i.i.d. sample $X_1, \ldots, X_n$ from $\text{Gamma}(\alpha, \theta)$, with density

$$f(x|\theta) = \frac{\theta^\alpha}{\Gamma(\alpha)} x^{\alpha-1} \exp(-\theta x) \tag{8.35}$$

   where $\alpha > 2$ is known. The prior is $\theta \sim \text{Gamma}(\alpha_0, \beta_0)$.
   (a) Calculate the posterior distribution.

(b) Consider the one point test problem $H_0 : \theta = \theta_0$ versus $H_1 : \theta \neq \theta_0$. Calculate the Bayes factor B_{01}.

(c) Suppose the Bayes factor $B_{10} = 200$. Which conclusion is possible?

3. Environmental scientists studied the accumulation of toxic elements in marine mammals. The mercury concentrations (microgram/gram) in the livers of 28 male striped dolphins (*Stenella coeruleoalba*) are given in the following table, taken from Augier et al. (1993). Set $m = 4$ and $n = 28$. Two

1.70	1.72	8.80	5.90	183.00	221.00
286.00	168.00	406.00	286.00	218.00	252.00
241.00	180.00	329.00	397.00	101.00	264.00
316.00	209.00	85.40	481.00	445.00	314.00
118.00	485.00	278.00	318.00		

different models are proposed,

$$\mathcal{P}_0 = \{N(\theta, \sigma_1^2)^{\otimes m} \otimes N(\mu, \sigma_2^2)^{\otimes(n-m)} : \theta \in \mathbb{R}\}, \qquad (8.36)$$

with parameters $\mu, \sigma_1^2, \sigma_2^2$ known, and

$$\mathcal{P}_1 = \{N(\theta, \sigma_1^2)^{\otimes m} \otimes N(a\theta, \sigma_2^2)^{\otimes(n-m)} : \theta \in \mathbb{R}\}, \qquad (8.37)$$

where the coefficient a is known and σ_1^2, σ_2^2 are known, and the same as in (8.36).

(a) Suggest a non-informative prior for each model.

(b) Derive the posterior distributions belonging to the related non-informative prior.

(c) Give the expressions of the fitted values $\widehat{x}_i^{(j)}$, $i = 1, \ldots, n$ and of $RSS^{(j)}$ in each model $j = 0, 1$.

(d) Derive the expression of the Bayes factor B_{01} for comparing both models.

(e) Set $\mu = 200$, $a = 50$ and $\sigma_1^2 = 10$, $\sigma_2^2 = 10000$. Compare $RSS^{(j)}$ $j = 0, 1$. Calculate $2\ln(B_{10})$. Draw the conclusion.

4. Let y_{ij} be the length of intensive care of Corona patient i in hospital j, where $i = 1, \ldots, 2n$, $j = 1, 2$. Further, the following information of each patient i is given: x_{1i} age, x_{2i} body mass index, x_{3i} chronic lung disease and x_{4i} serious heart conditions, where $x_{3i} \in \{0, 1\}$, $x_{4i} \in \{0, 1\}$, and $x_{3i} = 1$, $x_{4i} = 1$ mean the existence of the respective chronic disease. The following linear model is proposed

$$y_{1i} = \mu_1 + \beta_1 x_{1i} + \beta_2 x_{2i} + \beta_3 x_{3i} + \beta_4 x_{4i} + \varepsilon_{1i}, \quad i = 1, \ldots, n$$
$$y_{2i} = \mu_2 + \beta_1 x_{1i} + \beta_2 x_{2i} + \beta_3 x_{3i} + \beta_4 x_{4i} + \varepsilon_{2i}, \quad i = n+1, \ldots, 2n,$$

where $\varepsilon_{ij} \sim N(0, \sigma^2)$, i.i.d. Define $\theta = (\mu_1, \mu_2, \beta_1, \ldots, \beta_4)$. Suppose σ^2 is known. The parameters θ_i, $i = 1, \ldots, 6$, are independent with $\theta_i \sim N(0, 1)$.

(a) Calculate the posterior distribution of θ.

(c) Calculate the posterior distribution of $\mu_1 - \mu_2$.

(d) Propose a Bayes test for

$$\mathsf{H}_0 : -0.2 \le \mu_1 - \mu_2 \le 0.2 \quad \mathsf{H}_1 : |\mu_1 - \mu_2| > 0.2.$$

5. This problem on parallel lines is related to Problem 8 in Chapter 6. Let

$$
\begin{aligned}
y_i &= \alpha_y + \beta_y x_i + \varepsilon_i, \quad i = 1, \ldots, n, \quad \varepsilon_i \sim \mathsf{N}(0, \sigma^2) \ i.i.d. \\
z_i &= \alpha_z + \beta_z x_i + \xi_i, \quad i = 1, \ldots, n, \quad \xi_i \sim \mathsf{N}(0, \sigma^2) \ i.i.d.,
\end{aligned}
\tag{8.38}
$$

where ε_i and ξ_i are mutually independent and $\sum_{i=1}^{n} x_i = 0$. The unknown parameter is $\theta = (\alpha_y, \alpha_z, \beta_y, \beta_z, \sigma^2)$. We are interested in testing the parallelism of lines

$$\mathsf{H}_0 : \beta_y = \beta_z = \beta, \quad \mathsf{H}_1 : \beta_y \ne \beta_z.$$

Under H_0 the prior of $\theta_{(0)} = (\alpha_y, \alpha_z, \beta, \sigma^2)$ is $\mathsf{NIG}(a, b, 0, \mathbf{I}_3)$ and under H_1 the prior of $\theta_{(1)} = (\alpha_y, \alpha_z, \beta_y, \beta_z, \sigma^2)$ is $\mathsf{NIG}(a, b, 0, \mathbf{I}_4)$.

(a) Formulate the Bayes model under H_0 and derive the posterior $\mathsf{NIG}(a_1, b_1^{(0)}, \gamma^{(0)}, \Gamma^{(0)})$.

(b) Formulate the Bayes model under H_1 and derive the posterior $\mathsf{NIG}(a_1, b_1^{(1)}, \gamma^{(1)}, \Gamma^{(1)})$.

(c) Show that $\frac{1}{n}(b_1^{(0)} - b_1^{(1)}) = \frac{n+1}{2n+1}(\widehat{\beta}_y - \widehat{\beta}_z)^2 + \mathrm{rest}(n)$, with $\lim_{n \to \infty} \mathrm{rest}(n) = 0$, where $\widehat{\beta}_y$ and $\widehat{\beta}_z$ are the Bayes estimates of β_y and β_z under the alternative.

(d) Calculate the Bayes factor B_{01}.

6. Variance test in linear regression. We assume a univariate model

$$\mathbf{y} = \mathbf{X}\beta + \epsilon, \quad \epsilon \sim \mathsf{N}(0, \sigma^2 \mathbf{I}_n),$$

with Jeffreys prior $\pi(\theta) \propto \sigma^{-2}$. We are interested in testing

$$\mathsf{H}_0 : \sigma^2 = \sigma_0^2, \quad \mathsf{H}_1 : \sigma^2 \ne \sigma_0^2.$$

(a) Formulate the Bayes model under H_0 and derive the posterior.

(b) Formulate the Bayes model under H_1 and derive the posterior.

(c) Derive the expression of the Bayes factor B_{01}.

(d) Set $\widehat{\sigma}^2 = \frac{1}{n-p}\mathrm{RSS}$. Discuss the behavior of $2\ln(\mathsf{B}_{10})$ as function of $\frac{\widehat{\sigma}^2}{\sigma_0^2}$.

7. Test on correlation between two regression lines. This problem is related to Problem 10 in Chapter 6. We assume the model (6.162), i.e.,

$$
\begin{aligned}
y_i &= \alpha_y + \beta_y x_i + \varepsilon_i, \quad i = 1, \ldots, n, \quad \varepsilon_i \sim \mathsf{N}(0, \sigma_1^2) \ i.i.d. \\
z_i &= \alpha_z + \beta_z x_i + \xi_i, \quad i = 1, \ldots, n, \quad \xi_i \sim \mathsf{N}(0, \sigma_2^2) \ i.i.d.,
\end{aligned}
\tag{8.39}
$$

where $\mathsf{Cov}(\varepsilon_i, \xi_i) = \sigma_{12}$ and $\sum_{i=1}^{n} x_i = 0$. The unknown parameter is $\theta = (\alpha_y, \alpha_z, \beta_y, \beta_z, \sigma_1^2, \sigma_2^2, \sigma_{12})$. We are interested in testing

$$\mathsf{H}_0 : \sigma_1^2 = \sigma_2^2, \ \sigma_{12} = 0, \quad \mathsf{H}_1 : \text{else}.$$

(a) Formulate the Bayes model under H_0 with prior $\mathsf{NIG}(2, 2, 0, \mathbf{I}_4)$ and derive the posterior.

(b) Formulate the Bayes model under H_1 with prior $\mathsf{NIW}(2, \mathbf{0}, \mathbf{I}_2, \mathbf{I}_2)$ and derive the posterior.

(c) Set

$$\Sigma = \begin{pmatrix} \sigma_1^2 & \sigma_{12} \\ \sigma_{12} & \sigma_2^2 \end{pmatrix}.$$

Derive the Bayes estimate $\widetilde{\Sigma}$ of Σ under H_0 and the Bayes estimate $\widehat{\Sigma}$ of Σ under H_1. Compare both.

(d) Derive the expression of the Bayes factor B_{01}.

(e) Show that

$$\frac{2}{n+1} \ln B_{10} = \ln |\widetilde{\Sigma}| - \ln |\widehat{\Sigma}| + o_P(1).$$

Chapter 9

Computational Techniques

Following the Bayesian inference principle, all what we need is the posterior distribution. In general however, we rarely find a known distribution family which has the same kernel as $\pi(\theta)\,\ell(\theta|\mathbf{x})$. In this case computational methods help.

In this chapter we present methods for

- computing integrals $\int h(\theta, \mathbf{x})\,\pi(\theta|\mathbf{x})\,d\theta$,
- generating an i.i.d. sample $\theta_1, \ldots, \theta_N$ from $\pi(\theta|\mathbf{x})$, and
- generating a Markov chain with stationary distribution $\pi(\theta|\mathbf{x})$.

The first item is important for computing Bayes estimators, Bayes factors, and predictive distributions. The other items are useful for MAP estimators, HPD–regions, studying the shape of the posterior distribution, or to estimate the posterior density by smoothing methods.

Nowadays we can derive the posterior for almost all combinations of $\pi(\theta)$ and $\ell(\theta|\mathbf{x})$. Furthermore we can also do it in cases where we do not have an expression of the likelihood function, but only a data generating procedure. Likewise, a closed form expression of the prior is not required either.

This opens broad fields of applications for Bayesian methods.

Our personal advice is:

> Always try to find an analytical expression first!

Even when these expressions require computational methods for special functions, such as the incomplete beta function or digamma function, properties of these functions are well studied and can help understand the problem better. It can also be useful to approximate the posterior by analytic expressions, and in this case we can study the properties of the posterior more generally.

Bayesian computation is an exciting field on its own. Our purpose of writing this textbook is to explain main principles and the new possibilities. For more

DOI: 10.1201/9781003221623-9

Figure 9.1: The lazy mathematician gives up and lets the computer do it.

details we refer to Albert (2005), Chen et al. (2002), Givens and Hoeting (2005), Robert and Casella (2010), Lui (2001), Zwanzig and Mahjani (2020, Chapter 2), and the literature therein.

9.1 Deterministic Methods

In this section we present two methods, which are not based on simulations. We start with the general method "brute-force", known from cryptography.

9.1.1 Brute-Force

Algorithm 9.1 Brute-force

1. Discretize $\Theta \approx \{\theta_1, \ldots, \theta_N\}$.
2. Calculate $\pi(\theta_j)\,\ell(\theta_j|\mathbf{x})$ for $j = 1, \ldots, N$.
3. Calculate $m_N(\mathbf{x}) = \frac{1}{N}\sum_{j=1}^{N}\pi(\theta_j)\,\ell(\theta_j|\mathbf{x})$.
4. Approximate $\pi(\theta_j|\mathbf{x}) \approx \frac{1}{m_N(\mathbf{x})}\pi(\theta_j)\,\ell(\theta_j|\mathbf{x})$.

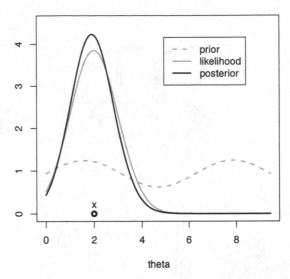

Figure 9.2: Example 9.1.

We illustrate this method by a toy example.

Example 9.1 (Brute-force)
Assume $\Theta = [0, 3\pi]$, $\pi(\theta) \propto \sin(\theta)/3 + 1$ and $X|\theta \sim \mathsf{N}(\theta, 1)$. We observe $x = 2$.
See Figure 9.2 and the following R code. The Bayes estimate is $\widehat{\theta} = 1.97$. ▢

R Code 9.1.10. Brute-force, Example 9.1.

```
theta<-seq(0,3*pi,0.001); L<-length(theta) # discretize
prior<-sin(theta)/3+1 # kernel of prior
x<-2; lik<-rep(0,L)
for(i in 1:L){lik[i]<-exp(-(x-theta[i])^2/2)} # likelihood
m<-mean(prior*lik); post<-prior*lik/m # posterior
theta.hat<-mean(theta*post) # estimate
```

9.1.2 Laplace Approximation

As second method we present an analytic approximation for $n \to \infty$, exploring
basic *Laplace approximation*,

$$\int b(\theta) \exp(-nh(\theta))\, d\theta = \sqrt{2\pi}\sigma n^{-1/2} \exp\left(-n\widehat{h}\right)\left(\widehat{b} + \frac{1}{n}\text{term}\right) + O(n^{-2}),$$

where $\widehat{\theta} = \arg\min h(\theta)$, $\widehat{b} = b(\widehat{\theta})$, $\widehat{h} = h(\widehat{\theta})$ and $\sigma^2 = h''(\widehat{\theta})^{-1}$; see Tierney
et al. (1989).

For better explanation of Laplace approximation we set $b(\theta) = 1$ and assume that $h(\theta)$ is smooth, has the unique minimum $\widehat{\theta} = \arg\min h(\theta)$, and $h'(\widehat{\theta}) = 0$. Applying Taylor expansion and the integral $\int \exp(-\frac{1}{2a}(x-b)^2)dx = \sqrt{2\pi a}$ we obtain

$$\int \exp(-nh(\theta))\,d\theta \approx \int \exp\left(-n\widehat{h} - n\widehat{h}'(\theta - \widehat{\theta}) - \frac{n}{2}\widehat{h}''(\theta - \widehat{\theta})^2\right)d\theta$$

$$= \int \exp\left(-n\widehat{h} - \frac{n}{2}\widehat{h}''(\theta - \widehat{\theta})^2\right)d\theta$$

$$= \exp\left(-n\widehat{h}\right)\int \exp\left(-\frac{n}{2}\widehat{h}''(\theta - \widehat{\theta})^2\right)d\theta$$

$$= \exp\left(-n\widehat{h}\right)\sqrt{2\pi}\sigma n^{-\frac{1}{2}},$$

where $\sigma^2 = (\widehat{h}'')^{-1}$.

Tierney et al. (1989) derived a second order analytic approximation for

$$\mu(\mathbf{x}) = \int g(\theta)\pi(\theta|\mathbf{x})\,d\theta = \frac{\int g(\theta)\ell(\theta|\mathbf{x})\pi(\theta)\,d\theta}{\int \ell(\theta|\mathbf{x})\pi(\theta)\,d\theta} = \frac{\int b_N(\theta)\exp(-nh_N(\theta))\,d\theta}{\int b_D(\theta)\exp(-nh_D(\theta))\,d\theta}.$$

They assume that $g(\theta)$ is smooth, and that n is sufficiently large such that the MAP estimator $\widehat{\theta}_{\text{MAP}}$ is unique. Note that, the second assumption is not very strong; see the results in Chapter 5. Further, they need that $\frac{1}{n}\ln(\ell(\theta|\mathbf{x}))$ is asymptotically independent of n.

Here we quote only one of their approximation expressions. The hats on b_N, h_N and their derivatives indicate evaluation at $\widehat{\theta}_N = \arg\min h_N(\theta)$; and respectively the hats on b_D, h_D and their derivatives indicate evaluation at $\widehat{\theta}_D = \arg\min h_D(\theta)$. Set $\sigma_N^2 = (\widehat{h}_N'')^{-1}$ and $\sigma_D^2 = (\widehat{h}_D'')^{-1}$. Note that, in general $\widehat{\theta}_N$ and $\widehat{\theta}_D$ are different, but in case of consistent posteriors they can converge to each other for $n \to \infty$; see Chapter 5.

$$\mu(\mathbf{x}) = \frac{\sigma_N \exp(-n\widehat{h}_N)}{\sigma_D \exp(-n\widehat{h}_D)} \times \left(\frac{\widehat{b}_N}{\widehat{b}_D} + \sigma_D^2 \frac{\widehat{b}_D\widehat{b}_N'' - \widehat{b}_N\widehat{b}_D''}{2n\widehat{b}_D^2} - \sigma_D^4 \widehat{h}_D'''\frac{\widehat{b}_D\widehat{b}_N' - \widehat{b}_N\widehat{b}_D'}{2n\widehat{b}_D^2}\right)$$
$$+ O(n^{-2}).$$

$$(9.1)$$

Depending on the specification of b_N, b_D we obtain from (9.1) different special cases. For positive functions $g(\theta) > 0$ we set $b_N = b_D$ and $h_N = h_D - \frac{1}{n}\ln g$, where $-nh_D = \ln(\ell(\theta|\mathbf{x}) + \ln(\pi(\theta))$ and obtain

$$\mu(\mathbf{x}) = \frac{\widehat{b}_N}{\widehat{b}_D}\frac{\sigma_N \exp(-n\widehat{h}_N)}{\sigma_D \exp(-n\widehat{h}_D)} + O(n^{-2}).$$

$$(9.2)$$

We illustrate the approximation (9.2) for $g(\theta) = \theta$ in the following example.

Example 9.2 (Binomial distribution and beta prior)
Set $X|\theta \sim \text{Bin}(n, \theta)$ and $\theta \sim \text{Beta}(\alpha_0, \beta_0)$. Then the posterior is $\text{Beta}(\alpha, \beta)$, with $\alpha = \alpha_0 + x$ and $\beta = \beta_0 + n - x$; see Example 2.11. We know that the expectation of $\text{Beta}(\alpha, \beta)$ is

$$\mu = \frac{\alpha}{\alpha + \beta}.$$

We set $b_N = b_D = 1$ and further

$$h_D(\theta) = -\frac{1}{n}\left((\alpha - 1)\ln(\theta) + (\beta - 1)\ln(1 - \theta)\right),$$

$$h_N(\theta) = -\frac{1}{n}\left(\alpha \ln(\theta) + (\beta - 1)\ln(1 - \theta)\right).$$

We calculate

$$h_D(\theta)' = -\frac{1}{n}\left(\frac{\alpha - 1}{\theta} - \frac{\beta - 1}{(1 - \theta)}\right), \quad h_D(\theta)'' = -\frac{1}{n}\left(-\frac{\alpha - 1}{\theta^2} - \frac{\beta - 1}{(1 - \theta)^2}\right).$$

Setting $h_D(\theta)' = 0$ we obtain the MAP estimator

$$\widehat{\theta}_D = \frac{\alpha - 1}{\alpha + \beta - 2}, \quad 1 - \widehat{\theta}_D = \frac{\beta - 1}{\alpha + \beta - 2}.$$

Analogously we obtain

$$\widehat{\theta}_N = \frac{\alpha}{\alpha + \beta - 1}, \quad 1 - \widehat{\theta}_N = \frac{\beta - 1}{\alpha + \beta - 1}.$$

Further, we get

$$\widehat{h}_D'' = \frac{1}{n}(\alpha + \beta - 2)^2 \left(\frac{1}{\alpha - 1} + \frac{1}{\beta - 1}\right), \quad \widehat{h}_N'' = \frac{1}{n}(\alpha + \beta - 1)^2 \left(\frac{1}{\alpha} + \frac{1}{\beta - 1}\right),$$

applying (9.2), and obtain the Laplace approximation

$$\mu = \left(\frac{\alpha + \beta - 2}{\alpha + \beta - 1}\right)^{\alpha + \beta + 0.5} \left(\frac{\alpha}{\alpha - 1}\right)^{\alpha + 0.5} \frac{\alpha - 1}{\alpha + \beta - 2} + O(n^{-2}). \tag{9.3}$$

For illustration we consider $\theta \sim U[0, 1]$. Then we have $\alpha + \beta - 2 = n$. Figure 9.3 shows the convergence in (9.3) for $n \to \infty$ under different values of $p = \frac{\alpha}{\alpha + \beta}$, such that $\alpha = p(n + 2)$, $\beta = (1 - p)(n + 2)$. □

9.2 Independent Monte Carlo Methods

For given data $\mathbf{x}$ we want to calculate the integral

$$\mu(\mathbf{x}) = \int h(\theta, \mathbf{x})\, \pi(\theta|\mathbf{x})\, d\theta = \frac{\int g(\theta, \mathbf{x})\, \ell(\theta|\mathbf{x})\pi(\theta)\, d\theta}{\int \ell(\theta|\mathbf{x})\pi(\theta)\, d\theta}. \tag{9.4}$$

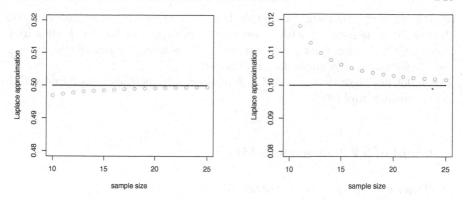

Figure 9.3: Laplace approximation in Example 9.2. Left: The symmetric case $\alpha = \beta$, $p = 0.5$. Right: $p = 0.1$. Note that, we need $n \geq 8$.

The striking idea of Monte Carlo methods (MC) is that, depending on the factorization,

$$h(\theta, \mathbf{x})\, \pi(\theta|\mathbf{x}) = m(\theta, \mathbf{x})\, p(\theta|\mathbf{x}), \tag{9.5}$$

the integral can be rewritten as an expected value

$$\mu(\mathbf{x}) = \int m(\theta, \mathbf{x})\, p(\theta|\mathbf{x})\, d\theta.$$

It can be estimated by a sample from $p(\theta|\mathbf{x})$. In case of an i.i.d. sample $\theta^{(1)}, \ldots, \theta^{(N)}$ the *independent Monte Carlo approximation* is given by

$$\widehat{\mu}(\mathbf{x}) = \frac{1}{N} \sum_{i=1}^{N} m(\theta^{(i)}, \mathbf{x}). \tag{9.6}$$

For

$$\sigma^2(\mathbf{x}) = \int \left(m(\theta, \mathbf{x}) - \mu(\mathbf{x}) \right)^2 p(\theta|\mathbf{x})\, d\theta < \infty, \tag{9.7}$$

the central limit theorem implies

$$\widehat{\mu}(\mathbf{x}) = \mu(\mathbf{x}) + \frac{1}{\sqrt{N}} \sigma(\mathbf{x}) O_{\mathrm{P}}(1).$$

Note that, the rate $\frac{1}{\sqrt{N}}$ cannot be improved for Monte Carlo Methods. The Monte Carlo approximation of the deterministic integral is a random variable, which for large N is close to the integral value with high probability.

The factorization ansatz (9.5) is essential. We need a distribution $p(\theta|\mathbf{x})$ with both a small variance and a good random number generator. The choice of the fraction in (9.4), the application of the MC on the denominator and numerator separately, and generating a sample from the prior $\pi(\theta)$ may not be good choices. Since non-informative priors can be improper or have at least a high variance. Otherwise, priors with small variances have the risk of dominating the likelihood; see Example 3.1 on page 30.

The following algorithm is called independent MC because it is based on an independent sample $\theta^{(1)}, \ldots, \theta^{(N)}$.

Algorithm 9.2 Independent MC

1. Draw $\theta^{(1)}, \ldots, \theta^{(N)}$ from distribution $p(\cdot|\mathbf{x})$.

2. Approximate $\mu(\mathbf{x}) = \int m(\theta, \mathbf{x}) p(\theta|\mathbf{x}) \, d\theta$ by

$$\widehat{\mu}(\mathbf{x}) = \frac{1}{N}(m(\theta^{(1)}, \mathbf{x})) + \ldots + m(\theta^{(N)}, \mathbf{x})).$$

We refer to Zwanzig and Mahjani (2020, Section 2.1), Albert (2005, Section 5.7), Robert and Casella (2010, Section 3.2). Here we give only an example for illustration of two MC methods, one based on sampling from posterior and the other on twice independent sampling from the prior for separate approximation of the numerator and dominator in the fraction (9.4).

Example 9.3 (Normal i.i.d. sample and inverse-gamma prior)
Recall Example 2.14 on page 19. The data consist of $n = 10$ observations $x_1, \ldots, x_{10}$ from $N(0, \sigma_0^2)$ given in R code 9.2.11. We set $\theta = \sigma^2$. The maximum likelihood estimate is

$$\widehat{\theta}_{\mathrm{MLE}} = \frac{1}{n} \sum_{i=1}^{n} x_i^2 = 2.05.$$

We assume $\theta \sim \mathsf{InvGamma}(\alpha, \beta)$, with $\alpha = 4$, $\beta = 12$, prior expectation 4, and prior variance 8. The posterior is $\theta|\mathbf{x} \sim \mathsf{InvGamma}(\alpha_1, \beta_1)$, with $\alpha_1 = 4 + n/2 = 9$ and $\beta_1 = \beta + \frac{n}{2}\widehat{\theta}_{MLE} = 22.25$; see Example 2.14. The expectation of the posterior is the Bayes estimate $\widehat{\theta}_{\mathrm{L}_2} = \frac{\beta_1}{\alpha_1 - 1} = 2.782$. The posterior variance is 1.106, essentially smaller than the prior variance. Two different methods are considered: sampling from the posterior and sampling from the prior. We set $N = 100$. The first method draws $\theta_1^{(1)}, \ldots, \theta_1^{(N)}$ from the posterior, then approximate

$$\widehat{\mu}(\mathbf{x}) = \frac{1}{N} \sum_{j=1}^{N} \theta_1^{(j)}.$$

Note that, for given data $\mathbf{x}$, $\widehat{\mu}(\mathbf{x})$ is a random approximation. We get $\widehat{\mu}(\mathbf{x}) = 2.803$. For the second method we draw $\theta_1^{(1)}, \ldots, \theta_1^{(N)}$ from the prior. Then we approximate the denominator integral

$$\mu_{\text{Den}}(\mathbf{x}) = \int_0^\infty \theta^{-(\alpha+\frac{n}{2}+1)} \exp(-\frac{\beta + \frac{n}{2}\widehat{\theta}_{\text{MLE}}}{\theta}) d\theta$$

by

$$\widehat{\mu}_{\text{Den}}(\mathbf{x}) = \frac{1}{N} \sum_{j=1}^N (\theta_1^{(j)})^{-\frac{n}{2}} \exp(-\frac{\frac{n}{2}\widehat{\theta}_{\text{MLE}}}{\theta_1^{(j)}}).$$

Further we draw $\theta_2^{(1)}, \ldots, \theta_2^{(N)}$ from the prior. Then we approximate the numerator integral

$$\mu_{\text{Num}}(\mathbf{x}) = \int_0^\infty \theta^{-(\alpha+\frac{n}{2})} \exp(-\frac{\beta + \frac{n}{2}\widehat{\theta}_{\text{MLE}}}{\theta}) d\theta$$

by

$$\widehat{\mu}_{\text{Num}}(\mathbf{x}) = \frac{1}{N} \sum_{j=1}^N (\theta_2^{(j)})^{-\frac{n}{2}-1} \exp(-\frac{\frac{n}{2}\widehat{\theta}_{\text{MLE}}}{\theta_2^{(j)}}).$$

The approximation of the Bayes estimate by the second method is given by the quotient

$$\widehat{\theta}_{L_2} = \frac{\mu_{\text{Num}}(x)}{\mu_{\text{Den}}(x)} \approx \frac{\widehat{\mu}_{\text{Num}}(x)}{\widehat{\mu}_{\text{Den}}(x)} = 2.716$$

Observe that, the true underlying parameter $\sigma_0^2 = 2.25$. In Figure 9.4 the approximation of $\widehat{\mu}(\mathbf{x})$ for increasing simulation size is shown. Both methods are compared in Figure 9.6.   □

R Code 9.2.11. Independent MC, Example 9.3.

```
x<-c(-0.06,2.43,1.55,1.76,1.27,0.94,0.48,1.16,1.32,1.81)# data
n<-length(x) # sample size
MLE<-sum(x**2)/n # maximum likelihood estimate
N<-100 # simulation size
library(invgamma)
a0<-4; b0<-12 # prior parameters
b0^2/((a0-1)^2*(a0-2)) # prior variance
a1<-a0+n/2; b1<-b0+MLE*n/2 # posterior parameters
b1^2/((a1-1)^2*(a1-2)) # posterior variance
b1/a1-1# Bayes estimate
# Sampling from the posterior (MC1)
MC1<-function(N,a1,b1)
    {mu<-mean(rinvgamma(N,a1,b1)); return(mu)}
MC1(N,a1,b1)# MC1 approximation of the Bayes estimate
# Sampling from the prior (MC2)
MC2<-function(N,a0,b0){
```

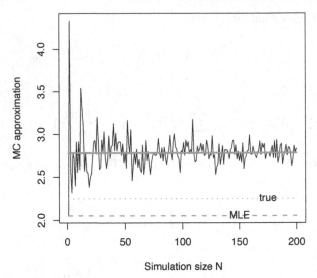

Figure 9.4: Example 9.3. The MC approximations converge to the Bayes estimate (straight line) with increasing simulation size N. But even for large N the approximations still have random deviations around the Bayes estimate. The Bayes estimate and the maximum likelihood estimate differ from the true parameter. For them, only a convergence for increasing sample size n is possible.

```
theta1<-rinvgamma(N,a0,b0) # simulate from the prior
lik1<-rep(0,length(theta1)) # likelihood function
for ( i in 1:length(theta1))
{lik1[i]<-theta1[i]**(-n/2)*exp(-n/2*MLE/theta1[i])}
m0<-mean(lik1); # MC approximation
theta2<-rinvgamma(N,a0,b0) # simulate from the prior
lik2<-rep(0,length(theta2)) # likelihood function
for ( i in 1:length(theta2))
{lik2[i]<-theta2[i]**(-n/2)*exp(-n/2*MLE/theta2[i])}
m1<-mean(theta2*lik2) # MC approximation
return(m1/m0)}
MC2(N,a0,b0)# MC2 approximation of Bayes estimate
```

9.2.1   Importance Sampling (IS)

IS is another approximation method for the integral

$$\mu(\mathbf{x}) = \int h(\theta, \mathbf{x})\, \pi(\theta|\mathbf{x})\, d\theta.$$

As independent MC the importance sampling method explores an independent sample and the law of large numbers.

For an efficient MC method, we search for a generating distribution $p(\theta|\mathbf{x})$ in (9.5) with small variance. The importance sampling method is based on another approach. The main idea is to sample from a *trial distribution* $g(\theta|\mathbf{x})$, and not from the *target* distribution, $\pi(\theta|\mathbf{x})$, and to correct the sampling from the wrong distribution by weights, called *importance*. The weights are the likelihood quotient of the target and the trial distribution.

The trial distribution g should be easy to sample, and its support must include the support of $h(\theta,\mathbf{x})\pi(\theta|\mathbf{x})$. We do not need a closed form expression of g, but we should be able to calculate the density at arbitrary points. The choice of the trial is essential for the performance of the procedure. High importance weights are a bad sign and imply a high variance of the approximation. The advice is to avoid this effect by choosing a heavier tail distribution as trial rather than the target. The trial $g(\theta|\mathbf{x})$ is also called *importance function* or *instrumental distribution*.

Importance sampling is particularly recommended for calculating posterior tail probabilities when the trial distribution is chosen such that more values are sampled from the tail region. The importance algorithm is given as follows.

Algorithm 9.3 Importance Sampling (IS)

1. Draw $\theta^{(1)},\ldots,\theta^{(j)},\ldots,\theta^{(N)}$ from the trial distribution $g(\cdot|\mathbf{x})$.

2. For $\pi(\theta|\mathbf{x}) \propto k(\theta|\mathbf{x})$, calculate the importance weights

$$w^{(j)} = w(\theta^{(j)}, \mathbf{x}) = \frac{k(\theta^{(j)}|\mathbf{x})}{g(\theta^{(j)}|\mathbf{x})}, \quad j = 1, \ldots, N.$$

3. Approximate $\mu(\mathbf{x}) = \int h(\theta,\mathbf{x})\,\pi(\theta|\mathbf{x})\,d\theta$ by

$$\widehat{\mu}(\mathbf{x}) = \frac{w^{(1)}h(\theta^{(1)},\mathbf{x}) + \ldots + w^{(N)}h(\theta^{(N)},\mathbf{x})}{w^{(1)} + \ldots + w^{(N)}}.$$

Note that, the constant of the posterior is not needed. Usually we apply $\pi(\theta)\ell(\theta|\mathbf{x}) \propto k(\theta|\mathbf{x})$. In contrast to independent MC, where the numerator integral and the denominator integral are approximated separately, we generate only one sample $\theta^{(1)},\ldots,\theta^{(N)}$.

The IS method is based on a weighted law of large numbers. For $\pi(\theta|\mathbf{x}) = c_0(\mathbf{x})\,k(\theta|\mathbf{x})$, we have

$$\frac{1}{N}\sum_{j=1}^{N} w^{(j)}h(\theta^{(j)},\mathbf{x}) \xrightarrow{P} \int h(\theta,\mathbf{x})w(\theta,\mathbf{x})\,g(\theta|\mathbf{x})\,d\theta = c_0(\mathbf{x})^{-1}\mu(\mathbf{x}),$$

since

$$\int \frac{h(\theta,\mathbf{x})k(\theta|\mathbf{x})}{g(\theta|\mathbf{x})}g(\theta|\mathbf{x})\,d\theta = \int h(\theta,\mathbf{x})k(\theta|\mathbf{x})\,d\theta = c_0(\mathbf{x})^{-1}\int h(\theta,\mathbf{x})\pi(\theta|\mathbf{x})\,d\theta,$$

and

$$\frac{1}{N}\sum_{j=1}^{N} w^{(j)} \xrightarrow{P} \int \frac{k(\theta|\mathbf{x})}{g(\theta|\mathbf{x})} g(\theta|\mathbf{x})\,d\theta = c_0(\mathbf{x})^{-1} \int \pi(\theta|\mathbf{x})\,d\theta = c_0(\mathbf{x})^{-1},$$

such that $\widehat{\mu}(\mathbf{x}) \xrightarrow{P} \mu(\mathbf{x})$.

The importance weights are quotients of two densities up to a constant. For this, we state the following theorem from Lui (2001, Theorem 2.5.1, page 31).

Theorem 9.1 Let $f(z_1, z_2)$ and $g(z_1, z_2)$ be two probability densities, where the support of f is a subset of the support of g. Then,

$$\mathsf{Var}_g\left(\frac{f(z_1, z_2)}{g(z_1, z_2)}\right) \geq \mathsf{Var}_g\left(\frac{f_1(z_1)}{g_1(z_1)}\right),$$

where $f_1(z_1)$ and $g_1(z_1)$ are the marginal densities.

PROOF: Using $f_1(z_1) = \int f(z_1, z_2)dz_2$, $g(z_1, z_2) = g_1(z_1)g_{2|1}(z_2|z_1)$, and $r(z_1, z_2) = \frac{f(z_1,z_2)}{g(z_1,z_2)}$ we obtain

$$r_1(z_1) = \frac{f_1(z_1)}{g_1(z_1)} = \int r(z_1, z_2)g_{2|1}(z_2|z_1)dz_2 = \mathsf{E}_g(r(z_1, z_2)|z_1).$$

The statement follows from

$$\mathsf{Var}(r(z_1, z_2)) = \mathsf{E}(\mathsf{Var}(r(z_1, z_2)|z_1)) + \mathsf{Var}(\mathsf{E}(r(z_1, z_2)|z_1)) \geq \mathsf{Var}(r_1(z_1)).$$

□

Theorem 9.1 is based on the Rao–Blackwell Theorem; see Liero and Zwanzig (2011, Theorem 4.5, page 104). Based on this result, Lui formulated a rule of thumb for Monte Carlo computation:

"One should carry out analytical computation as much as possible."

We refer to Zwanzig and Mahjani (2020, Section 2.1.1), Albert (2005, Section 5.9), Robert and Casella (2010, Section 3.3), Chen et al. (2002, Chapter 5) and Lui (2001, Section 2.5).

Example 9.4 (Normal distribution and inverse-gamma prior)
We continue with the data and estimation problem given in Example 9.3. We apply importance sampling using the trial distribution $\mathsf{Gamma}(6, 2)$ with density

$$g(\theta) = \frac{2^6}{\Gamma(6)}\,\theta^5 \exp(-2\,\theta).$$

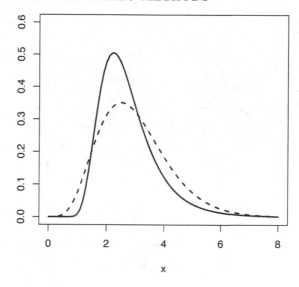

Figure 9.5: Example 9.3. The continuous line, inverse-gamma distribution, is the target, and the broken line, the gamma distribution, is the trial.

In Figure 9.5 the posterior and the trial distribution are plotted together. We set $N = 100$ and generate observations from the trial, given by $\theta^{(1)}, \ldots, \theta^{(N)}$. Using $\pi(\theta)\ell(\theta|\mathbf{x}) \propto k(\theta|\mathbf{x})$, with

$$k(\theta|\mathbf{x}) = \left(\frac{1}{\theta}\right)^{\alpha+1+\frac{n}{2}} \exp(-\frac{\beta + \frac{n}{2}\widehat{\theta}_{\mathrm{MLE}}}{\theta}),$$

the importance weights are calculated by $w^{(j)} = \frac{k(\theta^{(j)}|\mathbf{x})}{g(\theta^{(j)})}$. We obtain $\sum_{j=1}^{m} w^{(j)} = 2.778851 \times 10^{-6}$, and $\sum_{j=1}^{m} w^{(j)}\theta(j) = 7.860973 \times 10^{-6}$. Thus the Bayes estimate approximated by importance sampling is 2.83. Recall that, the Bayes estimate is 2.78 and $\sigma_0^2 = 2.25$. See also the following R code. $\square$

R Code 9.2.12. Important sampling, Example 9.4.

```
N<-100
a0<-4; b0<-12 # prior parameters
a<-6; b<-2 # trial parameters
IS<-function(N,a0,b0,MLE,a,b)
{
  theta<-rgamma(N,a,b); k<-theta^{-(a0+6)}*exp(-(b0+5*MLE)/theta);
  w<-k/dgamma(theta,a,b);
  return(sum(w*theta)/sum(w))
}
```

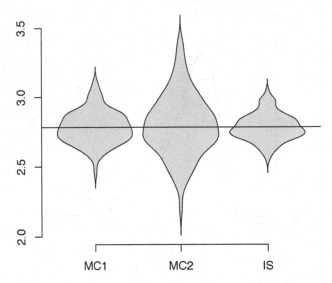

Figure 9.6: Comparison of methods in Examples 9.3 and 9.4 by `violinplots`. The straight line is the Bayes estimate.

MC approximations deliver random numbers. For comparison of methods, we have to compare distributions. For continuous distributions we recommend the R function `violinplot` which plots the density estimation upright and normalized by a standard hight. To carry out this, we repeat each method several times. In Examples 9.3 and 9.4 the importance sampling delivers better results; see the following R code and Figure 9.6.

R Code 9.2.13. Violinplot, Examples 9.3 and 9.4.

```
library(invgamma);library(UsingR)
M1<-rep(NA,200);M2<-rep(NA,200);M3<-rep(NA,200)
for (i in 1:200){
M1[i]<-MC1(N,a1,b1); M2[i]<-MC2(N,a0,b0)
M3[i]<-IS(N,a0,b0,MLE,a,b)} # run each method 200 times
Mcomb<-c(M1,M2,M3); comb<-c(rep(1,200),rep(2,200),rep(3,200))
violinplot(Mcomb~comb,col=grey(0.8),names=c("MC1","MC2","IS"))
```

9.3 Sampling from the Posterior

In this section we describe two methods for sampling from the posterior (target) with the help of a trial distribution. The methods are based on different approaches. First, we consider the *sampling importance resampling* (SIR), introduced by Rubins in 1987, which corrects sampling from the trial by weights.

This method is based on a weighted bootstrap procedure and delivers a sample, approximatively distributed like the target. As other method, we describe the *rejection algorithm*, already introduced by von Neumann in 1951. This method corrects sampling from the wrong distribution by a test. The updated values are independent and are exactly distributed like the target.

9.3.1 Sampling Importance Resampling (SIR)

The method consists of two sampling steps. The first step samples from the trial and calculate the importance weights. The second step is a weighted bootstrap step; it resamples from the first sample using the importance weights. The algorithm is given as follows.

Algorithm 9.4 Sampling Importance Resampling (SIR)

1. Draw $\vartheta^{(1)}, \ldots, \vartheta^{(m)}$ from the trial distribution $g(\cdot|\mathbf{x})$.
2. For $\pi(\theta|\mathbf{x}) \propto k(\theta|\mathbf{x})$, calculate and standardize the importance weights

$$w^{(j)} = \frac{k(\vartheta^{(j)}|\mathbf{x})}{g(\vartheta^{(j)}|\mathbf{x})}, \quad w_s^{(j)} = \frac{w^{(j)}}{\sum_{i=1}^{m} w^{(i)}}, \quad j = 1, \ldots, m.$$

3. Resample $\theta^{(1)}, \ldots, \theta^{(N)}$ from $\vartheta^{(1)}, \ldots, \vartheta^{(m)}$ with replacement, using probabilities $w_s^{(1)}, \ldots, w_s^{(m)}$.

Thus the target distribution is approximated by the discrete distribution defined on first sample with mass function equal to the importance weights. For distributional convergence we require $N/m \to 0$. We refer to Albert (2005, Section 5.10), Robert and Casella (2010, Section 3.3.2), and Givens and Hoeting (2005, Section 6.2.4). Here we continue with Examples 9.3 and 9.4.

Example 9.5 (Normal distribution and inverse-gamma prior)
The data and estimation problem are given in Example 9.3. The goal is to sample from the posterior $\mathsf{InvGamma}(a_1, b_1)$, where $a_1 = 9$ and $b_1 = 22.26$. We apply the trial distribution $\mathsf{Gamma}(6, 2)$ with density $g(\theta)$. We set $m = 1000$ and generate $\vartheta^{(1)}, \ldots, \vartheta^{(m)}$ from the trial. The importance weights are calculated as $w^{(j)} = \frac{\pi(\vartheta^{(j)}|\mathbf{x})}{g(\vartheta^{(j)})}$. Set $N = 100$ and resample $\theta^{(1)}, \ldots, \theta^{(m)}$ from $\vartheta^{(1)}, \ldots, \vartheta^{(N)}$. The distribution $\mathsf{P}_{\mathsf{sim}}$ of $\theta^{(1)}, \ldots, \theta^{(N)}$ is tested by a one–sample Kolmogorov–Smirnov test for $\mathsf{H}_0 : \mathsf{P}_{\mathsf{sim}} = \mathsf{InvGamma}(a_1, b_1)$; we get p-value $= 0.275$. Further a third sample is generated directly from $\mathsf{InvGamma}(a_1, b_1)$ and a two–sample Kolmogorov–Smirnov test is carried out to compare this third sample with $\theta^{(1)}, \ldots, \theta^{(N)}$, which gives p-value $= 0.699$.

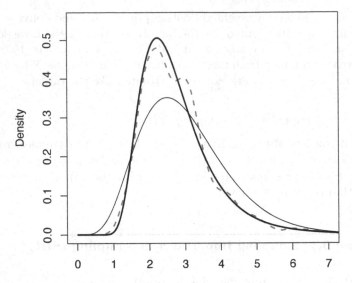

Figure 9.7: Example 9.5. The thick continuous line is the target, the thin line is the trial and the broken line is the density estimation of SIR simulated sample.

Thus we conclude that the SIR method works. See also the following R code and Figure 9.7.   □

R Code 9.3.14. SIR, Examples 9.3 and 9.4.

```
library(invgamma)
a0<-4; b0<-12; n<-10; MLE<-2.051
a1<-a0+n/2; b1<-b0+n/2*MLE #posterior parameters
a<-6; b<-2 # trial parameters
m<-1000; y<-rgamma(m,a,b) # step 1
w<-dinvgamma(y,a1,b1)/dgamma(y,a,b); ws<-w/sum(w) # weights
N<-100; yy<-sample(y,N,replace=TRUE,prob=ws)# step 2
ks.test(yy,pinvgamma(a1,b1)) # one sample ks test
z<-rinvgamma(N,a1,b1); ks.test(yy,z) # two sample ks.test
```

9.3.2   Rejection Algorithm

The rejection algorithm consists of two steps. In the first step a random value is drawn from the trial distribution; in the second step a test is carried out, which can reject this value.

The trial distribution $g(\theta|\mathbf{x})$ cannot be chosen arbitrarily. Assume that the target distribution $\pi(\theta|\mathbf{x})$ has kernel $k(\theta|\mathbf{x})$; then it is required that there exists $M(\mathbf{x})$ such that

$$k(\theta|\mathbf{x}) \leq M(\mathbf{x})g(\cdot|\mathbf{x}), \quad \text{for all } \theta \in \Theta. \tag{9.8}$$

Note that, the constant of the target distribution is not needed. The condition (9.8) implies that the trial has heavier tails than the target. We will see that the algorithm rejects less, when the trial distribution is close to the target and $M(\mathbf{x})$ is as small as possible. The rejection algorithm generates an independent sample $\theta^{(1)}, \ldots, \theta^{(j)}, \ldots, \theta^{(N)}$ from the target. It is given as following.

Algorithm 9.5 Rejection algorithm

Given a current state $\theta^{(j-1)}$.

1. Draw θ from $g(\theta|\mathbf{x})$ and compute the ratio

$$r(\theta, \mathbf{x}) = \frac{k(\theta|\mathbf{x})}{M(\mathbf{x})\,g(\theta|\mathbf{x})}.$$

2. Draw independently u from $\mathsf{U}[0,1]$

$$\theta^{(j)} = \begin{cases} \theta & \text{if } u \leq r(\theta, \mathbf{x}) \\ \text{new trial} & \text{otherwise} \end{cases}.$$

Let us explain why the rejection algorithm works. Set $\mathsf{I} = 1$ if θ is accepted and $\mathsf{I} = 0$ otherwise. An updated θ has the distribution $g(\theta|\mathbf{x}, \mathsf{I} = 1)$. For $\pi(\theta|\mathbf{x}) = c_0(\mathbf{x})k(\theta|\mathbf{x})$, we have

$$P(\mathsf{I} = 1) = \int r(\theta, \mathbf{x})g(\theta|\mathbf{x})\,d\theta = \frac{1}{M(\mathbf{x})}\int k(\theta|\mathbf{x})\,d\theta$$

$$= \frac{1}{c_0(\mathbf{x})M(\mathbf{x})}\int \pi(\theta|\mathbf{x})\,d\theta = \frac{1}{c_0(\mathbf{x})M(\mathbf{x})},$$

so that

$$g(\theta|\mathbf{x}, \mathsf{I} = 1) = \frac{P(\mathsf{I} = 1|\theta)}{P(\mathsf{I} = 1)} = r(\theta, \mathbf{x})g(\theta|\mathbf{x})c_0(\mathbf{x})M(\mathbf{x})$$

$$= \frac{k(\theta|\mathbf{x})}{g(\theta|\mathbf{x})M(\mathbf{x})}M(\mathbf{x})g(\theta|\mathbf{x})c_0(\mathbf{x})$$

$$= k(\theta|\mathbf{x})c_0(\mathbf{x}) = \pi(\theta|\mathbf{x}).$$

One criticism to rejection method is that it generates "useless" simulations when rejecting. The probability of the proposal to be accepted is

Figure 9.8: Rejection algorithm. The girl changes the fortune area on the wheel depending on the chosen ball.

$(c_0(\mathbf{x})M(\mathbf{x}))^{-1}$, that is why the choice of a small $M(x)$ fulfilling (9.8) is essential for an effective algorithm.

We refer to Zwanzig and Mahjani (2020, Section 1.3), Lui (2001, Section 2.2), Robert and Casella (2010, Section 2.3), and Givens and Hoeting (2005, Section 6.2.3). Here we continue with Examples 9.3 and 9.4.

Example 9.6 (Rejection Algorithm)

The data and estimation problem are given in Example 9.3. The goal is to sample from the posterior $\mathsf{InvGamma}(a_1, b_1)$, where $a_1 = 9$ and $b_1 = 22.26$. For comparison, we apply the same trial as in Example 9.5, $\mathsf{Gamma}(6, 2)$ with density $g(\theta)$. Condition (9.8) is fulfilled for $M = 1.6$, which is relatively high, since target and trial have different modes; see Figure 9.9. The calculations are given in the following R code, where the acceptance probability is calculated by $r(\theta, \mathbf{x}) = \frac{\pi(\theta|\mathbf{x})}{1.6\,g(\theta)}$. We generate 1000 observations by the rejection algorithm. On the right side of Figure 9.9 we can see that it is a sample from the target $\mathsf{InvGamma}(a_1, b_1)$ and not from the trial. $\square$

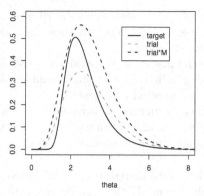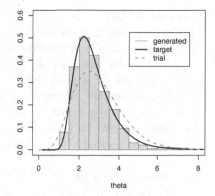

Figure 9.9: Example 9.6. The trial distribution is Gamma(6, 2). The target is InvGamma(9, 22.26). Left: The target is covered with $M = 1.6$, which is large since trial and target have different modes. Right: The histogram of the random numbers generated by the rejection algorithm and the density of the target and trial. The generated sample belongs to the target.

R Code 9.3.15. Rejection Algorithm, Example 9.6.

```
library(invgamma)
n<-10; MLE<-2.051922 # sample size, sufficient statistic
a1<-a0+n/2; b1<-b0+MLE*n/2 # posterior parameters
rand.rej<-function(N,M)
{rand<-rep(0,N)
 for( i in 1:N)
  {
 L<-TRUE
 while(L){
    rand[i]<-rgamma(1,6,2)
    r<-dinvgamma(rand[i],a1,b1)/(dgamma(rand[i],6,2)*M)
    if(runif(1)<r){L<-FALSE}
           }
  }
return(rand)
}
R<-rand.rej(1000,1.6)
```

9.4 Markov Chain Monte Carlo (MCMC)

Generating an i.i.d sample from the posterior can be an inefficient task; especially when the parameter is high dimensional. The way out is to generate a sequence where each member is based on the previous member, i.e., to generate a Markov chain with posterior $\pi(\theta|\mathbf{x})$ as stationary distribution. In this section we explain the main principle of two types of these methods: *Metropolis–Hastings algorithm* and *Gibbs sampling*.

The Metropolis–Hastings algorithm has the posterior as target and generates new proposals by a trial distribution depending on previous member. The simulation from the wrong distribution is corrected by a test.

The Gibbs sampler explores the hierarchical structure of Bayes models, where the posterior can be factorized by conditional distributions. The new value is generated recursively conditioned on the previous member. A test is not needed.

The common principle of MCMC is briefly formulated as follows: instead of the law of large numbers for i.i.d. samples we apply the ergodic properties of Markov chains.

We consider the problem of determining

$$\mu(\mathbf{x}) = \int h(\theta, \mathbf{x})\pi(\theta|\mathbf{x})\, d\theta. \tag{9.9}$$

The general principle of Markov Chain Monte Carlo Methods can be formulated as follows.

Algorithm 9.6 General MCMC

Start with any configuration $\theta^{(0)}$.

1. Sample a Markov Chain
$$\theta^{(1)}, \ldots, \theta^{(N)},$$

   with stationary distribution $\pi(\theta|\mathbf{x})$ and with actual transition probability

$$A(\vartheta, \theta) = \mathsf{P}\left(\theta^{(j)} = \theta \mid \theta^{(j-1)} = \vartheta\right),$$

   such that
$$\pi(\theta|\mathbf{x}) = \int A(\vartheta, \theta)\pi(\vartheta|\mathbf{x})\, d\vartheta. \tag{9.10}$$

2. After a burn-in period of length m approximate $\mu(\mathbf{x})$ by

$$\widehat{\mu}(\mathbf{x}) = \frac{1}{N-m} \sum_{j=m+1}^{N} h(\theta^{(j)}, \mathbf{x}).$$

Let us explain, why MCMC works; for more details see Lui (2001, p. 249 and pp. 269). Suppose the Markov chain $\theta^{(1)}, \ldots, \theta^{(N)}$ is irreducible and aperiodic. Then for any initial distribution of $\theta^{(0)}$ it holds that $\widehat{\mu}(\mathbf{x}) \xrightarrow{\mathsf{P}} \mu(\mathbf{x})$ for $N \to \infty$

and furthermore

$$\frac{\sqrt{N}\,(\widehat{\mu}(\mathbf{x}) - \mu(\mathbf{x}))}{\sigma_h(\mathbf{x})} \to \mathsf{N}(0,1), \quad \text{where } \sigma_h^2(\mathbf{x}) = \sigma^2(\mathbf{x})(1 + 2\sum_{j=1}^{\infty} \rho_j(\mathbf{x})),$$

with

$$\sigma^2(\mathbf{x}) = \mathsf{Var}(h(\theta^{(1)}, \mathbf{x})), \quad \rho_j(\mathbf{x})\sigma^2(\mathbf{x}) = \mathsf{Cov}(h(\theta^{(1)}, \mathbf{x}), h(\theta^{(j+1)}, \mathbf{x})).$$

Thus

$$\widehat{\mu}(\mathbf{x}) = \mu(\mathbf{x}) + \frac{1}{\sqrt{N}}\sigma_h(\mathbf{x})O_\mathsf{P}(1). \tag{9.11}$$

The approximation in (9.11) is better for smaller $\sigma_h(\mathbf{x})$. We search for a Markov chain which can be easily generated by using the previous members, but has still sufficiently low dependence structure.

9.4.1 Metropolis–Hastings Algorithms

Note that, in literature the name MCMC is often used for Metropolis–Hastings algorithms only. The goal is to generate a Markov chain with stationary distribution $\pi(\theta|\mathbf{x})$. This problem is solved by the *Metropolis algorithm*, first published in Metropolis et al. (1953), later generalized by Hastings (1970).

The *Metropolis–Hastings algorithm* is a trial and error procedure. The trial step generates a proposal ϑ from $T(\theta^{(j-1)}, \vartheta) = p(\vartheta|\theta^{(j-1)})$. The error step carries out a test, to decide whether ϑ fits in the Markov chain or not. In case it fits, the new member of the chain is the proposal, $\theta^{(j)} = \vartheta$; otherwise the chain gets stuck and we set $\theta^{(j)} = \theta^{(j-1)}$.

Set $\pi(\theta|\mathbf{x}) = c_0(\mathbf{x})k(\theta|\mathbf{x})$. The $c_0(\mathbf{x})$ is not needed in the algorithm; the test depends on the trial distribution and on the kernel $k(\theta|\mathbf{x})$ only.

The main steps of Metropolis–Hastings algorithm are given as follows.

Algorithm 9.7 Metropolis–Hastings
Given current state $\theta^{(j-1)}$.

1. Draw ϑ from $T(\theta^{(j-1)}, \cdot)$.

2. Calculate the *Metropolis–Hastings ratio*

$$\mathsf{R}(\theta^{(j-1)}, \vartheta) = \frac{k(\vartheta|\mathbf{x})}{k(\theta^{(j-1)}|\mathbf{x})} \frac{T(\vartheta, \theta^{(j-1)})}{T(\theta^{(j-1)}, \vartheta)}.$$

3. Generate u from $\mathsf{U}[0,1]$. Update

$$\theta^{(j)} = \begin{cases} \vartheta & \text{if } u \le \min(1, \mathsf{R}(\theta^{(j-1)}, \vartheta)) \\ \theta^{(j-1)} & \text{otherwise} \end{cases}.$$

Let us explain, why this algorithm produces a Markov chain with $A(\vartheta, \vartheta')$, such that (9.10) holds. Note that, the actual transition function is not equivalent to the trial distribution. It is sufficient for (9.10) to show the *balance equality*, i.e.,

$$\pi(\vartheta|\mathbf{x})A(\vartheta, \vartheta') = \pi(\vartheta'|\mathbf{x})A(\vartheta', \vartheta), \qquad (9.12)$$

since $\int A(\vartheta', \vartheta)d\vartheta = 1$ and

$$\int \pi(\vartheta|\mathbf{x})A(\vartheta, \vartheta')\,d\vartheta = \int \pi(\vartheta'|\mathbf{x})A(\vartheta', \vartheta)d\vartheta$$
$$= \pi(\vartheta'|\mathbf{x})\int A(\vartheta', \vartheta)d\vartheta$$
$$= \pi(\vartheta'|\mathbf{x}).$$

It remains to show (9.12). We assume that the trial function gives always a new proposal: $T(\vartheta, \vartheta) = 0$. That implies that the chain only gets stuck when the proposal is rejected. Let $\delta_\vartheta(\vartheta')$ be the Dirac mass, i.e., $\delta_\vartheta(\vartheta') = 1$ for $\vartheta = \vartheta'$ and $\delta_\vartheta(\vartheta') = 0$ otherwise. The proposal is accepted with probability $r(\vartheta, \vartheta') = \min(1, \mathsf{R}(\vartheta, \vartheta'))$. Thus

$$
\begin{aligned}
A(\vartheta, \vartheta') &= \mathsf{P}(\text{``coming from } \vartheta \text{ to } \vartheta'''\text{''}), \\
&= \mathsf{P}(\text{``coming from } \vartheta \text{ to } \vartheta'''\text{''})(1 - \delta_\vartheta(\vartheta')) \\
&\quad + \mathsf{P}(\text{``coming from } \vartheta \text{ to } \vartheta'''\text{''})\delta_\vartheta(\vartheta') \\
&= \mathsf{P}(\text{``}\vartheta' \text{ is proposed and accepted''})(1 - \delta_\vartheta(\vartheta')) \\
&\quad + \mathsf{P}(\text{``all rejected proposals''})\delta_\vartheta(\vartheta').
\end{aligned}
$$

The proposal is independent of the test, such that

$$
\begin{aligned}
\mathsf{P}(\text{``}\vartheta' \text{ is proposed and accepted''}) &= \mathsf{P}(\text{``}\vartheta' \text{ is proposed''})\mathsf{P}(\text{``}\vartheta' \text{ is accepted''}) \\
&= T(\vartheta, \vartheta')r(\vartheta, \vartheta').
\end{aligned}
$$

Using $\int T(\vartheta, \vartheta')d\vartheta' = 1$ and setting $r(\vartheta) = \int T(\vartheta, \vartheta')r(\vartheta, \vartheta')\,d\vartheta'$, we get

$$\mathsf{P}(\text{``all not accepted proposals''}) = \int T(\vartheta, \vartheta')(1 - r(\vartheta, \vartheta'))\,d\vartheta' = 1 - r(\vartheta).$$

Thus

$$A(\vartheta, \vartheta') = T(\vartheta, \vartheta')r(\vartheta, \vartheta')(1 - \delta_\vartheta(\vartheta')) + (1 - r(\vartheta))\delta_\vartheta(\vartheta').$$

Since $T(\vartheta, \vartheta')\delta_\vartheta(\vartheta') = T(\vartheta, \vartheta) = 0$, we have

$$A(\vartheta, \vartheta') = T(\vartheta, \vartheta')r(\vartheta, \vartheta') + (1 - r(\vartheta))\delta_\vartheta(\vartheta') = A_1(\vartheta, \vartheta') + A_2(\vartheta, \vartheta').$$

We get the symmetry in ϑ' and ϑ, since

$$
\begin{aligned}
\pi(\vartheta|\mathbf{x})A_1(\vartheta,\vartheta') &= \pi(\vartheta|\mathbf{x})T(\vartheta,\vartheta')r(\vartheta,\vartheta') \\
&= \pi(\vartheta|\mathbf{x})T(\vartheta,\vartheta')\min(1,\frac{\pi(\vartheta'|\mathbf{x})}{\pi(\vartheta|\mathbf{x})}\frac{T(\vartheta',\vartheta)}{T(\vartheta,\vartheta')}) \\
&= \min(\pi(\vartheta|\mathbf{x})T(\vartheta,\vartheta'),\pi(\vartheta'|\mathbf{x})T(\vartheta',\vartheta)) \\
&= \pi(\vartheta'|\mathbf{x})A_1(\vartheta',\vartheta)
\end{aligned}
$$

and

$$
\pi(\vartheta|\mathbf{x})A_2(\vartheta,\vartheta') = \pi(\vartheta|\mathbf{x})(1-r(\vartheta))\delta_\vartheta(\vartheta') = \pi(\vartheta'|\mathbf{x})A_2(\vartheta',\vartheta).
$$

Hence (9.12) holds.

The Markov chains have a "burn-in" time m, the waiting time until the equilibrium is reached. To determine m, it is useful to run the algorithm with different starting values and determine the time when the chains reach the same area; see Figure 9.11.

The key point is to find a good proposal distribution. A good balance is needed between low correlation between the consecutive members and an easy way of calculating the new member from the previous one, because high correlation can give a bad approximation; see (9.11). Also, often stuck up chains are not useful; see Figure 9.12.

There are two methods which can be helpful for solving this conflict: *annealing* and *thinning*. An annealing procedure works as follows. We carry out Metropolis–Hastings algorithms with several trial distributions and select the procedure with best properties; for instance let the variances of the trial stepwise increase. The thinning method takes a subsequence of the generated chain, then the correlation between members of the new chain is less.

There exists a huge literature on MCMC; we recommend Zwanzig and Mahjani (2020, Section 1.3), Albert (2005, Chapter 6), Chen et al. (2002, Chapter 2), Lui (2001, Chapter 5), Robert and Casella (2010, Chapter 6), Givens and Hoeting (2005, Section 7.1), and the references therein. The number of methods is rapidly increasing. Generally, our advice is:

> Be careful in implementing an MCMC procedure!

Here we present only the algorithm, where the proposal distribution is a random walk. The previous member is disturbed by a random variable with symmetric distribution around zero. The step length from one member of the chain to the following is regulated by the variance of the disturbance; see Figure 9.12.

Algorithm 9.8   Random-walk Metropolis
Given current state $\theta^{(j-1)}$.

1. Draw ε from g_σ, where g_σ is symmetric around zero and σ is a scaling parameter. Set

$$\vartheta = \theta^{(j-1)} + \varepsilon.$$

2. Draw u from $\mathsf{U}[0,1]$. Update

$$\theta^{(j)} = \begin{cases} \vartheta & \text{if}\quad u \leq \min(1, \frac{k(\vartheta|\mathbf{x})}{k(\theta^{(j-1)}|\mathbf{x})}) \\ \theta^{(j-1)} & \text{otherwise} \end{cases}.$$

The random-walk algorithm is a special case of the Metropolis–Hastings algorithm. For instance set $\varepsilon \sim \mathsf{N}_p(0, \sigma^2\mathbf{I}_p)$, then $\vartheta|\theta^{(j-1)} \sim \mathsf{N}(\theta^{(j-1)}, \sigma^2\mathbf{I}_p)$ and $\theta^{(j-1)}|\vartheta \sim \mathsf{N}(\vartheta, \sigma^2\mathbf{I}_p)$. Both trials have the same density, $T(\vartheta, \vartheta') = T(\vartheta', \vartheta)$, thus the trial distribution is reduced in the Metropolis–Hastings ratio. Note that, the original Metropolis algorithm requires the symmetry of $T(\vartheta, \vartheta')$; Hastings impact was the generalization to non-symmetrical trials, hence the random-walk method above is a Metropolis algorithm.

The following example is just an illustration of a random-walk Metropolis algorithm with the uniform disturbance distribution $\mathsf{U}[-d, d] \times \mathsf{U}[-d, d]$ in a simple linear regression with normally distributed errors, known error variance and Cauchy priors for the intercept and slope.

Example 9.7 (Random-Walk metropolis)
Consider a simple linear regression $y_i = \theta_1 + x_i\theta_2 + \varepsilon_i$ with $\varepsilon_i \sim \mathsf{N}(0, 0.5)$ and $n = 8$. The design points x_i, $i = 1, \ldots, 8$, are equidistant between 0 and 1.4. The data points are plotted in Figure 9.10. We have $\theta = (\theta_1, \theta_2)$. As Cauchy priors we set $\theta_1 \sim \mathsf{C}(0, 1.5)$ and $\theta_2 \sim \mathsf{C}(1, 2)$ independently, plotted in Figure 9.10. The trial distribution is a random walk. Suppose the current state is $\theta^{(j-1)} = (\theta_1^{(j-1)}, \theta_2^{(j-1)})$. Then the proposal $\vartheta = (\vartheta_1, \vartheta_2)$ is generated by

$$\vartheta_1 = \theta_1^{(j-1)} + u_1 \quad \text{and} \quad \vartheta_2 = \theta_2^{(j-1)} + u_2$$

where u_1 and u_2 are independently $\mathsf{U}[-d, d]$ distributed. The tuning parameter is d; see Figure 9.12 for different choices. The Metropolis–Hastings ratio is

$$\mathsf{R}(\vartheta, \theta^{(j-1)}) = \min(1, \mathsf{LRT}(\vartheta, \theta^{(j-1)})\mathsf{PRT}(\vartheta, \theta^{(j-1)})),$$

where LRT is the likelihood ratio and PRT is the prior ratio, here the quotient of Cauchy densities

$$\mathsf{LRT}(\vartheta, \theta) = \frac{\ell(\vartheta|\mathbf{y})}{\ell(\theta|\mathbf{y})}, \quad \mathsf{PRT}(\vartheta, \theta) = \frac{\pi(\vartheta)}{\pi(\theta)}.$$

We see in Figure 9.11 that after $m = 120$ the equilibrium is reached. We take $d = 0.8$ and $N = 1000$, and obtain for the Bayes estimates

$$\widehat{\theta}_{1,\text{Bayes}} = 0.726 \text{ and } \widehat{\theta}_{2,\text{Bayes}} = 2.363.$$

The least squares estimates are $\widehat{\theta}_{1,\text{lse}} = 0.338$ and $\widehat{\theta}_{2,\text{lse}} = 2.969$. The true values are $\theta = (1, 2)$. Observe, that the sample size $n = 8$ is quite small. The following R code gives the calculations. □

R Code 9.4.16. Random-Walk Metropolis, Example 9.7.

```
MCMC<-function(d,seed1,seed2,N)
{
 rand1<-rep(0,N); rand2<-rep(0,N)
 LRT<-rep(1,N); PRT<-rep(1,N); R<-rep(1,N)
 rand1[1]<-seed1; rand2[1]<-seed2
for ( i in 2:N)
    {
    rand1[i]<-rand1[i-1]+runif(1,-d,d)# proposal intercept
    rand2[i]<-rand2[i-1]+rnorm(1,-d,d)# proposal slope
    A<-prod(dnorm(Y,rand1[i]+rand2[i]*x,sqrt(0.5)))# likelihood
    B<-prod(dnorm(Y,rand1[i-1]+rand2[i-1]*x,sqrt(0.5)))
    LRT[i]<-A/B # new/old # likelihood ratio
    p1<-dcauchy(rand1[i],1,2)*dcauchy(rand1[i],0,1.5)
    p2<-dcauchy(rand1[i-1],1,2)*dcauchy(rand1[i],0,1.5)
    PRT[i]<-p1/p2 # new/old # prior ratio
    R[i]<- LRT[i]*PRT[i] # Metropolis ratio
    r<-min(1,R[i])
    u<-runif(1)
    if (u<r){rand1[i]<-rand1[i]}else{rand1[i]<-rand1[i-1]}
    if (u<r){rand2[i]<-rand2[i]}else{rand2[i]<-rand2[i-1]}
    }
return(data.frame(rand1,rand2,LRT,PRT,R))
}
MM<-MCMC(0.8,0,0,1000) # carry out MCMC
 mean(MM$rand1[121:1000]) # Bayes estimate intercept
 mean(MM$rand2[121:1000]) # Bayes estimate slope
```

9.4.2 Gibbs Sampling

Example 2.19 on page 25 illustrates hierarchical Bayes modelling. The authors in Dupuis (1995) apply *Gibbs sampling*.
We explain the general principle. Suppose that the parameter can be written as

$$\theta = (\theta_1, \ldots, \theta_p),$$

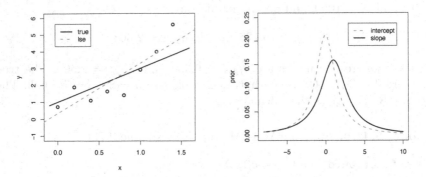

Figure 9.10: Example 9.7. Left: Generated data points and the true regression line. The broken line is the least squares fit. Right: The prior distribution for the intercept is a Cauchy distribution with location parameter 0, and scale parameter 1.5; the prior for the slope is a Cauchy distribution with location parameter 1 and scale parameter 2.

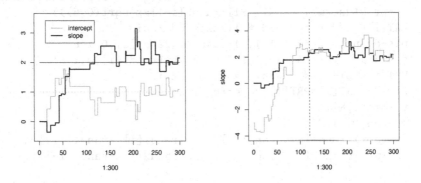

Figure 9.11: Example 9.7. Left: The generated Markov chains with $d = 0.8$ and seed $= (0.0)$. Right: Different start values for studying the burn-in time.

where the θ_i's are either one- or multidimensional. Moreover, suppose that we can simulate the corresponding conditional posterior distributions $\pi_1, \ldots, \pi_p$; that is we can simulate

$$\theta_i \mid (\mathbf{x}, \theta_{[-i]}) \sim \pi_i \left(. \mid \mathbf{x}, \theta_{[-i]}\right), \quad \text{where } \theta_{[-i]} = (\theta_1, \ldots, \theta_{i-1}, \theta_{i+1}, \ldots, \theta_p).$$

Thus Gibbs sampling reduces the generation of high-dimensional values to low-dimensional ones. The densities $\pi_1, \ldots, \pi_p$ are called full conditionals - only they are used for simulation.

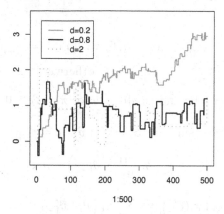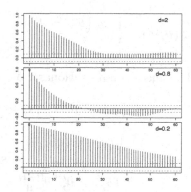

Figure 9.12: Example 9.7. Right: The generated Markov chain for the intercept with $d = 0.2$ which is too small and with $d = 2$ which is too large. Left: The autocorrelation functions for different turning parameter d.

First we consider the random-scan algorithm, where in every step one component is randomly chosen and updated, while the other components remain unchanged.

Algorithm 9.9 Random-Scan Gibbs Sampler
Given current state $\theta^{(t)} = (\theta_i^{(t)})_{i=1,\dots,p}$.

1. Randomly select a coordinate i from $1, \dots, p$ according to the probability distribution $p_1, \dots, p_p$ on $1, \dots, p$.

2. Draw $\theta_i^{(t+1)}$ from $\pi_i\left(. \mid \mathbf{x}, \theta_{[-i]}^{(t)}\right)$.

3. Set $\theta_{[-i]}^{(t+1)} = \theta_{[-i]}^{(t)}$.

Alternatively, we can also change the components one after the other.

Algorithm 9.10 Systematic-Scan Gibbs Sampler
Given current state $\theta^{(t)} = (\theta_i^{(t)})_{i=1,\dots,p}$. For $i = 1, \dots, p$, draw $\theta_i^{(t+1)}$ from

$$\pi_i\left(. \mid \mathbf{x}, \theta_1^{(t+1)}, \dots, \theta_{i-1}^{(t+1)}, \theta_{i+1}^{(t)}, \dots, \theta_p^{(t)}\right).$$

Let us explain why the Gibbs sampler produces a "right" Markov chain. Let $i = i_t$ be the index chosen at step t. Then the actual transition probability from ϑ to ϑ' at step t is

$$
A_t(\vartheta, \vartheta') = \begin{cases} \pi_i\left(\theta_i^{(t)} = \vartheta_i' \mid \theta_{[-i]}^{(t-1)} = \vartheta_{[-i]}\right) & \text{for} \quad \vartheta, \vartheta', \text{ with } \vartheta_{[-i]} = \vartheta_{[-i]}' \\ 0 & \text{otherwise} \end{cases}.
$$

The chain is non-stationary, but aperiodic and irreducible for $\pi(\theta|\mathbf{x}) > 0$ for all θ.

Moreover $\sum_\vartheta A_t(\vartheta, \vartheta')\pi(\vartheta|\mathbf{x}) = \pi(\vartheta'|\mathbf{x})$, since for $S_i = \{\vartheta : \vartheta_{[-i]} = \vartheta_{[-i]}'\}$

$$
\sum_\vartheta A_t(\vartheta, \vartheta')\pi(\vartheta|\mathbf{x}) = \sum_{\vartheta \in S_i} \pi_i\left(\theta_i^{(t)} = \vartheta_i' \mid \theta_{[-i]}^{(t-1)} = \vartheta_{[-i]}\right)\pi(\vartheta|\mathbf{x})
$$

$$
= \pi_i\left(\theta_i^{(t)} = \vartheta_i' \mid \theta_{[-i]}^{(t-1)} = \vartheta_{[-i]}'\right)\pi_i(\vartheta_{[-i]}'|\mathbf{x}) \sum_{\vartheta \in S_i} \pi_i\left(\vartheta_i \mid \vartheta_{[-i]}'\right)
$$

$$
= \pi_i\left(\theta_i^{(t)} = \vartheta_i' \mid \theta_{[-i]}^{(t-1)} = \vartheta_{[-i]}'\right)\pi_i(\vartheta_{[-i]}'|\mathbf{x}) = \pi(\vartheta'|\mathbf{x}).
$$

For more details see Zwanzig and Mahjani (2020, Section 2.2.5), Albert (2005, Chapter 10), Chen et al. (2002, Section 2.1), Lui (2001, Chapter 6), Robert and Casella (2010, Chapter 7), Givens and Hoeting (2005, Section 7.2), and the references therein.

Here we give two examples for illustration. The first example explores the simulated data from Example 9.7.

Example 9.8 (Systematic-Scan Gibbs sampler)

The data is given in Example 9.7 and plotted in Figure 9.10. In contrast to Example 9.7 we suppose that the variance is unknown, such that $\theta = (\theta_1, \theta_2, \theta_3)$, where the first component is the intercept, the second the slope, and the third the variance. We assume the conjugate prior, $\theta \sim \mathsf{NIG}(a_0, b_0, \beta_0, \Sigma_0)$ with $a_0 = 2$, $b_0 = 2$ $\beta_0 = (0, 4)^T$, and $\Sigma_0 = \text{diag}(1, 2)$. Applying Theorem 6.5 on page 150, and using the R code (see below) we obtain the posterior, $\theta|\mathbf{y} \sim \mathsf{NIG}(a_1, b_1, \beta_1, \Sigma_1)$, with $a_1 = 10$, $b_1 = 6.995$ and

$$
\beta_1 = \begin{bmatrix} 0.128 \\ 3.246 \end{bmatrix}, \quad \Sigma_1 = \begin{bmatrix} 0.259 & -0.237 \\ -0.237 & 0.382 \end{bmatrix}.
$$

The goal is to generate $\theta^{(1)}, \ldots, \theta^{(N)}$ with approximatively $\theta^{(j)} \sim \mathsf{NIG}(a_1, b_1, \beta_1, \Sigma_1)$ by using Gibbs sampling. The complete conditionals follow as

$$
\pi(\theta|\mathbf{y}) = \pi_1((\theta_1, \theta_2)|\mathbf{y}, \theta_3)\pi_2(\theta_3|\mathbf{y}) = \pi_{11}(\theta_1|\mathbf{y}, \theta_2, \theta_3)\,\pi_{12}(\theta_2|\mathbf{y}, \theta_3)\pi_2(\theta_3|\mathbf{y}).
$$

It holds for bivariate normal distribution, $(\theta_1, \theta_2)|\mathbf{y}, \theta_3 \sim \mathsf{N}_2(\beta_1, \theta_3\Sigma_1)$, that

$$
\theta_1|\mathbf{y}, \theta_2, \theta_3 \sim \mathsf{N}(\beta_{1,1} + \frac{\sigma_{12}}{\sigma_{22}}(\theta_2 - \beta_{2,1}), \theta_3(\sigma_{11} - \frac{\sigma_{12}^2}{\sigma_{22}})), \ \theta_2|\mathbf{y}, \theta_3 \sim \mathsf{N}(\beta_{2,1}, \sigma_{22})
$$

where $\beta_1 = (\beta_{1,1}, \beta_{2,1})$ and $\Sigma_1 = (\sigma_{ij})_{i=1,2;j=1,2}$. Further $\theta_3|\mathbf{y} \sim$ InvGamma$(a_1/2, b_1/2)$. Here $\pi_{11}(\theta_1|\mathbf{y}, \theta_2, \theta_3)$ is $N(2.147-0.622\theta_2, \theta_3\,0.111)$ and $\pi_{12}(\theta_2|\mathbf{y}, \theta_3)$ is $N(3.246, \theta_3\,0.382)$. Then Gibbs sampling is carried out as follows:

1. Generate $\theta_3^{(j)}$ from InvGamma$(5, 3.498)$.

2. Generate $\theta_2^{(j)}$ from $N(3.246, \theta_3^{(j)}\,0.382)$.

3. Generate $\theta_1^{(j)}$ from $N(2.147 - 0.622\theta_2^{(j)}, \theta_3^{(j)}\,0.111)$.

After 1000 runs we take the average and obtain as approximation for the Bayes estimates $\widehat{\alpha} = 0.129$, $\widehat{\beta} = 3.25$ and $\widehat{\sigma}^2 = 0.839$. The Bayes estimates in this set up are $\widehat{\alpha}_{\text{Bayes}} = 0.128$, $\widehat{\beta}_{\text{Bayes}} = 3.247$, $\widehat{\sigma}^2_{\text{Bayes}} = 0.874$. $\qquad\square$

R Code 9.4.17. Systematic-Scan Gibbs Sampler, Example 9.8.

```
xx2<-seq(0,1.5,0.2); n<-length(xx2) # data
Y<-c(0.727, 1.898, 1.110, 1.648, 1.420, 2.919, 3.993, 5.613)
xx1<-rep(1,n); X11<-sum(xx1*xx1); X12<-sum(xx2*x1)
X22<-sum(xx2*xx2); XX<-matrix(c(X11,X12,X12,X22),ncol=2)
S0<-matrix(c(1,0,0,2),ncol=2); B0<-c(0,4); a0<-2; b0<-2 #prior
SS<-solve(XX+solve(S0)) # posterior
bb<-as.numeric(coef(M))# LSE
B1<-SS%*%(XX%*%bb+solve(S0)%*%B0) # Bayes estimator
a1<-a0+n # posterior
r<-Y-B1[1]*xx1-xx2*B1[2] # Bayes residuum
b1<-b0+sum(r^2)+t(B1-B0)%*%solve(S0)%*%(B1-B0)# posterior
b1/(a1-2)# Bayes variance estimate
### Gibbs sampler
v11<-SS[1,1]-SS[1,2]*SS[1,2]/SS[2,2]
aa1<-B1[1]-SS[1,2]/SS[2,2]*B1[2]
bb1<-SS[1,2]/SS[2,2]
library(invgamma)
gibbs<-function(N)
{
aa<-rep(0,N)# intercept
bb<-rep(0,N)# slope
ss<-rep(1,N) # variance
for( i in 1:N)
{
 ss[i]<-rinvgamma(1,a1/2,b1/2)
 bb[i]<-rnorm(1,B1[2],sqrt(ss[i]*SS[2,2]))
 s<-ss[i]*v11; aa[i]<-rnorm(1,aa1+bb1*bb[i],sqrt(s))
}
return(data.frame(aa,bb,ss))
}
G<-gibbs(1000) # Carry out the sampler
```

```
mean(G$aa[100:1000]) # estimate intercept
mean(G$bb[100:1000]) # estimate slope
mean(G$ss[100:1000]) # estimate variance
```

The next example gives an idea for *data augmentation*. We are interested in generating random numbers from the beta-binomial distribution, $X \sim$ BetaBin(n, α, β). We apply a Bayes model and obtain the beta-binomial distribution as marginal distribution of X from (X, θ).

Example 9.9 (Beta-binomial distribution)
Consider $X|\theta \sim$ Bin(n, θ) and $\theta \sim$ Beta(α, β), then $\theta|x \sim$ Beta$(\alpha+x, \beta+n-x)$; see Example 2.11 on page 16. Also $X \sim$ BetaBin(α, β); see Example 3.6 on page 35.

Here we generate the joint distribution of (X, θ) using the conditional distribution of $\theta|X$ and of $X|\theta$. The marginal distribution of θ is the prior distribution Beta(α, β), while the marginal distribution of X is a beta-binomial distribution with

$$P(X = k) = \binom{n}{k} \frac{B(\alpha + k, \beta + n - k)}{B(\alpha, \beta)}.$$

Algorithm: Given the current state $(x^{(t)}, \theta^{(t)})$.

1. Draw $\theta^{(t+1)}$ from Beta$(\alpha + x^{(t)}, n - x^{(t)} + \beta)$.

2. Draw $x^{(t+1)}$ from Bin$(n, \theta^{(t+1)})$.

Using R code 9.4.18, $N = 5000$ values are simulated and compared with the true distributions for $\alpha = \beta = 0.5$ in Figure 9.13.   □

R Code 9.4.18. Gibbs sampler of beta-binomial distribution, Example 9.9

```
gibbs.beta<-function(a,b,n,N)
{
  theta=rep(0,N);x=rep(0,N);
  theta[1]<-rbeta(1,a,b);
  x[1]<-rbinom(1,n,theta[1]);
  for(i in 2:N){
      theta[i]<-rbeta(1,a+x[i-1],b+n-x[i-1]);
      x[i]<-rbinom(1,n,theta[i])
      }
return(data.frame(x,theta))
}
```

9.5   Approximative Bayesian Computation (ABC)

The aim of ABC is to generate an i.i.d. sample $\theta^{(1)}, \ldots, \theta^{(N)}$ from the posterior distribution $\pi(\theta|\mathbf{x})$. The ABC methods are recommended when the likelihood

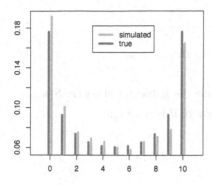 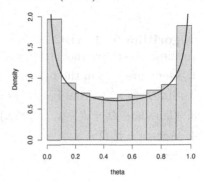

Figure 9.13: Example 9.9. The values are generated by the Gibbs sampler, using R code 18. Left: BetaBin$(10, 0.5, 0.5)$ Right: Beta$(0.5, 0.5)$

is intractable, i.e., there is no closed form likelihood expression or it is computationally expensive. Theoretically it is possible to apply ABC in all situations where we are able to generate new data for a given parameter.

With ABC we have a very general tool which allows Bayesian calculations for almost all combinations of likelihood and prior. Nevertheless we have to

- check that the used model is reasonable and

- test that the algorithm is well-implemented.

Both items are not easy to fulfill. We explain here only the main underlying idea and give two illustrative examples.

The main idea is intuitive and convincing.

1. Generate θ_{new} from the prior $\pi(\cdot)$.

2. Simulate new data $\mathbf{x}_{new}$ from $p(\cdot|\theta_{new})$.

3. Compare the new data $\mathbf{x}_{new}$ with the observed data $\mathbf{x}$. If there is "no difference", accept the parameter θ_{new}; otherwise reject and go back to Step 1.

The accepted θ's are independent random variables and have approximately the distribution $\pi(\theta|\mathbf{x})$. The crucial point is the test of "no difference". The acceptance rate of ABC can be very low.

Assume that $S = (S_1, S_2, ..., S_p)$ is a sufficient statistic in model $\mathcal{P} = \{\mathsf{P}_\theta : \theta \in \Theta\}$, where P_θ is the data generating distribution.

Algorithm 9.11 ABC

Assume the Bayes model $\{\mathcal{P}, \pi\}$. Given the data $\mathbf{x}$.

1. Generate θ from the prior π.

2. Simulate $\mathbf{x}_{\text{new}}$ from P_θ and compute the sufficient statistic $S(\mathbf{x}_{\text{new}})$.

3. Calculate $D = d(S(\mathbf{x}_{\text{new}}), S(\mathbf{x}))$, where d is a metric.

4. Accept θ if $D \leq q$.

5. Return to 1.

The problem is to determine the distance $d(S(\mathbf{x}_{\text{new}}), S(\mathbf{x}))$ and the threshold q. For small threshold q the rejection rate is high and the algorithm is slow. On the other hand for large thresholds the generated sample follows the prior more than the posterior; see Figure 9.14.

In the following we present an example for illustration. It is the same set up as in Example 9.3, on page 246, with data $\mathbf{x}$ are given in R code 11. This example has the advantage that we can compare the ABC simulated sample with the known posterior distribution.

Example 9.10 (ABC) The data $\mathbf{x}$ consists of $n = 10$ i.i.d. observations from $\mathsf{N}(0, \sigma^2)$. The parameter of interest is $\theta = \sigma^2$. A sufficient statistic is $S(\mathbf{x}) = \sum_{i=1}^{n} x_i^2 = 2.05$. The prior is $\mathsf{InvGamma}(\alpha_0, \beta_0)$ with $\alpha_0 = 4$ and $\beta_0 = 12$. The posterior distribution is $\mathsf{InvGamma}(\alpha_1, \beta_1)$ with $\alpha_1 = 9$ and $\beta_1 = 22.25$; see Example 9.3.

1. Generate θ from $\mathsf{InvGamma}(4, 12)$.

2. Generate an i.i.d. sample $\mathbf{z} = (z_1, \ldots, z_{10})$ from $\mathsf{N}(0, \theta)$.

3. Calculate $S(\mathbf{z})$ and $D = |S(\mathbf{z}) - S(\mathbf{x})|$

4. For $D < q$ we accept θ; otherwise we go back to Step 1.

The following R code gives the procedure. The results for $N = 1000$ and different thresholds q are presented in Figure 9.14.   □

R Code 9.5.19. ABC, Example 9.10.

```
x<-c(-0.06,2.43,1.55,1.76,1.27,0.94,0.48,1.16,1.32,1.81)# data
S<-sum(x*x) # maximum likelihood estimate, sufficient statistic
library(invgamma)
a0<-4; b0<-12 # prior parameters
a1<-a0+n/2; b1<-b0+S/2 # posterior parameters
## ABC ##
```

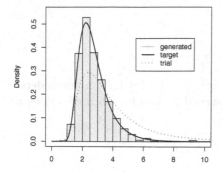 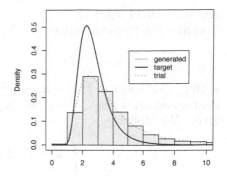

Figure 9.14: Example 9.10. The histogram of the random numbers generated by the ABC algorithm. The continuous line is the target, the posterior density InvGamma(9, 22.25); the dotted line is the trial, the prior InvGamma(4, 12). Left: For the threshold $q = 0.1$ the generated sample is approximatively distributed like the target. Right: For the large threshold $q = 10$ the generated sample is approximatively distributed like the trial.

```
ABC.dist<-function(N,q,S)
{
rand<-rep(0,N)
for( i in 1:N)
  {
L<-TRUE
while(L){
   rand[i]<-rinvgamma(1,a0,b0) # draw theta
   xx<-rnorm(n,0,sqrt(rand[i])) # new data
   SS<-sum(xx*xx) # new sufficient statistics
   D<-abs(SS-S) # metric
   if(D<q){L<-FALSE}
       }
  }
return(rand)
}
C<-ABC.dist(1000,0.1,2.05) # carry out ABC
# Figure
hist(C,freq=FALSE,xlab="",ylim=c(0,0.6),main="",nclass=14)
xx<-seq(0.01,15,0.1); lines(xx,dinvgamma(xx,a1,b1),lty=2,lwd=2)
box(lty=1,col=1)
```

The following example is an illustration of the ABC method when the likelihood function is not given in a closed form and a data generating algorithm is used.

Example 9.11 (ABC)

Consider the following latent model

$$y_i = f(x_i, \theta) + \varepsilon_i, \; i = 1, \ldots, n$$

with $\theta = (a, b, c)$ and $f(x) = ax + bx^2 + cx^3$, and ε_i, $i = 1, \ldots, n$ are independent standard normal. The y_i are unobserved and only truncated values are given. We observe

$$z_i = \begin{cases} x_i + 1 & \text{if } y_i > x_i + 1, \\ y_i & \text{if } x_i - 1 \leq y_i \leq x_i + 1 \\ x_i - 1 & \text{if } y_i < x_i - 1, \end{cases}$$

The z_i have no standard probability distribution, but for given θ we can generate new data points. In Step 3 of the algorithm we use the squared distance between the observed data $\mathbf{z} = (z_1, \ldots, z_n)$ and the proposed data points $\mathbf{z} = (z_1', \ldots, z_n')$, i.e.,

$$d(\mathbf{z}, \mathbf{z}') = \sum_{i=1}^{n} (z_i - z_i')^2.$$

The priors are $a \sim N(-1, 1)$, $b \sim N(1, 1)$, and $c \sim N(0, 1)$, mutually independent. The fitted Bayes curve $\tilde{f}(x)$ is calculated from $\theta^{(1)}, \ldots, \theta^{(N)}$ by

$$\tilde{\theta} = \frac{1}{N} \sum_{j=1}^{N} \theta^{(j)} = (\tilde{a}, \tilde{b}, \tilde{c}), \quad \tilde{f}(x) = \tilde{a}x + \tilde{b}x^2 + \tilde{c}x^3.$$

The left side of Figure 9.15 shows the true curve and the non-truncated points. On the right side, data points together with filter region are plotted. In Figure 9.16 the true and a fitted Bayes curves are compared. The algorithm with $q = 3$ delivers a good fit but the choice $q = 13$ gives bad fit.   □

R Code 9.5.20. ABC, Example 9.11

```
# data generation procedure
Data<-function(a,b,c,){
xx<-seq(0,5,0.5); ff<-a*xx+b*xx^2+c*xx^3 # regression curve
ynew<-ff+rnorm(n,0,1)# new data
for( i in 1:n){if(ynew[i]>xxx[i]+1){ynew[i]<-xxx[i]+1}}
for( i in 1:n){if(ynew[i]<xxx[i]-1){ynew[i]<-xxx[i]-1}}
return(ynew)
}
# ABC algorithm
ABC<-function(N,q){
```

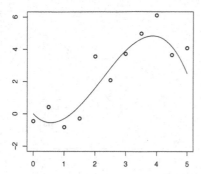
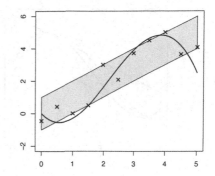

Figure 9.15: Example 9.11. Left: The latent model. The observations are simulated with $a = -2$, $b = 2$ and $c = 0.3$. Right: The observed model. Observations inside the gray area are unchanged; otherwise the point on the borderline is taken.

```
randa<-rep(0,N); randb<-rep(0,N);randc<-rep(0,N)
for( i in 1:N)
  {
   L<-TRUE
    while(L){
      a<-rnorm(1,-1,1) # priors
      b<-rnorm(1,1,1); c<-rnorm(1,0,1)
      D<-Data(a,b,c)
      Diff<-sum((yobs-D)*(yobs-D))
      if(Diff<q){randa[i]<-a;randb[i]<-b;randc[i]<-c;L<-FALSE}
      }
  }
rand<-data.frame(randa,randb,randc)
return(rand)
}
AA<-ABC(100,3) # carry out ABC
ma<-mean(AA$randa); mb<-mean(AA$randb) # Bayes estimates
mc<-mean(AA$randc)
```

There are many different ABC methods. We refer to Beaumont (2019) and the references therein.

We mention only one more. The combination of MCMC and ABC avoids generation from the prior; instead MCMC steps are included. Note that, this algorithm produces a Markov chain so that the elements in the sequence are no longer independent.

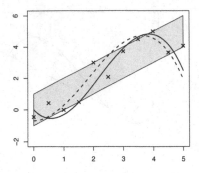 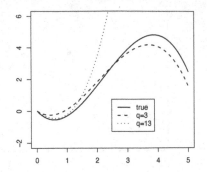

Figure 9.16: Example 9.11 with $N = 100$. Left: The broken line is the Bayes curve, calculated with $q = 3$. Right: The ABC procedure delivers a reasonable fit for $q = 3$ and a bad fit for $q = 13$.

Algorithm 9.12 ABC–MCMC

Given current parameter $\theta^{(j-1)}$.

1. Generate ϑ from the trial distribution $T(\theta^{(j-1)}, \cdot)$.

2. Simulate $\mathbf{x}'$ from $p(\cdot | \vartheta)$.

3. Compare $\mathbf{x}'$ with $\mathbf{x}$. In case of a good fit continue, otherwise go back to Step 1.

4. Calculate the Metropolis–Hastings ratio

$$R(\theta^{(j-1)}, \vartheta) = \frac{\pi(\vartheta) T(\vartheta, \theta^{(j-1)})}{\pi(\theta^{(j-1)}) T(\theta^{(j-1)}, \vartheta)}.$$

5. Generate u from $\mathrm{U}[0, 1]$. Update

$$\theta^{(j)} = \begin{cases} \vartheta & \text{if } u \leq \min\left(1, R(\theta^{(j-1)}, \vartheta)\right) \\ \theta^{(j-1)} & \text{otherwise} \end{cases}$$

9.6 Variational Inference (VI)

Given a Bayes model $\{\mathcal{P}, \pi\}$, the family of posterior distributions is determined, by $\pi(\theta | \mathbf{x}) \propto \pi(\theta) \ell(\theta | \mathbf{x})$. In case this posterior family is difficult to handle, it can be useful to approximate it by another family. In this context, a model $\mathcal{D}$ for the posterior is posed and the best approximation of $\pi(\theta | \mathbf{x})$ in $\mathcal{D}$ is determined. This is the idea of variational Bayes inference.

It is especially applicable when the likelihood does not belong to an exponential family, where we have the same family for prior and posterior and only need to update the parameter; see Section 3.2, page 38.

The Kullback–Leibler divergence $K(Q|P)$ measures how much Q deviates from P. It is defined by

$$K(Q|P) = \int q(\theta) \ln \left(\frac{q(\theta)}{p(\theta)} \right) d\theta, \qquad (9.13)$$

where $K(Q|P) \geq 0$, with $=$ only for $Q = P$. Using the Kullback–Leibler divergence as measure for proximity we want $K(Q|P)$ to be small. We call $q^*(\theta|\mathbf{x})$ *variational density* iff

$$q^*(\theta|\mathbf{x}) = \arg \min_{q \in \mathcal{D}} K(q(\theta|\mathbf{x})|\pi(\theta|\mathbf{x})). \qquad (9.14)$$

The model $\mathcal{D}$ is named *variational family*. The idea is to use $q^*(\theta|\mathbf{x})$ instead of $\pi(\theta|\mathbf{x})$ and to explore properties of $\mathcal{D}$. The challenge is to find a compromise between a simple model $\mathcal{D}$, which gives a bad approximation and a complex model $\mathcal{D}$ with high quality approximation; otherwise, for complex models it is hard to find $q^*(\theta|\mathbf{x})$.

We refer to Lee (2022) and Bishop (2006) for more details. Here we explain only the main approach.

Set $\Theta = \Theta_1 \times \Theta_j \times \Theta_m \subset \mathbb{R}^p$. We focus on the *mean-field variational* family, $\mathcal{D}_{\mathrm{MF}}$, which requires that the components of θ are independent:

$$q(\theta|\mathbf{x}) = \prod_{j=1}^{m} q_j(\theta_j|\mathbf{x}), \quad \theta_j \in \Theta_j, \qquad (9.15)$$

where $q_j(\theta_j|\mathbf{x})$ is called the jth *variational factor*.

We discuss the criterion (9.14) and give an illustrative example where we can compute both $q^*(\theta|\mathbf{x})$ and $\pi(\theta|\mathbf{x})$.

First, we reformulate criterion (9.14). Note that, $p(\theta, \mathbf{x}) = p(\mathbf{x})\pi(\theta|\mathbf{x})$, such that

$$
\begin{aligned}
K(q(\theta|\mathbf{x})|\pi(\theta|\mathbf{x})) &= \int q(\theta|\mathbf{x}) \ln \left(\frac{q(\theta|\mathbf{x})}{\pi(\theta|\mathbf{x})} \right) d\theta \\
&= \int q(\theta|\mathbf{x}) \ln \left(\frac{q(\theta|\mathbf{x})p(\mathbf{x})}{p(\theta, \mathbf{x})} \right) d\theta \\
&= \int q(\theta|\mathbf{x}) \ln(p(\mathbf{x})) \, d\theta + \int q(\theta|\mathbf{x}) \ln \left(\frac{q(\theta|\mathbf{x})}{p(\theta, \mathbf{x})} \right) d\theta \\
&= \ln(p(\mathbf{x})) - \mathrm{ELBO}(q)
\end{aligned}
$$

where

$$\mathrm{ELBO}(q) = \int q(\theta|\mathbf{x}) \ln(p(\theta, \mathbf{x})) \, d\theta - \int q(\theta|\mathbf{x}) \ln(q(\theta|\mathbf{x})) \, d\theta.$$

Since $K(q(\theta|\mathbf{x})|\pi(\theta|\mathbf{x}) \geq 0$ it holds that

$$\mathsf{ELBO}(q) \leq \ln(p(\mathbf{x})).$$

This explains the notation ELBO, it stands for *evidence lower bound*. Minimization of the Kullback–Leibler divergence means maximization of ELBO. Let us illustrate the concept of variational inference for a linear model with normal errors and known error variance.

Example 9.12 (Variational inference in linear Bayes model)
We assume

$$\mathbf{y}|\beta \sim \mathsf{N}_n(\mathbf{X}\beta, \Sigma) \quad \text{and} \quad \beta \sim \mathsf{N}_p(\gamma_0, \Gamma_0).$$

Applying Corollary 6.1 on page 135, we have the posterior $\beta|\mathbf{y} \sim \mathsf{N}_p(\gamma_1, \Gamma_1)$ with

$$\gamma_1 = \Gamma_1(\mathbf{X}^T \Sigma^{-1} \mathbf{y} + \Gamma_0^{-1}\gamma),$$
$$\Gamma_1 = (\Gamma_0^{-1} + \mathbf{X}^T \Sigma^{-1} \mathbf{X})^{-1}.$$

Assume for the posterior the following mean-field variational family,

$$\mathcal{D} = \{\mathsf{N}_p(\gamma, \Gamma) : \gamma \in \mathbb{R}^p, \Gamma = \mathrm{diag}(\lambda_1, \ldots, \lambda_p); \lambda_i > 0\}.$$

The Kullback–Leibler divergence for multivariate normal distributions is

$$K(\mathsf{N}_p(\gamma, \Gamma)|\mathsf{N}_p(\gamma_1, \Gamma_1))$$
$$= \frac{1}{2}\left(\mathrm{tr}(\Gamma_1^{-1}\Gamma) - p + (\gamma_1 - \gamma)^T \Gamma_1^{-1}(\gamma_1 - \gamma) + \ln(\frac{\det(\Gamma_1)}{\det(\Gamma)})\right). \quad (9.16)$$

For $\gamma_1 = \gamma$ we have

$$K(\mathsf{N}_p(\gamma, \Gamma)|\mathsf{N}_p(\gamma, \Gamma_1)) = \frac{1}{2}\left(\mathrm{tr}(\Gamma_1^{-1}\Gamma) - p + \ln(\frac{\det(\Gamma_1)}{\det(\Gamma)})\right).$$

Furthermore, for $\Gamma_1 = \Gamma$, $\mathrm{tr}(\Gamma_1^{-1}\Gamma) = \mathrm{tr}(\mathbf{I}_p) = p$ and $\det(\Gamma_1) = \det(\Gamma)$, such that $K(\mathsf{N}_p(\gamma, \Gamma)|\mathsf{N}_p(\gamma, \Gamma)) = 0$. Thus for orthogonal design $\mathbf{X}^T\mathbf{X} = \mathbf{I}_p$ and for $\Sigma = \mathbf{I}_n$ we have $\min_{\gamma,\Gamma} K(\mathsf{N}_p(\gamma, \Gamma)|\mathsf{N}_p(\gamma_1, \Gamma_1)) = 0$. The posterior belongs to $\mathcal{D}$ and $q^*(\beta|\mathbf{y}) = \pi(\beta|\mathbf{y})$. In the general case we choose $\gamma = \gamma_1$ but we still have to solve

$$\min_{\{\lambda_1, \ldots, \lambda_p; \lambda_i > 0\}} \left(\mathrm{tr}(\Gamma_1^{-1}\mathrm{diag}(\lambda_1, \ldots, \lambda_p)) - \ln(\prod_{i=1}^{p}\lambda_i)\right). \quad (9.17)$$

Denote the diagonal elements of $\Gamma_0^{-1} + \mathbf{X}^T \Sigma^{-1} \mathbf{X}$ by a_{ii}. Then

$$\min_{\{\lambda_1, \ldots, \lambda_p; \lambda_i > 0\}} \left(\sum_{i=1}^{p} a_{ii}\lambda_i - \sum_{i=1}^{p} \ln(\lambda_i)\right) = \sum_{i=1}^{p} \min_{\lambda_i}(a_{ii}\lambda_i - \ln(\lambda_i))$$
$$= \sum_{i=1}^{p}(a_{ii}\lambda_i^* - \ln(\lambda_i^*)),$$

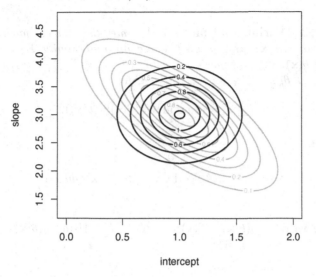

Figure 9.17: Example 9.12. The grey contours are from the posterior $N(\gamma_1, \Gamma_1)$, the black contours belong to the variational distribution $N(\gamma_1, \Gamma^*)$. Observe that, the components of the variational distribution are independent, but have different variances.

where $\lambda_i^* = a_{ii}^{-1}$. We get $N(\gamma_1, \Gamma^*)$ with $\Gamma^* = \mathrm{diag}(\lambda_1^*, \ldots, \lambda_p^*)$ as variational distribution $q^*(\theta|\mathbf{x})$.

Consider simple linear regression with 11 equidistant design points $0, 0.1, 0.2, \ldots, 1$, independent standard normal errors, and prior $N(\mathbf{0}, \mathbf{I}_2)$. Then (see Figure 9.17)

$$\Gamma_1^{-1} = \begin{pmatrix} 12 & 5.5 \\ 5.5 & 4.85 \end{pmatrix}, \quad \Gamma^* = \begin{pmatrix} 0.083 & 0 \\ 0 & 0.206 \end{pmatrix}.$$

$\square$

The following result on partial optimization gives the theoretical background for Algorithm 9.13, which approximates the variational distribution $q^*(\theta|\mathbf{x})$ by stepwise conditioning.

Theorem 9.2 (Variational factor) *Assume the Bayes model* $\{\mathcal{P}, \pi\}$ *with posterior* $\pi(\theta|\mathbf{x})$ *and mean-field variational family* $\mathcal{D}_{MF}$. *Set, for* $q \in \mathcal{D}_{MF}$, $q(\theta|\mathbf{x}) = q_k(\theta_k|\mathbf{x})q(\theta_{[-k]})$ *with* $q(\theta_{[-k]}) = \prod_{j=1, j \neq k}^{m} q_j(\theta_j|\mathbf{x})$ *and* $\theta_{[-k]} = (\theta_1, \ldots, \theta_{k-1}, \theta_{k+1}, \ldots, \theta_m)$. *Then, for*

$$q_k^*(\theta_k|\mathbf{x}) = \arg\min_{q_k} \mathsf{K}(q(\theta|\mathbf{x})|\pi(\theta|\mathbf{x})),$$

it holds that

$$q_k^*(\theta_k|\mathbf{x}) \propto \exp\left(\int q(\theta_{[-k]}) \ln(\pi(\theta_k|\theta_{[-k]}, \mathbf{x})) \, d\theta_{[-k]} \right). \qquad (9.18)$$

PROOF: We write $\mathsf{K}(q(\theta|\mathbf{x})|\pi(\theta|\mathbf{x})) = K_1 - K_2$. Using $q(\theta|\mathbf{x}) = q_k(\theta_k|\mathbf{x})$ $q(\theta_{[-k]})$,

$$K_1 = \int q(\theta|\mathbf{x}) \ln(q(\theta|\mathbf{x})) \, d\theta$$

$$= \int q_k(\theta_k|\mathbf{x})q(\theta_{[-k]}) \ln(q_k(\theta_k|\mathbf{x})q(\theta_{[-k]})) \, d\theta$$

$$= \int q_k(\theta_k|\mathbf{x})q(\theta_{[-k]}) \ln(q_k(\theta_k|\mathbf{x})) \, d\theta + \int q_k(\theta_k|\mathbf{x})q(\theta_{[-k]}) \ln(q(\theta_{[-k]})) \, d\theta$$

$$= \int q_k(\theta_k|\mathbf{x}) \ln(q_k(\theta_k|\mathbf{x})) \, d\theta_k + \int q(\theta_{[-k]}) \ln(q(\theta_{[-k]})) \, d\theta_{[-k]}$$

$$= \int q_k(\theta_k|\mathbf{x}) \ln(q_k(\theta_k|\mathbf{x})) \, d\theta_k + \mathrm{const}_{K_1}$$

Using additionally $\pi(\theta|\mathbf{x}) = \pi(\theta_k|\theta_{[-k]}, \mathbf{x}) \, \pi(\theta_{[-k]}|\mathbf{x})$ and setting

$$\ln(v_k(\theta_k)) = \int q(\theta_{[-k]}) \ln(\pi(\theta_k|\theta_{[-k]}, \mathbf{x})) d\theta_{[-k]}, \quad c_v = \int v_k(\theta_k) d\theta_k,$$

we have

$$K_2 = \int q_k(\theta_k|\mathbf{x})q(\theta_{[-k]}) \ln(\pi(\theta|\mathbf{x})) \, d\theta$$

$$= \int q_k(\theta_k|\mathbf{x})q(\theta_{[-k]}) \ln(\pi(\theta_k|\theta_{[-k]}, \mathbf{x})) d\theta + \int q_k(\theta_k|\mathbf{x})q(\theta_{[-k]}) \ln(\pi(\theta_{[-k]}|\mathbf{x})) d\theta$$

$$= \int q_k(\theta_k|\mathbf{x}) \ln(v_k(\theta_k)) \, d\theta_k + \int q(\theta_{[-k]}) \ln(\pi(\theta_{[-k]}|\mathbf{x})) \, d\theta_{[-k]}$$

$$= \int q_k(\theta_k|\mathbf{x}) \ln(\frac{v_k(\theta_k)}{c_v}) \, d\theta_k + \mathrm{const}_{K_2}$$

Summarizing

$$\mathsf{K}(q(\theta|\mathbf{x})|\pi(\theta|\mathbf{x})) = \mathsf{K}(q_k(\theta_k|\mathbf{x})|\frac{v_k(\theta_k)}{c_v}) + \mathrm{const}.$$

Since $K(Q|P) = 0$ for $Q = P$, the minimizer is $\frac{v_k(\theta_k)}{c_v}$.

$\square$

Applying Theorem 9.2 we formulate the *coordinate ascent variational inference* (CAVI) algorithm.

Algorithm 9.13 CAVI
Given the current state $\widehat{q}(\theta|\mathbf{x})^{(t)} = \prod_{j=1}^{m} \widehat{q}_j(\theta_j|\mathbf{x})^{(t)}$. For $k = 1, \ldots, m$, calculate

$$\widehat{q}_k(\theta_k|\mathbf{x})^{(t+1)} \propto \exp\left(\int q_{[-j]}(\theta|\mathbf{x}) \ln(\pi(\theta_k|\theta_{[-k]}, \mathbf{x})) \, d\theta_{[-k]} \right)$$

where

$$q_{[-j]}(\theta|\mathbf{x}) = \prod_{j=1}^{k-1} \widehat{q}_j(\theta_j|\mathbf{x})^{(t+1)} \prod_{j=k+1}^{m} \widehat{q}_j(\theta_j|\mathbf{x})^{(t)}.$$

We specify the CAVI algorithm for an i.i.d sample $x_1, \ldots, x_n$ from $N(\mu, \sigma^2)$.

Example 9.13 (CAVI algorithm)
We apply the set up of Example 7.3 on page 185, where we have $\theta = (\mu, \sigma^2)$, the prior $NIG(\alpha_0, \beta_0, \mu_0, \sigma_0^2)$, and the posterior $NIG(\alpha_1, \beta_1, \mu_1, \sigma_1^2)$, given in (7.1), with

$$\sigma_1^{-2} = \sigma_0^{-2} + n, \quad \mu_1 = \sigma_1^2(n\bar{x} + \sigma_0^{-2}\mu_0),$$

$$\alpha_1 = \alpha_0 + n, \quad \beta_1 = \beta_0 + (n-1)s^2 + \frac{1}{\sigma_0^2}\mu_0^2 + n\bar{x}^2 - \frac{1}{\sigma_1^2}\mu_1^2.$$

$InvGamma(\alpha_1, \beta_1, \mu_1, \sigma_1^2)$ implies the factorization

$$\mu|\sigma^2, \mathbf{x} \sim N(\mu_1, \sigma^2 \sigma_1^2), \quad \sigma^2|\mathbf{x} \sim InvGamma(\frac{\alpha_1}{2}, \frac{\beta_1}{2}).$$

Applying Lemma 6.5 on page 149, we get $\pi(\mu|\mathbf{x})$. The distribution of $\pi(\sigma^2|\mu, \mathbf{x})$ is obtained by following arguments. Define $z_i = x_i - \mu$. Then $z_1, \ldots, z_n$ are i.i.d. from $N(0, \sigma^2)$ with prior $\sigma^2 \sim InvGamma(\frac{\alpha_0}{2}, \frac{\beta_0}{2})$. In Example 2.14, on page 19, we derived $\sigma^2|\mathbf{z} \sim InvGamma(\frac{\alpha_1}{2}, \frac{\beta_1(\mu)}{2})$, with $\beta_1(\mu) = \sum_{i=1}^{n} z_i^2 + \beta_0$. Thus we have the factorization

$$\mu|\mathbf{x} \sim t(\alpha_1, \mu_1, \frac{\alpha_1}{\beta_1}\sigma_1^2), \quad \sigma^2|\mu, \mathbf{x} \sim InvGamma(\frac{\alpha_1}{2}, \frac{\beta_1(\mu)}{2}),$$

where $\beta_1(\mu) = \sum_{i=1}^{n}(x_i - \mu)^2 + \beta_0$. Applying Theorem 9.2, we obtain the algorithm: Given the current state $q(\mu|\mathbf{x})^{(t)} q(\sigma^2|\mathbf{x})^{(t)}$ calculate

$$q(\mu|\mathbf{x})^{(t+1)} \propto \exp\left(\int q(\sigma^2|\mathbf{x})^{(t)} \ln(\pi(\mu|\sigma^2, \mathbf{x})) \, d\sigma^2 \right)$$

$$q(\sigma^2|\mathbf{x})^{(t+1)} \propto \exp\left(\int q(\mu|\mathbf{x})^{(t+1)} \ln(\pi(\sigma^2|\mu, \mathbf{x}))\, d\mu\right).$$

Especially a, we have with $\sigma^2(t) = (\int q(\sigma^2|\mathbf{x})^{(t)}(\sigma^2)^{-1}d\sigma^2)^{-1}$, that $q(\mu|\mathbf{x})^{(t+1)}$ is $\mathsf{N}(\mu_1, \sigma_1^2\sigma^2(t))$. Further, for

$$\beta_1(t) = \int q(\mu|\mathbf{x})^{(t)}\beta_1(\mu)\, d\mu = \sum_{i=1}^{n}(x_i - \mu_1)^2 + n\sigma_1^2\sigma^2(t) + \beta_0,$$

we have that $q(\sigma^2|\mathbf{x})^{(t)}$ is $\mathsf{InvGamma}(\frac{\alpha_1}{2}, \frac{\beta_1(t)}{2})$. It holds because

$$\int q(\sigma^2|\mathbf{x})^{(t)} \ln(\pi(\mu|\sigma^2, \mathbf{x}))\, d\sigma^2 = \int q(\sigma^2|\mathbf{x})^{(t)}(-\frac{1}{2\sigma^2\sigma_1^2}(\mu - \mu_1)^2)\, d\sigma^2 + \text{const}$$

and

$$\int q(\mu|\mathbf{x})^{(t)} \ln(\pi(\sigma|\mu, \mathbf{x}))\, d\mu = \int q(\mu|\mathbf{x})^{(t)}(\frac{\beta_1(\mu)}{\sigma^2})\, d\mu - (\alpha_1 + 1)\ln(\sigma^2) + \text{const}.$$

$\square$

9.7 List of Problems

1. Consider the stroke data given in the following table. We assume that the treatment and control samples are independent.

	success	size
treatment	119	11037
control	98	11034

Let p_1 be the probability of success (to get a stroke) under the treatment and p_2 the probability of success in the control group. We are interested in $\theta = \frac{p_1}{p_2}$, especially in testing $H_0 : \theta \leq 1$.

 (a) Which distribution model is valid?
 (b) Give the Jeffreys prior for p_1 and p_2.
 (c) Give the least favourable prior for p_1 and p_2.
 (d) Derive the posterior distributions for p_1 and p_2 in (b) and (c).
 (e) Give an R code for generating a sample of size N from the posterior distributions of θ.
 (f) In Figure 9.18 the density estimators of the posterior distributions of θ are plotted. Can we reject $H_0 : \theta \leq 1$?

2. Here are R codes for three different methods for calculating the same integral. Method 1

density.default(x = odds.ratio)

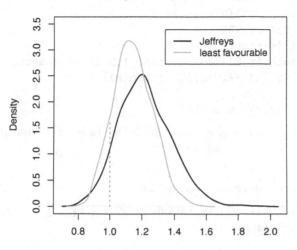

Figure 9.18: Problem 1(f).

```
x1<-runif(10000);
sum(dbeta(x1,1,20)*dbinom(15,1000,x1))/10000
```
Method 2
```
x2<-rbeta(10000,1,20); sum(dbinom(15,1000,x2))/10000
```
Method 3
```
f1<-function(x){dbinom(15,1000,x)*dbeta(x,1,20)}
integrate(f1,0,1)
```
Following results are obtained:

Method	1	2	3
Result	0.01228	0.01397	0.01476

(a) Which integral is calculated?

(b) Give the name of each method.

(c) Give the steps of the algorithm of Method 2.

(d) Why are the results so different? Which result would you prefer the most, and why?

(e) Compare the accuracies of Methods 1 and 2.

3. Consider the following R code:

```
rand<-function(N,M){R<-rep(NA,N);
for(i in 1:N){ L<-TRUE;
while(L){R[i]<-rcauchy(1,C);
r<-prod(dcauchy(x,R[i]))*dnorm(R[i],2,1)/(M*dcauchy(R[i],C));
if(runif(1)<r){L<-FALSE}}}; return(R)}
M<-5; C<-3
R1<-rand(10000,M); mean(R1)
```

(a) Which Bayesian model is considered? Determine $\pi(\theta)$ and $p(x|\theta)$.

(b) Which method is carried out?

(c) What is generated?

(d) Give the main steps of the algorithm.

4. Consider n independent observations $(\mathbf{x}, \mathbf{y}) = \{(x_1, y_1), \ldots, (x_n, y_n)\}$ from a logistic regression model

$$P(y_i = 1|\theta) = \frac{\exp(\theta x_i)}{1 + \exp(\theta x_i)}, \quad i = 1, \ldots, n.$$

The prior $\pi(\theta)$ is a Cauchy distribution. Given the data set $(\mathbf{x}, \mathbf{y})$:

x	0.2	0.6	0.8	1	1.5	2	2.3	3	3.2	3.5
y	1	0	0	1	0	1	0	1	1	1

In order to carry out a test the user is interested in the integral

$$\mu = \int_{\theta < 0.2} \pi(\theta|\mathbf{x}, \mathbf{y}) \, d\theta.$$

(a) Apply a random-walk Metropolis–Hastings algorithm to calculate μ. Consider the walk

$$\vartheta = \theta^j + a \times u, \quad u \sim \mathsf{U}[-1, 1].$$

Specify $\pi(\theta)$, $\ell(\theta|\mathbf{x}, \mathbf{y})$ and the trial distribution $T(\theta^j, \theta)$.

(b) Write an R code for this MCMC random-walk procedure.

(c) Make a reasonable choice for a and for the burn-time.

(d) Carry out the procedure and estimate the integral.

5. We are interested in a Gibbs sampler from

$$(X, Y) \sim \mathsf{N}_2 \left(0, \begin{pmatrix} 1 & \rho \\ \rho & 1 \end{pmatrix} \right). \tag{9.19}$$

(a) Give the main steps of a systematic-scan Gibbs sampler.

(b) Give an R code.

(c) Is (9.19) the stationary distribution of the generated Markov chain?

6. Consider the following R code:

```
library(MASS); library(tmvtnorm); library(truncnorm)
data(iris); # Iris data set,
V1<-iris[1:50,1] # length of sepal of Setosa
V2<-iris[1:50,2] # width of sepal of Setosa
data<-data.frame(V1,V2);
m_1<-mean(V1);m_2<-mean(V2);
abc<-function(N,tol)
{
rand1<-rep(NA,N);rand2<-rep(NA,N);rand3<-rep(NA,N);
rand4<-rep(NA,N);rand5<-rep(NA,N);
for(i in 1:N)
  { L<-TRUE;
    while(L)
    {rand1[i]<-rtruncnorm(1,0,Inf,3,1);
     rand2[i]<-rtruncnorm(1,0,Inf,2,1);
     rand3[i]<-runif(1,0.02,2); rand4[i]<-runif(1,0.02,2);
     rand5[i]<-runif(1,0,0.9);
     S11<-rand3[i]^2; S12<-rand5[i]*rand3[i]*rand4[i];
     S22>-rand4[i]^2;
     S<-matrix(c(S11,S12,S12,S22),2,2);
     a<-c(0,0); b<-c(Inf,Inf);
     Z<-rtmvnorm(50,c(rand1[i],rand2[i]),S,a,b);
     D<-(mean(Z[,1])-m1)^2+(mean(Z[,2])-m2)^2;
     if(D<tol){L<-FALSE}
      }
   };
 D<-data.frame(rand1,rand2,rand3,rand4,rand5);return(D)
 }
```

(a) Formulate the parametric statistical model used for generating the data.

(b) Which priors are applied?

(c) What random sample is generated?

(d) Give the main steps of the algorithm.

Chapter 10

Solutions

This chapter provides solutions to the problems listed at the end of each chapter. In some cases, main arguments or sketch of solution are given.

10.1 Solutions for Chapter 2

1. *Students exam results*: The posterior probabilities $\pi(\theta|x)$ are given as follows.

x	D	C	B	A
$\theta = 0$	2/5	1/34	1/28	0
$\theta = 1$	2/5	30/34	12/28	1/3
$\theta = 2$	0	3/34	15/28	2/3

2. *Sufficient statistic*: To show that $\pi(\theta|x) = \pi(\theta|T(x))$ iff $T(x)$ is a sufficient statistic. It holds that $T(x)$ is a sufficient statistic iff the factorization criterion holds; see Liero and Zwanzig (2011, page 54), i.e., $p(x|\theta) = g(T(x)|\theta)h(x)$. We get

$$\pi(\theta|x) \propto p(x|\theta)\pi(\theta)$$
$$\propto g(T(x)|\theta)h(x)\pi(\theta)$$
$$\propto g(T(x)|\theta)\pi(\theta) \propto \pi(\theta|T(x)).$$

3. *Find a prior*: It should hold a.s. for all $x \in \mathcal{X}$ and all $\theta \in \Theta$, that the prior is independent of x, $\pi(\theta) \propto \frac{\pi(\theta|x)}{\ell(\theta|x)}$. Here $\theta = n$ and

$$\ell(\theta, x) \propto \binom{\theta}{x}(\frac{1}{2})^\theta \text{ and } \pi(\theta|x) \propto \binom{\theta+x+1}{x}(\frac{1}{2})^{\theta+x}.$$

It holds that

$$\frac{\pi(\theta|x)}{\ell(\theta|x)} \propto \frac{\binom{\theta+x+1}{x}}{\binom{\theta}{x}}.$$

This quotient is not independent of x. Set $\theta = 1$. Especially, for $x = 0$ the quotient equals 1, but for $x = 1$ the quotient is 3. There exists no prior such that the posterior of $\theta = n$ is negative binomial.

DOI: 10.1201/9781003221623-10

4. *Posterior for precision parameter*: In Example 2.13, set $n = 4$ and $\alpha = \beta = 2$. We obtain

$$\tau | \mathbf{x} \sim \text{Gamma}(4, 2 + \frac{1}{2} \sum_{i=1}^{4} x_i^2).$$

5. *Two alternative models*:

(a) Posterior in model $\mathcal{M}_0$: $p | (X = 15) \sim \text{Beta}(16, 1005)$; see Example 2.11. Posterior in model $\mathcal{M}_1$: $\lambda | (X = 15) \sim \text{Gamma}(35, 2)$, since

$$\pi(\lambda | X = k) \propto \lambda^{\alpha-1} \exp(-\lambda\beta) \frac{\lambda^k}{k!} \exp(-\lambda)$$

$$\propto \lambda^{\alpha+k-1} \exp(-\lambda(\beta + 1)).$$

(b) Following the law of rare events the difference between $\mathsf{P}_0 = \text{Bin}(1000, 0.05)$ and $\mathsf{P}_1 = \text{Poi}(20)$ is relatively small, since we set $\mathsf{E}p = \frac{1}{20} = 0.05$ in the first model and $\mathsf{E}\lambda = 20$ in the second. The first two prior moments in model $\mathcal{M}_0$ are $\mathsf{E}p = \frac{1}{20} = 0.05$, $\text{Var}p = 0.002$ the posterior moments are $\mathsf{E}(p|X = 15) = 0.016$, $\text{Var}(p|X = 15) \approx 0$. The first two prior moments in model $\mathcal{M}_1$ are $\mathsf{E}\lambda = 20$, $\text{Var}\lambda = 20$ the posterior moments are $\mathsf{E}(\lambda|X = 15) = 17.5$, $\text{Var}(\lambda|X = 15) = 8.75$. Have in mind that the parameter of interest is $\theta = p$ in $\mathcal{M}_0$ and $\theta = \lambda/1000$ in $\mathcal{M}_1$ the posterior expectations of θ are 0.016 and 0.017. The posterior variances of θ are very small in both models.

(c) In both models the prior variances of θ are very small, hence there is a risk of dominating priors. A clear recommendation is not possible.

6. *Linear regression*: We calculate the posterior

$$\pi(\beta|\mathbf{y}$$
$$\propto \pi(\beta)\ell(\beta|\mathbf{y})$$
$$\propto \exp(-\frac{1}{2} \sum_{k=0}^{2}(\beta_k - 1)^2) \exp(-2 \sum_{i=1}^{n}(y_i - \beta_0 - \beta_1 x_i - \beta_2 z_i)^2)$$
$$\propto \exp(-\frac{1}{2} \sum_{k=0}^{2}(\beta_k - 1)^2) - 2n\left((\beta_0 - \bar{y})^2 + (\beta_1 - \overline{xy})^2 + (\beta_2 - \overline{zy})^2\right))$$

with $\bar{y} = \frac{1}{n}\sum_{i=1}^{n} y_i$, $\overline{xy} = \frac{1}{n}\sum_{i=1}^{n} x_i y_i$, $\overline{zy} = \frac{1}{n}\sum_{i=1}^{n} z_i y_i$. Completing the squares

$$c(x - a)^2 + (x - b)^2 = (c+1)(x - \frac{ac+b}{c+1})^2 - \frac{(ac+b)^2}{c+1} + ca^2 + b^2,$$

we obtain

$$\pi(\beta|\mathbf{y}) \propto \exp(-\frac{4n+1}{2}\left((\beta_0 - b_0)^2 + (\beta_1 - b_1)^2 + (\beta_2 - b_2)^2\right))$$

with

$$b_0 = \frac{4n\bar{y}+1}{4n+1}, \quad b_1 = \frac{4n\overline{xy}+1}{4n+1}, \quad b_2 = \frac{4n\overline{zy}+1}{4n+1}.$$

Thus posteriors are independent normal distributions:

$$\beta_0|\mathbf{y} \sim \mathsf{N}(b_0, \frac{1}{4n+1}), \quad \beta_1|\mathbf{y} \sim \mathsf{N}(b_1, \frac{1}{4n+1}), \quad \beta_2|\mathbf{y} \sim \mathsf{N}(b_2, \frac{1}{4n+1}).$$

The independence is a consequence of $\sum_{i=1}^{n} x_i = 0$, $\sum_{i=1}^{n} z_i = 0$ and $\sum_{i=1}^{n} x_i z_i = 0$.

7. *Hierarchical model*:

(a) The joint distribution of (θ, μ) is normal, thus the marginal distribution of θ is $\mathsf{N}_n(\mathsf{E}(\theta), \mathsf{Cov}(\theta))$ with $\mathsf{E}(\theta) = \mathsf{E}\,(\mathsf{E}(\theta|\mu)) = 0$. Set $\mathbf{1}_n = (1, \ldots, 1)^T$, for an $n \times 1$ vector of 1s. Then $\mathsf{E}\theta = \mu\mathbf{1}$. Applying (2.5),

$$\mathsf{Cov}(\theta) = \mathsf{E}_\mu(\mathsf{Cov}(\theta|\mu)) + \mathsf{Cov}_\mu(\mathsf{E}(\theta|\mu)) = \mathbf{I}_n + \mathbf{1}_n\mathbf{1}_n^T.$$

(b) Recall that the posterior depends on the sufficient statistics only. The sufficient statistic is

$$\overline{\mathbf{X}} = (\frac{1}{k}\sum_{j=1}^{k} X_{1j}, \ldots, \frac{1}{k}\sum_{j=1}^{k} X_{nj}), \quad \overline{\mathbf{X}}|\theta \sim \mathsf{N}_n(\theta, \frac{1}{k}\mathbf{I}_n)$$

Thus $\pi(\theta|\mathbf{x}) = \pi(\theta|\overline{\mathbf{x}})$, where

$$\pi(\theta|\mathbf{x}) \propto \exp\left(-\frac{1}{2}\theta^T(\mathbf{I}_n + \mathbf{1}_n\mathbf{1}_n^T)^{-1}\theta\right) \exp\left(-\frac{k}{2}(\overline{\mathbf{x}} - \theta)^T(\overline{\mathbf{x}} - \theta)\right).$$

Set $\mathbf{b}_1 = (\mathbf{A} + \mathbf{B})^{-1}\mathbf{B}\mathbf{b}$. Using

$$\theta^T\mathbf{A}\theta + (\theta - \mathbf{b})^T\mathbf{B}(\theta - \mathbf{b}) = (\theta - \mathbf{b}_1)^T(\mathbf{A} + \mathbf{B})(\theta - \mathbf{b}_1) + \text{const}$$

with $\mathbf{A} = (\mathbf{I}_n + \mathbf{1}_n\mathbf{1}_n^T)^{-1}$, $\mathbf{B} = k\mathbf{I}_n$, and $\mathbf{b} = \overline{\mathbf{X}}$, and using the relation

$$(\mathbf{I}_n + c\mathbf{1}_n\mathbf{1}_n^T)^{-1} = (\mathbf{I}_n - \frac{c}{cn+1}\mathbf{1}_n\mathbf{1}_n^T)$$

we get

$$\mathbf{A} + \mathbf{B} = (k+1)\mathbf{I}_n - \frac{1}{n+1}\mathbf{1}_n\mathbf{1}_n^T$$

and

$$(\mathbf{A} + \mathbf{B})^{-1} = \frac{1}{k+1}(\mathbf{I}_n + \frac{1}{k(n+1)+1}\mathbf{1}_n\mathbf{1}_n^T).$$

Summarizing, the posterior is $\mathsf{N}_n(\mathbf{b}_1, \Sigma)$ with $\mathbf{b}_1 = k\Sigma\overline{\mathbf{x}}$ and

$$\Sigma = \frac{1}{k+1}(\mathbf{I}_n + \frac{1}{k(n+1)+1}\mathbf{1}_n\mathbf{1}_n^T).$$

(c) We have $\bar{\theta} = \frac{1}{n}\mathbf{1}_n^T\theta$ thus

$$\bar{\theta}|\mathbf{x} \sim \mathsf{N}(m, v^2), \quad m = \frac{n+1}{k(n+1)+1}\frac{1}{nk}\sum_{i=1}^{n}\sum_{j=1}^{k} x_{ij}, \quad v^2 = \frac{1}{n}\frac{n+1}{k(n+1)+1}.$$

10.2   Solutions for Chapter 3

1. *Costumer satisfaction*:

   (a) Set the prior $\pi(0) = 0.3, \pi(1) = 0.5, \pi(2) = 0.2$. The posterior probabilities $\pi(\theta|x)$ are given as follows.

x	$+$	$+/-$	$-$	0
$\theta = 2$	12/17	1/9	0	20/121
$\theta = 1$	5/17	5/9	50/287	100/121
$\theta = 0$	0	3/9	237/287	1/121

   (b) Highest entropy prior is $\pi(0) = \pi(1) = \pi(2) = 1/3$. The posterior is the normalized likelihood.

   (c) Set $\pi(0) = p, \pi(1) = 2p, \pi(2) = 1 - 3p$. Using (3.25),

   $$H(p) = -p\ln(p) - 2p\ln(2p) - (1-3p)\ln(1-3p), \quad \arg\max_p H(p) = \frac{1}{3 + 2^{\frac{2}{3}}}.$$

   (d) Set $\pi(0) = \pi(1) = p, \pi(2) = 1 - 2p$ and $E\theta = p + 2(1-2p) = 1$. Thus $p = 11/30$.

2. *Log-normal distribution*:

   (a) Cauchy distribution $C(m, \gamma)$ is symmetric around m. Take $m = 3$.

   (b) Distribution function of $C(3, \gamma)$ is

   $$F(\theta) = \frac{1}{2} + \frac{1}{\pi}\arctan\left(\frac{\theta - 3}{\gamma}\right), \quad F(10) = 0.7, \quad \gamma = \frac{7}{\tan(0.2\pi)} = 9.63.$$

   (c) The posterior is not Cauchy distribution since the posterior moments exist.

   (d) Fisher information is $\frac{1}{\sigma^2}$. Thus $\pi_{\text{Jeff}}(\theta) \propto 1$.

3. *Pareto distribution*:

   (a) No. The support depends on μ.

   (b) No. The support depends on μ.

   (c) Yes, with natural parameter α and

   $$T(\mathbf{x}) = -\sum_{i=1}^{n} x_i, \quad \Psi(\alpha) = -\ln(\alpha), \quad \text{because } f(x, \alpha) = \alpha\frac{1}{x}\exp(-\alpha x).$$

   (d) By Theorem 3.3, the conjugate prior is

   $$\pi(\alpha|\mu, \lambda) \propto \exp(\alpha\mu + \lambda\ln(\alpha)).$$

   Set $\mu = -b$. Then $\alpha \sim \text{Gamma}(\lambda + 1, b), b > 0, \lambda > 0$.

4. *Geometric distribution*:

(a) It belongs to an exponential family with natural parameter $\eta = \ln(1-\theta)$,

$$P(X = k|\theta) = \exp\left(k\ln(1-\theta) + \ln(\theta)\right),$$

$$T(k) = k, \quad \Psi(\eta) = -\ln(1 - \exp(\eta)).$$

(b) By Theorem 3.3, we get the conjugate prior for η as

$$\pi_\eta(\eta|\mu, \lambda) \propto \exp\left(\eta\mu - \lambda\Psi(\eta)\right).$$

Transform $\eta = h(\theta) = \ln(1-\theta)$. Then

$$\pi_\theta(\theta|\mu, \lambda) = \pi_\eta(h(\theta)|\mu, \lambda)|h'(\theta)|$$

$$\propto \exp\left(\ln(1-\theta)\mu + \lambda\ln(\theta)\right)\frac{1}{1-\theta} \propto (1-\theta)^{\mu-1}\theta^\lambda.$$

(c) $\theta \sim \text{Beta}(\lambda + 1, \mu)$

(d) $\theta \mid x_1, \ldots, x_n \sim \text{Beta}(\lambda + n + 1, \mu + \sum_{i=1}^n x_i)$

(e) Fisher information: $\ln(p(k|\theta)) = k\ln(1-\theta) + \ln(\theta)$ and

$$\frac{d}{d\theta}\ln(p(k|\theta)) = -k\frac{1}{1-\theta} + \frac{1}{\theta}, \quad I(\theta) = \text{Var}(\frac{k}{1-\theta}) = \frac{1}{(1-\theta)\theta^2}.$$

(f) $\pi_{\text{Jeff}}(\theta) \propto \theta^{-1}(1-\theta)^{-\frac{1}{2}}$

(g) Jeffreys prior is improper.

5. *Gamma distribution*:

The likelihood is

$$\ell(\alpha, \beta|\mathbf{x}) = \prod_{i=1}^n \frac{\beta^\alpha}{\Gamma(\alpha)}x_i^{\alpha-1}\exp(-\beta x_i)$$

$$= \frac{\beta^{n\alpha}}{\Gamma(\alpha)^n}\prod_{i=1}^n x_i^{-1}\exp(\alpha\sum_{i=1}^n \ln(x_i) - \beta\sum_{i=1}^n x_i). \tag{10.1}$$

(a) The sample is generated by a 2-dimensional exponential family of form the (3.17) with natural parameter $\theta = (\alpha, \beta)$ and

$$T(x) = (\ln(x), -x), \quad \Psi(\theta) = \ln(\Gamma(\alpha)) - \alpha\ln(\beta).$$

(b) Applying Theorem 3.3 a conjugate prior is $\pi(\theta) \propto \exp(\theta^T\mu - \lambda\Psi(\theta))$.

(c) The posterior is $\pi(\theta|\mathbf{x}) \propto \exp(\theta^T(T(\mathbf{x}) + \mu) - (\lambda + 1)\Psi(\theta))$.

(d) Consider the conjugate family, $\pi(\theta) = \pi(\beta|\alpha)\pi(\alpha)$. Setting $b = -\mu_2 > 0$,

$$\pi(\beta|\alpha) \propto \exp(-\beta b + \lambda\alpha\ln(\beta)) \propto \beta^{\alpha\lambda}\exp(-\beta b),$$

i.e., $\beta|\alpha \sim \text{Gamma}(\lambda\alpha + 1, b)$. The marginal distribution of α,

$$\pi(\alpha) \propto \Gamma(\alpha)^{-\lambda}\exp(\alpha\mu_1),$$

does not belong to an exponential family. It is not a "well known" distribution either.

6. *Binomial distribution* $\mathrm{Bin}(n, \theta)$ *and odds ratio* $\eta = \frac{\theta}{1-\theta}$:

(a) Apply $\pi(\theta) = \pi_\eta(h(\theta)) \left| \frac{h(\theta)}{d\theta} \right|$ with $h(\theta) = \frac{\theta}{1-\theta}$. We get

$$\pi(\theta) \propto \frac{1}{h(\theta)} \frac{1}{(1-\theta)^2} \propto \frac{1-\theta}{\theta} \frac{1}{(1-\theta)^2} \propto \frac{1}{\theta(1-\theta)}.$$

(b) The distribution in (a) is called *Haldane* distribution. It is improper. For $x \in \{1, \ldots, n-1\}$,

$$\pi(\theta|x) \propto \theta^{-1}(1-\theta)^{-1}\theta^x(1-\theta)^{n-x} \propto \theta^{x-1}(1-\theta)^{n-x-1} \propto \theta^a(1-\theta)^b,$$

with $a \geq 0$, $b \geq 0$ is proper.

(c) Applying $\pi(\xi) = \pi_\theta(h(\xi)) \left| \frac{h(\xi)}{d\xi} \right|$ with

$$\theta = h(\xi) = \frac{\xi}{c+\xi}, \quad \frac{dh(\xi)}{d\xi} = \frac{1}{(c+\xi)^2}, \quad c = \frac{b}{a},$$

gives

$$\pi(\xi) \propto \left(\frac{\xi}{c+\xi} \right)^{a-1} \left(\frac{c}{c+\xi} \right)^{b-1} \frac{1}{(c+\xi)^2} \propto \frac{\xi^{a-1}}{(b + a\xi)^{a+b}}.$$

It is the kernel of an F-distribution with $(2a, 2b)$ degrees of freedom.

7. *Binomial distribution* $\mathrm{Bin}(n, \theta)$ *and natural parameter*:

(a) Yes, it is. Natural parameter: $\vartheta = \ln(\eta) = \ln(\frac{\theta}{1-\theta})$ and

$$p(x|\theta) \propto \theta^x(1-\theta)^{n-x} \propto \exp(x \ln(\theta) + (n-x)\ln(1-\theta)) \propto \exp(\vartheta x - \Psi(\vartheta))$$

with

$$\theta = \frac{\exp(\vartheta)}{1+\exp(\vartheta)}, \quad \Psi(\vartheta) = -n\ln(1-\theta) = -n\ln\left(\frac{1}{1+\exp(\vartheta)} \right).$$

(b) Applying Theorem 3.3, for $\lambda > 0, \mu \in \mathbb{R}$, the conjugate prior is

$$\pi(\vartheta) \propto \exp(\vartheta\mu - \lambda n \ln(1+\exp(\vartheta)))$$
$$\propto \frac{\exp(\vartheta)^\nu}{(1+\exp(\vartheta))^{n\lambda}} \propto \frac{\exp(\vartheta)^a}{(1+\exp(\vartheta))^{a+b}}$$

with $a = \mu$ and $b = n\lambda - \mu$.

(c) Applying Theorem 3.3,

$$\pi(\vartheta|x) \propto \frac{\exp(\vartheta)^{\nu+x}}{(1+\exp(\vartheta))^{n(\lambda+1)}} \propto \frac{\exp(\vartheta)^{a_1}}{(1+\exp(\vartheta))^{a_1+b_1}}$$

with $a_1 = a + x$, $b_1 = b + n - x$.

(d) We shift the natural parameter to obtain a known family. Set

$$z = \frac{1}{2}\ln\left(\frac{b}{a}\frac{\theta}{1-\theta}\right) = C + \frac{1}{2}\vartheta, \quad \vartheta = 2z - 2C = h(z), \quad C = \ln(\frac{b}{a}).$$

Applying $\pi(z) = \pi_\vartheta(h(z))|\frac{h(z)}{dz}|$ with $\frac{d\,h(z)}{dz} = 2$, $\exp(h(z)) = \frac{a}{b}\exp(2z)$,

$$\pi(z) \propto \frac{\exp(z)^{2a}}{(\frac{a}{b}\exp(2z)+1)^{a+b}} \propto \frac{\exp(z)^{2a}}{(2a\exp(2z)+2b)^{a+b}}.$$

It is the kernel of *Fisher's z-distribution* with $f_1 = 2a$, $f_2 = 2b$ degrees of freedom. This distribution is defined as $\frac{1}{2}\ln(F)$, where $F \sim F_{f_1,f_2}$. See the solution of Problem 6 above.

8. *Multinomial distribution:* Jeffreys prior for $\theta = (p_1, p_2)$ is given in (3.45).

$$\pi_{\text{Jeff}}(\theta) \propto \frac{1}{(\theta_1\theta_2)^{1/2}}, \quad \theta_j = \frac{\eta_j}{1+\eta_1+\eta_2} = h_j(\eta), \quad j = 1,2$$

$$J = \left[\frac{\partial h_i(\eta)}{\partial \eta_j}\right]_{i=1,2,j=1,2} = \frac{1}{(1+\eta_1+\eta_2)^2}\begin{pmatrix} 1+\eta_2 & -1 \\ -1 & 1+\eta_1 \end{pmatrix}.$$

Applying (3.27)

$$\pi_{\text{Jeff}}(\eta) = \pi_{\text{Jeff}}(h(\eta))|J| \propto \frac{\eta_1+\eta_2+\eta_1\eta_2}{(\eta_1\eta_2(1+\eta_1+\eta_2)^5)^{1/2}}. \tag{10.2}$$

9. *Hardy–Weinberg model with natural parametrization:*

(a) Yes. Set $I_k(x) = 1$ for $x = k$ and zero else. Set $x = 0$ for aa, $x = 1$ for Aa, $x = 2$ for AA. Set $I_k(x) = 1$ for $x = k$ and zero else, $I_0(x)+I_1(x)+I_2(x) = 1$. For $x \in \{0,1,2\}$ and $T(x) = 2I_0(x) + I_1(x) = 2 - I_1(x) + 2I_2(x)$.

$$\begin{aligned} P(x|\theta) &= \theta^{2I_0(x)}(2\theta(1-\theta))^{I_1(x)}((1-\theta))^{2I_2(x)} \\ &= \theta^{2I_0(x)+I_1(x)}(1-\theta)^{I_1(x)+2I_2(x)}2^{I_1(x)} \\ &= \exp\left(T(x)\ln(\frac{\theta}{1-\theta}) + 2\ln(1-\theta)\right)2^{I_1(x)}. \end{aligned}$$

(b) Natural parameter: $\eta = \ln(\frac{\theta}{1-\theta})$, sufficient statistic: $T(x) = 2I_0(x) + I_1(x)$, and $\Psi_0(\theta) = -2\ln(1-\theta)$. It has form (3.17).

(c) We have $\theta = h(\eta) = \frac{\exp(\eta)}{1+\exp(\eta)}$. Apply Theorem 3.3 with $\Psi(\eta) = \Psi_0(h(\theta)) = 2\ln(1+\exp(\eta))$. A conjugate prior is

$$\pi(\eta|\mu,\lambda) \propto \exp(\eta\mu - 2\lambda\ln(1+\exp(\eta)) \propto \frac{\exp(\eta)^\mu}{(1+\exp(\eta))^{2\lambda}}.$$

(d) Theorem 3.3 delivers the posterior

$$\pi(\eta|x) = \pi(\eta|\mu + T(x), \lambda + 1) \propto \frac{\exp(\eta)^{\mu+T(x)}}{(1+\exp(\eta))^{2(\lambda+1)}}.$$

(e) Calculate the Fisher information $I(\eta) = \mathsf{Var}(V(\eta|x))$, with score function $V(\eta|x) = \ln(\mathsf{P}(x|\eta))$, as

$$V(\eta|x) = T(x) - 2\frac{\exp(\eta)}{1+\exp(\eta)}, \quad \mathsf{Var}(T(x)) = 2\frac{\exp(\eta)}{(1+\exp(\eta))^2}.$$

Jeffreys prior is given as

$$\pi_{\text{Jeff}}(\eta) \propto |I(\eta)|^{\frac{1}{2}} \propto \frac{\exp(\eta)^{\frac{1}{2}}}{1+\exp(\eta)}.$$

(f) Jeffreys prior belongs to the conjugate family with $\mu = \lambda = \frac{1}{2}$.

10. *Reference prior*:

(a) See Examples 3.21 and 3.24: $\pi_{\text{Jeff}}(\theta) \propto$ const and

$$\pi_{\text{Jeff}}(\theta|x) = \frac{1}{\sqrt{2\pi}} \exp(-\frac{1}{2}(x-\theta)^2).$$

(b) For $\theta \in [-k, k]$, $\pi_{\text{Jeff}}^k(\theta) = \frac{1}{2k}$, the posterior is the truncated normal distribution,

$$\pi_{\text{Jeff}}^k(\theta|x) = c(k)^{-1}\frac{1}{\sqrt{2\pi}} \exp(-\frac{1}{2}(x-\theta)^2), \quad c(k) = \Phi(k-x) - \Phi(-k-x).$$

(c) The Kullback–Leibler divergence $K(\pi_{\text{Jeff}}(.|x)|\pi_{\text{Jeff}}^k(.|x))$ equals

$$\int_{-k}^{k} c(k)^{-1}\frac{1}{\sqrt{2\pi}} \exp(-\frac{1}{2}(x-\theta)^2)(-\log(c(k)))d\theta = -\ln(c(k))$$

so that

$$\lim_{k\to\infty} K(\pi(.|x)|\pi^k(.|x)) = -\ln(\lim_{k\to\infty} c(k)) = -\ln(1) = 0.$$

(d) The prior $\pi_{\text{Jeff}}^k(\theta)$ is independent of θ, thus $p_0(\theta) = 1$.

11. *Shannon entropy of $N(\mu, \sigma^2)$*: We have

$$H(N(\mu,\sigma^2)) = -\int \varphi_{\mu,\sigma^2}(x) \ln(\frac{1}{\sqrt{2\pi}\sigma})dx + \frac{1}{2\sigma^2}\int \varphi_{\mu,\sigma^2}(x)(x-\mu)^2 dx$$

$$= \ln(\sqrt{2\pi}\sigma) + \frac{1}{2\sigma^2}\sigma^2 = \frac{1}{2}(\ln(2\pi\sigma)+1) = \frac{1}{2}\ln(2\pi\sigma^2 e).$$

12. *Compare two normal priors:*

(a) A sufficient statistic is $\bar{x} \sim N(\theta, \frac{\sigma^2}{n})$; see Example 2.12. For $j = 1, 2$, the posterior distributions related to π_j are $N(\mu_{j,1}, \sigma_{j,1}^2)$ with

$$\sigma_{1,1}^2 = \frac{\sigma_0^2 \sigma^2}{n\sigma_0^2 + \sigma^2}, \quad \sigma_{2,1}^2 = \frac{\lambda\sigma_0^2 \sigma^2}{n\lambda\sigma_0^2 + \sigma^2}.$$

(b) Use (3.49) and the result of Problem 11 above. Note that the Shannon entropy of $N(\mu, \sigma^2)$ is independent of the mean.

$$I(\mathcal{P}^n, \pi_1) = \frac{1}{2}\ln(2\pi\sigma_0^2 e) - \frac{1}{2}\ln(2\pi\sigma_{1,1}^2 e) = \frac{1}{2}\ln(n\sigma_0^2 + \sigma^2) - \frac{1}{2}\ln(\sigma^2)$$

and

$$I(\mathcal{P}^n, \pi_2) = \frac{1}{2}\ln(n\lambda\sigma_0^2 + \sigma^2) - \frac{1}{2}\ln(\sigma^2).$$

(c) We have

$$I(\mathcal{P}^n, \pi_1) - I(\mathcal{P}^n, \pi_2) = \frac{1}{2}\ln\left(\frac{n\sigma_0^2 + \sigma^2}{n\lambda\sigma_0^2 + \sigma^2}\right),$$

$$\lim_{n\to\infty} \frac{n\sigma_0^2 + \sigma^2}{n\lambda\sigma_0^2 + \sigma^2} = \lim_{n\to\infty} \frac{\sigma_0^2 + \sigma^2/n}{\lambda\sigma_0^2 + \sigma^2/n} = \lambda^{-1},$$

$$\lim_{n\to\infty} (I(\mathcal{P}^n, \pi_1) - I(\mathcal{P}^n, \pi_2)) = -\frac{1}{2}\ln(\lambda).$$

(d) For $\lambda > 1$ and every μ_0 the second prior is less informative.

10.3 Solutions for Chapter 4

1. *Query:*

(a) From

$$\pi(\theta = 0|x) = \frac{\pi(0)\,p(x|0)}{\pi(0)\,p(x|0) + \pi(1)p(x|1)} = 1 - \pi(\theta = 1|x),$$

it follows that

$$\pi(\theta = 0|x = -1) = 1, \quad \pi(\theta = 0|x = 0) \propto 0.3\,p, \quad \pi(\theta = 0|x = 1) \propto 0.2\,p.$$

(b) Applying Theorem 4.5 we have three different classes of priors.

		x	-1	0	1
$0 \le p \le \frac{2}{7}$	$\delta_1(x)$		0	0	0
$\frac{2}{7} < p \le \frac{3}{8}$	$\delta_2(x)$		0	0	1
$\frac{3}{8} < p \le 1$	$\delta_3(x)$		0	1	1

(c) The frequentist risk is calculated by $R(0, \delta^\pi) = P(C_1|0)$, $R(1, \delta^\pi) = 1 - P(C_1|1)$, where $C_1 = \{x : \delta^\pi = 1\}$. It is given as

θ	0	1
$0 \leq p \leq \frac{2}{7}$	0	1
$\frac{2}{7} < p \leq \frac{3}{8}$	0.2	0.5
$\frac{3}{8} < p \leq 1$	0.5	0

(d) The Bayes risk $r(\pi) = (1-p)R(0, \delta^\pi) + p\,R(1, \delta^\pi)$, given as function of p,

case	risk	behavior	sup
$0 \leq p \leq \frac{2}{7}$	p	linear increasing	0.2857
$\frac{2}{7} < p \leq \frac{3}{8}$	$0.2 + 0.3\,p$	linear increasing	0.3125
$\frac{3}{8} < p \leq 1$	$0.5 - 0.5\,p$	linear decreasing	0.3125

(e) It holds that $\max_p r(\pi) = \max_p(0.2 + 0.3\,p) = 0.3125$. The least favourable prior is $\pi_0(1) = \frac{3}{8}$.

2. *Binomial model*:

(a) Applying Theorem 4.3 the Bayes estimator is the $\frac{k_2}{k_1+k_2}$ fractile of the posterior $\mathsf{Beta}(\alpha + x, \beta + n - x)$.

(b) See Example 4.9. For $k_1 = k_2$ the Bayes estimator is the median of $\mathsf{Beta}(\alpha + x, \beta + n - x)$, approximated by (4.19).

(c) The asymmetric approach is of interest, when the underestimation of the success probability is more dangerous than overestimation, i.e., when the success means the risk of a traffic accident.

(d) Set $n = 1$ and $k_1 = k_2$. The frequentist risk is

$$R(\theta, \delta^\pi) = (1 - \theta)L(\theta, \delta^\pi(0)) + \theta L(\theta, \delta^\pi(1)).$$

Exploring the inequalities for the median of $\mathsf{Beta}(\alpha, \beta)$,

$$\frac{\alpha - 1}{\alpha + \beta - 2} \leq \text{median} \leq \frac{\alpha}{\alpha + \beta}, \quad \text{for } 1 < \alpha < \beta$$

$$\frac{\alpha - 1}{\alpha + \beta - 2} \geq \text{median} \geq \frac{\alpha}{\alpha + \beta}, \quad \text{for } 1 < \beta < \alpha$$

we get $\delta^\pi(0) < \delta^\pi(1)$. Calculating $R(\theta, \delta^\pi)$ we have three different cases:

$$\theta < \delta^\pi(0) < \delta^\pi(1), \quad \delta^\pi(0) < \theta < \delta^\pi(1), \quad \delta^\pi(0) < \delta^\pi(1) < \theta$$

In each case it is impossible to find a beta prior such that the frequentist risk is constant. We consider the first case only.

$$R(\theta, \delta^\pi) = \delta^\pi(0) + \theta(1 + \delta^\pi(1) - \delta^\pi(0)).$$

But $1 + \delta^\pi(1) - \delta^\pi(0) > 0$ for all beta priors with $\alpha > 1, \beta > 1$. Thus inside $\{\text{Beta}(\alpha, \beta) : \alpha > 1, \beta > 1\}$ we cannot find a prior, which fulfills the sufficient condition for a minimax estimator given in Theorem 4.13.

3. *Gamma distribution $\theta = \beta$, Jeffreys prior:*

   (a) The loglikelihood $\ln \ell(\theta|x)$ and the score function $V(\theta|x) = \frac{d}{d\theta} \ln \ell(\theta|x)$ are

   $$\ln \ell(\theta|x) = -\ln(\Gamma(\alpha)) + \alpha \ln(\theta) + (\alpha - 1)\ln(x) - \theta x, \quad V(\theta|x) = \frac{\alpha}{\theta} - x.$$

   Using $\text{Var}(X|\theta) = \frac{\alpha}{\theta^2}$ we obtain the Fisher information and Jeffreys prior

   $$I(\theta) = \text{Var}(V(\theta|x)) = \frac{\alpha}{\theta^2}, \quad \pi_{\text{Jeff}}(\theta) \propto I(\theta)^{\frac{1}{2}} \propto \frac{1}{\theta}.$$

   See also Example 3.13. Using (2.9), the posterior distribution is Gamma(α, x).

   (b) Applying Theorem 4.2, $\delta^\pi(x) = \mathsf{E}(\theta|x) = \frac{\alpha}{x}$.

   (c) Frequentist risk: For $\delta^\pi(x) = \mathsf{E}(\theta|x)$ and quadratic loss,

   $$R(\theta, \delta^\pi) = \text{Var}(\delta^\pi(X)|\theta) + (\theta - \mathsf{E}(\delta^\pi(X)|\theta))^2.$$

   Note that, if $X \sim \text{Gamma}(\alpha, \beta)$ then $X^{-1} \sim \text{InvGamma}(\alpha, \beta)$. We apply the moments of inverse-gamma distribution; see Section 11.2.

   $$\mathsf{E}(\delta^\pi(X)|\theta) = \frac{\alpha\theta}{\alpha - 1}, \quad \text{Var}(\delta^\pi(X)|\theta) = \frac{\alpha^2\theta^2}{(\alpha - 1)^2(\alpha - 2)},$$

   $$R(\theta, \delta^\pi) = \theta^2 d(\alpha), \quad d(\alpha) = \frac{\alpha^2}{(\alpha - 1)^2(\alpha - 2)} + \frac{1}{(\alpha - 1)^2}.$$

   Posterior expected risk:

   $$\rho(\pi, \delta^\pi(x)|x) = \mathsf{E}((\theta - \delta^\pi(x))^2|x) = \text{Var}(\theta|x) = \frac{\alpha}{x^2}$$

   Bayes risk:

   $$r(\pi) = \int_0^\infty R(\theta, \delta^\pi)\frac{1}{\theta}\, d\theta = d(\alpha)\int_0^\infty \theta\, d\theta = \infty.$$

   (d) We can neither apply Theorem 4.9 nor Theorem 4.10. Since δ^π is unique, by Theorem 4.11, δ^π is admissible.

4. *Gamma distribution $\theta = \frac{1}{\beta}$, conjugate prior*:

(a) See Example 3.13. Natural parameter $\eta = -\beta = -\frac{1}{\theta}$ and $T(x) = x$.

(b) See Liero and Zwanzig (2011, Theorem 4.3). Since $\mathsf{E}(T(X)) = \frac{\alpha}{\beta} = \theta\alpha$, the efficient estimator is $\delta^*(x) = \frac{x}{\alpha}$.

(c) In Example 3.13 the conjugate prior for β is $\mathsf{Gamma}(\alpha_0, \beta_0)$ and the posterior is $\mathsf{Gamma}(\alpha_0 + \alpha, \beta_0 + x)$. Thus the conjugate prior of $\theta = \frac{1}{\beta}$ is the inverse-gamma distribution $\mathsf{InvGamma}(\alpha_0, \beta_0)$ and the posterior is $\mathsf{InvGamma}(\alpha_0 + \alpha, \beta_0 + x)$.

(d) Applying Theorem 4.2 and the expectation of $\mathsf{InvGamma}(\alpha_0+\alpha, \beta_0+x)$,
$\delta^\pi(x) = \frac{\beta_0+x}{\alpha_0+\alpha-1}$.

(e) Frequentist risk of δ^π and δ^*:

$$R(\theta, \delta^\pi) = \mathsf{Var}(\delta^\pi(X)|\theta) + (\theta - \mathsf{E}(\delta^\pi(X)|\theta))^2$$

$$= \frac{1}{(\alpha_0 + \alpha - 1)^2}\left(\theta^2(\alpha + \alpha_0) - 2\theta\alpha_0(\beta_0 + 1) + (\beta_0^2 + 1)^2\right),$$

$$R(\theta, \delta^*) = \mathsf{Var}(\delta^*(X)|\theta) = \frac{1}{\alpha^2}\frac{\alpha}{\beta^2} = \frac{\theta^2}{\alpha}.$$

Posterior expected risk:

$$\rho(\pi, \delta^\pi(x)|x) = \mathsf{E}((\theta - \delta^\pi(x))^2|x) = \mathsf{Var}(\theta|x) = \frac{(\beta_0 + x)^2}{(\alpha - 1)^2(\alpha - 2)}.$$

Bayes risk: The frequentist risk is quadratic in θ. For $\alpha_0 > 2$ the second prior moment exists, such that $r(\pi) < \infty$. Particularly

$$r(\pi) = \frac{1}{(\alpha_0 + \alpha + 1)^2}\left(\frac{-2}{\alpha_0 - 2}\beta_0^2 + \frac{\alpha_0 - 2}{\alpha_0 - 1}\beta_0 + 1\right).$$

(f) Applying Theorem 4.9, 4.10 or 4.11 we note that under $\alpha_0 > 2$ the Bayes estimate is admissible.

(g) Due to admissibility of the Bayes estimator, there exists a θ' with $R(\theta', \delta^*) > R(\theta', \delta^\pi)$, which is no contradiction to the efficiency of δ^* since the Bayes estimator is biased.

5. *Multinomial distribution, minimax estimator*:

(a) See Example 3.8. Set $x = (n_1, n_2, n_3)$. The posterior is $\theta|x \sim \mathsf{Dir}(\alpha_1 + n_1, \alpha_2 + n_2, \alpha_3 + n_3)$.

(b) Set $\alpha_0 = \alpha_1 + \alpha_2 + \alpha_3$ and $\alpha = (\alpha_1, \alpha_2, \alpha_3)$. Then, by Theorem 4.12,

$$\delta^\pi(x) = \frac{1}{\alpha_0 + n}(\alpha + x) \in \Theta.$$

(c) Frequentist risk of δ^π: Using $\theta_1+\theta_2+\theta_3 = 1$ and the mean and covariance of $X|\theta \sim \mathsf{Mult}(n, \theta_1, \theta_2, \theta_3)$, with

$$E(X|\theta) = n\theta, \quad \mathsf{Cov}(X|\theta) = n \left(\mathrm{diag}(\theta) - \theta\theta^T\right)$$

we obtain bias $= \mathsf{E}(\delta^\pi(X)|\theta) - \theta = \frac{1}{\alpha_0+n}(\alpha - \alpha_0\theta)$ and

$$R(\theta, \delta^\pi) = \mathrm{tr}\,\mathsf{Cov}(\delta^\pi(X)|\theta) + \mathrm{bias}^T\mathrm{bias}$$

$$= \frac{1}{(\alpha_0 + n)^2} \left((\alpha_0^2 - n) \sum_{i=1}^{3} \theta_i^2 - 2\alpha_0 \sum_{i=1}^{3} \alpha_i\theta_i + \sum_{i=1}^{3} \alpha_i^2 + n \right).$$

(d) Bayes risk, $r(\pi) = \mathsf{E}^\pi R(\theta, \delta^\pi)$, is finite since the moments of $\mathsf{Dir}(\alpha_1, \alpha_2, \alpha_3)$ are finite.

(e) Least favourable prior: Applying Theorems 4.7 and 4.12 we search for $\alpha_1, \alpha_2, \alpha_3$ such that $R(\theta, \delta^\pi) = \mathrm{const}$. As $R(\theta, \delta^\pi)$ is permutation invariant with respect to $\alpha_1, \alpha_2, \alpha_3$, we set $\alpha_1 = \alpha_2 = \alpha_3$. For $\alpha_i = \frac{1}{3}\sqrt{n}$, $R(\theta, \delta^\pi) = \frac{2}{3} \frac{n}{(\sqrt{n}+n)^2}$.

(f) Applying Theorem 4.7 the minimax estimate is

$$\widehat{\theta}_{\mathrm{minimax},i} = \delta^{\pi_0}(x)_i = \frac{\frac{1}{3}\sqrt{n} + x_i}{\sqrt{n} + n}.$$

10.4   Solutions for Chapter 5

1. *Consistent estimation from an i.i.d. sample of* $\mathsf{Exp}\,(\theta)$:

(a) The posterior is $\mathsf{Gamma}(\alpha + n, \beta + n\bar{x})$, with $\bar{x} = \frac{1}{n}\sum_{i=1}^{n} x_i$, since

$$\pi(\theta|\mathbf{x}_{(n)}) \propto \theta^{\alpha-1} \exp(-\theta\beta) \prod_{i=1}^{n} (\theta \exp(-\theta x_i)) \propto \theta^{\alpha+n-1} \exp(-\theta(\beta+n\bar{x})).$$

(b) The Bayes estimator is $\delta(\mathbf{x}_{(n)}) = \mathsf{E}(\theta|\mathbf{x}_{(n)}) = \frac{\alpha+n}{\beta+n\bar{x}}$.

(c) Applying the law of large numbers, $\bar{x} \to \mathsf{E}X, a.s.$ with $\mathsf{E}X = \frac{1}{\theta}$, so that

$$\delta(\mathbf{x}_{(n)}) = \frac{\frac{\alpha}{n} + 1}{\frac{\beta}{n} + \bar{x}} \to \theta \ a.s.$$

2. *Kullback–Leibler divergence in exponential families*:

(a) Apply (3.11) and (3.28),

$$\mathsf{K}(\mathsf{P}_{\theta_0}|\mathsf{P}_\theta) = \int p(x|\theta_0) \ln(\frac{p(x|\theta_0)}{p(x|\theta)})dx$$

$$= \int p(x|\theta_0) \ln \left(\frac{C(\theta_0) \exp\left(\sum_{j=1}^{k} \zeta_j(\theta_0)T_j(x)\right)}{C(\theta) \exp\left(\sum_{j=1}^{k} \zeta_j(\theta)T_j(x)\right)} \right) dx$$

$$= \ln \left(\frac{C(\theta_0)}{C(\theta)} \right) + \sum_{j=1}^{k} \mathsf{E}_{\theta_0} T_j(X) \left(\zeta_j(\theta_0) - \zeta_j(\theta)\right).$$

(b) Gamma$(\alpha, \beta$ belongs to a two parametric exponential family with natural parameters $\eta = (\alpha, \beta)$, sufficient statistics $T(x) = (\ln(x), -x)$, and $C(\alpha, \beta) = \frac{\beta^\alpha}{\Gamma(\alpha)}$.

Applying $\mathsf{E}X = \frac{\alpha}{\beta}$ and $\mathsf{E}\ln(X) = \Psi(\alpha) - \ln(\beta)$, where $\Psi(\alpha)$ is the digamma function, we get for $\mathsf{K}(\mathsf{Gamma}(\alpha_0, \beta_0)|\mathsf{Gamma}(\alpha, \beta))$

$$\ln\left(\frac{\Gamma(\alpha)}{\Gamma(\alpha_0)}\right) + \alpha \ln\left(\frac{\beta_0}{\beta}\right) + \Psi(\alpha_0)(\alpha_0 - \alpha) - \frac{\alpha_0}{\beta_0}(\beta_0 - \beta). \qquad (10.3)$$

3. *Kullback–Leibler divergence for uniform distributions*: It holds that

$$\mathsf{K}(\mathsf{U}(0, \theta_0)|\mathsf{U}(0, \theta)) = \int_0^{\theta_0} \frac{1}{\theta_0} \ln\left(\frac{\theta I_{(0,\theta_0)}(x)}{\theta_0 I_{(0,\theta)}(x)}\right) dx.$$

(a) For $\theta < \theta_0$,

$$\mathsf{K}(\mathsf{U}(0, \theta_0)|\mathsf{U}(0, \theta)) = \int_0^\theta \frac{1}{\theta_0} \ln\left(\frac{\theta}{\theta_0}\right) dx + \int_\theta^{\theta_0} \frac{1}{\theta_0} \ln\left(\frac{1}{0}\right) dx = \infty.$$

For $\theta > \theta_0$,

$$\mathsf{K}(\mathsf{U}(0, \theta_0)|\mathsf{U}(0, \theta)) = \int_0^{\theta_0} \frac{1}{\theta_0} \ln\left(\frac{\theta}{\theta_0}\right) dx = \ln\left(\frac{\theta}{\theta_0}\right).$$

(b) Further $\mathsf{K}(\mathsf{P}|\mathsf{P}) = 0$. Applying the results above, the Kullback–Leibler divergence is only one-sided continuous.

4. *Comparison of posteriors related to different priors*
The posteriors are $\mathsf{P}_i = \mathsf{P}_n^{\pi_i}(.|\mathbf{x}_{(n)}) = \mathsf{Gamma}(n\nu + \alpha_i, n\bar{x} + \beta)$, $i = 1, 2$.
Applying (10.3),

$$\mathsf{K}(\mathsf{Gamma}(\alpha_1, \beta)|\mathsf{Gamma}(\alpha_2, \beta)) = \ln\left(\frac{\Gamma(\alpha_2)}{\Gamma(\alpha_1)}\right) + \Psi(\alpha_1)(\alpha_1 - \alpha_2).$$

Using the asymptotic relations for $\Gamma(.)$ and digamma function $\Psi(.)$

$$\lim_{n \to \infty} \frac{\Gamma(n + z)}{\Gamma(n) z^n} = 1, \quad \lim_{z \to \infty} \frac{\Psi(z)}{\ln(z) - \frac{1}{2z}} = 1,$$

we have

$$\mathsf{K}(\mathsf{P}_1|\mathsf{P}_2) = \left(\ln\left(\frac{n\nu + \alpha_1}{n\nu}\right) - \frac{1}{2(n\nu + \alpha_1)}\right)(\alpha_1 - \alpha_2) + o(1) = o(1).$$

5. *Conditions of Schwartz' Theorem*: First condition (5.14): The prior density of $\mathsf{N}(\mu, \sigma_0^2)$ is positive for all $\theta \in \Theta = \mathbb{R}$. Furthermore from Example 4.12, on page 96, we know that

$$\mathcal{K}_\varepsilon(\theta_0) = \{\theta : \mathsf{K}(\mathsf{P}_{\theta_0}|\mathsf{P}_\theta) < \varepsilon\} = \{\theta : (\theta_0 - \theta)^2 < 2\varepsilon\}.$$

Second condition (5.15): Let a, b, $O = (a, b)$ and $d > 0$ be such that $a < \theta_0 < b$, and $[\theta_0 - d, \theta_0 + d] \subset (a, b)$. Using $p_n(\mathbf{x}_{(n)}|\theta)\pi(\theta) = \pi(\theta|\mathbf{x}_{(n)})p(\mathbf{x}_{(n)})$ and the posterior $N(\mu_1, \sigma_1^2)$ given in Example 2.12, on page 16, we have for the density of P_{n,O^c}

$$q(\mathbf{x}_{(n)}) = \int_{O^c} p_n(\mathbf{x}_{(n)}|\theta)\frac{\pi(\theta)}{P^\pi(O^c)}\, d\theta = \frac{p_n(\mathbf{x}_{(n)})}{P^\pi(O^c)}\left(\Phi(a_1) + 1 - \Phi(b_1)\right),$$

where $a_1 = \frac{a-\mu_1}{\sigma_1}$ and $b_1 = \frac{b-\mu_1}{\sigma_1}$. For sufficiently large n, $\bar{x} \in [\theta_0 - d, \theta_0 + d] \subset (a, b)$, there exist positive constants c_1, c_2 such that $a_1 < -c_1\sqrt{n}$ and $b_1 > c_2\sqrt{n}$. Using $\bar{x} - \theta_0 \sim N(0, \frac{\sigma^2}{n})$ and

$$\Phi(-t) \le \frac{1}{2}\exp(-\frac{t^2}{2}), \quad 1 - \Phi(t) \le \frac{1}{2}\exp(-\frac{t^2}{2}), \quad \text{for } t > 0,$$

there exists a positive constant c_0 such that $\Phi(a_1) + 1 - \Phi(b_1) \le \exp(-c_0 n)$. Hence for $P^\pi(O^c) > C^{-1}$, we have $q(\mathbf{x}_{(n)}) \le p(\mathbf{x}_{(n)})C\exp(-c_0 n)$. Set $H = H_{\frac{1}{2}}(P_{\theta_0}^{\otimes n}, P_{n,O^c})$ and $p_0(\mathbf{x}_{(n)}) = p_n(\mathbf{x}_{(n)}|\theta_0)$ and split the integration region in $B_d = \{\mathbf{x}_{(n)} : |\bar{x} - \theta_0| < d\}$ and B_d^c. Then

$$H = \int_{B_d} \sqrt{p_0(\mathbf{x}_{(n)})q(\mathbf{x}_{(n)})}\, d\mathbf{x}_{(n)} + \int_{B_d^c} \sqrt{p_0(\mathbf{x}_{(n)})q(\mathbf{x}_{(n)})}\, d\mathbf{x}_{(n)} = H_1 + H_2,$$

where

$$H_1 = \int_{B_d} \sqrt{p_0(\mathbf{x}_{(n)})p(\mathbf{x}_{(n)})}\, d\mathbf{x}_{(n)} < C\exp(-c_0 n).$$

Applying Cauchy–Schwarz inequality and $P_{\theta_0}(B_d^c) \le \exp(-\frac{nd^2}{2\sigma^2})$,

$$H_2 \le \left(\int_{B_d^c} p_0(\mathbf{x}_{(n)})\, d\mathbf{x}_{(n)}\right)^{\frac{1}{2}} \left(\int_{B_d^c} q(\mathbf{x}_{(n)})\, d\mathbf{x}_{(n)}\right)^{\frac{1}{2}} \le P_{\theta_0}(B_d^c)^{\frac{1}{2}}$$

For $D_0 > C + 1$ and $q_0 = \exp(-\min(\frac{d^2}{4\sigma^2}, c_0))$, (5.15) follows.

6. *Counterexample to Theorem 5.3:*

(a) Bayes estimate: Applying Theorem 6.2 with $\mathbf{X} = 1_2^T = (1, 1)$ the posterior is $N(\gamma_1, \Gamma_1)$ with $\Gamma_1 = \mathbf{I}_2 - \lambda_n 11^T$, $\gamma_1 = \lambda_n 1_2\bar{x}$, and $\lambda_n = \frac{1}{\frac{1}{n}+2}$. The Bayes estimator equals $\gamma_1 = (\gamma_{1,1}, \gamma_{1,2})$. We have $\bar{x} \to a + b$, $a.s.$ and $\lambda_n \to \frac{1}{2}$. For $a > 0$, $b > 0$ and $a \ne b$ it holds that

$$\lim_{n\to\infty} \gamma_{1,1} = \frac{1}{2}(a + b) \ne a, \quad \lim_{n\to\infty} \gamma_{1,2} = \frac{1}{2}(a + b) \ne b.$$

(b) Inconsistency of the posterior: Take $\theta_0 = (a_0, b_0)$ with $a_0 > 0$, $b_0 > 0$ and $a_0 \neq b_0$. Set $O = \{\theta : \|\theta - \theta_0\| < c\}$ and $A = \{\theta : |a - a_0| < c, b \in \mathbb{R}\}$. Then $O \subset A$ and

$$P^\pi(O|\mathbf{x}_{(n)}) < P^\pi(A|\mathbf{x}_{(n)}) = \Phi(U_n) - \Phi(L_n) \to \Phi(c_1) - \Phi(c_2) = c_0 < 1$$

where

$$L_n = \frac{-c - \gamma_{11} + a_0}{\sqrt{(1 - \lambda_n)}} \to \frac{1}{\sqrt{2}}(-c + \frac{1}{2}(a_0 - b_0)) = c_1 \ \ a.s.$$

$$U_n = \frac{c - \gamma_{11} + a_0}{\sqrt{(1 - \lambda_n)}} \to \frac{1}{\sqrt{2}}(c + \frac{1}{2}(b_0 - a_0)) = c_2 \ \ a.s..$$

(c) Check condition (5.15): Let O be the open set defined above. Applying $p_n(\mathbf{x}_{(n)}|\theta)\pi(\theta) = \pi(\theta|\mathbf{x}_{(n)})p(\mathbf{x}_{(n)})$, and setting $C = \frac{1-c_0}{2P^\pi(O^c)}$,

$$q(\mathbf{x}_{(n)}) = \frac{p_n(\mathbf{x}_{(n)})}{P^\pi(O^c)} P^\pi(O^c|\mathbf{x}_{(n)}) < Cp_n(\mathbf{x}_{(n)}).$$

Thus

$$H_{\frac{1}{2}}(P_{\theta_0}^{\otimes n}, P_{n,O^c}) > C H_{\frac{1}{2}}(P_{\theta_0}^{\otimes n}, P^{\mathbf{x}_{(n)}}).$$

$P_{\theta_0}^{\otimes n}$ is $N_n(\mu_1, \Sigma_1)$ and $P^{\mathbf{x}_{(n)}}$ is $N_n(\mu_2, \Sigma_2)$ (see Theorem 6.2 on page 134), with $\mu_1 = \mathbf{1}_n \mathbf{1}_2^T \theta_0$, $\Sigma_1 = \mathbf{I}_n$, and $\mu_2 = \mathbf{1}_n \mathbf{1}_2^T \gamma_1$, $\Sigma_2 = \mathbf{I}_n + 2\mathbf{1}_n \mathbf{1}_n^T$. The Hellinger transform $H_n = H_{\frac{1}{2}}(N_n(\mu_1, \Sigma_1), N_n(\mu_2, \Sigma_2))$ of two multivariate normal distributions is given as

$$H_n = \left(\frac{\det(\Sigma_1)\det(\Sigma_2)}{\det(\Sigma)^2}\right)^{\frac{1}{4}} \exp\left(-\frac{1}{8}(\mu_1 - \mu_2)^T \Sigma^{-1}(\mu_1 - \mu_2)\right)$$

where $\Sigma = \frac{1}{2}(\Sigma_1 + \Sigma_2)$. Applying

$$\det(\mathbf{I}_n + \mathbf{a}\mathbf{a}^T) = 1 + \mathbf{a}^T\mathbf{a}, \quad (\mathbf{I}_n + \mathbf{a}\mathbf{a}^T)^{-1} = \mathbf{I}_n - \frac{1}{1 + \mathbf{a}^T\mathbf{a}}\mathbf{a}\mathbf{a}^T,$$

we obtain

$$H_n = \left(\frac{1 + 2n}{(1 + n)^2}\right)^{\frac{1}{4}} \exp\left(-\frac{1}{8}\frac{n}{\sigma^2(1 + n)}(a_0 + b_0)^2\right).$$

H_n converges to zero, but not with the required rate. It holds for all $q_0 < 1$, that $\lim_{n\to\infty} q_0^{-n} H_n = \lim_{n\to\infty} q_0^{-n} n^{-\frac{1}{4}} \text{const} = \infty$.

10.5   Solutions for Chapter 6

1. *Specify (6.23) and (6.24):*

   (a) Matrix form is given by $\mathbf{y} = \mathbf{X}\beta + \epsilon$ with

$$
\mathbf{y} = \begin{pmatrix} y_1 \\ y_2 \\ \vdots \\ y_n \end{pmatrix}, \quad
\mathbf{X} = \begin{pmatrix} 1 & x_1 & z_1 \\ 1 & x_2 & z_2 \\ \vdots & \vdots & \vdots \\ 1 & x_n & z_n \end{pmatrix}, \quad
\beta = \begin{pmatrix} \beta_1 \\ \beta_2 \\ \beta_3 \end{pmatrix}, \quad
\epsilon = \begin{pmatrix} \varepsilon_1 \\ \varepsilon_2 \\ \vdots \\ \varepsilon_n \end{pmatrix}.
$$

   (b) We have $\mathbf{X}^T\Sigma\mathbf{X} = n\mathbf{I}_3$. Applying

$$
\begin{pmatrix} a & b & 0 \\ b & c & 0 \\ 0 & 0 & d \end{pmatrix}^{-1} = \frac{1}{(ac-b^2)}\begin{pmatrix} c & -b & 0 \\ -b & a & 0 \\ 0 & 0 & \frac{ac-b^2}{d} \end{pmatrix}
$$

   with $n\overline{xy} = \sum_{i=1}^{n} x_i y_i$, $n\,\overline{zy} = \sum_{i=1}^{n} z_i y_i$ we get

$$
\Gamma^{-1} = \begin{pmatrix} \frac{4}{3} & -\frac{2}{3} & 0 \\ -\frac{2}{3} & \frac{4}{3} & 0 \\ 0 & 0 & 1 \end{pmatrix}, \quad
\Gamma_1 = \frac{1}{d}\begin{pmatrix} \frac{4}{3}+n & \frac{2}{3} & 0 \\ \frac{2}{3} & \frac{4}{3}+n & 0 \\ 0 & 0 & \frac{d}{1+n} \end{pmatrix},
$$

$$
\gamma_1 = \frac{2}{3d}\begin{pmatrix} n\bar{y}(2+\frac{3n}{2}) + n\overline{xy} + n + 2 \\ n\bar{y} + (2+\frac{3n}{2})n\overline{xy} + n + 2 \\ \frac{3}{2}d(n+1) - 1(n\overline{zy}+1) \end{pmatrix}
$$

   where $d = (\frac{4}{3}+n)^2 - \frac{4}{9}$.

2. *Simple linear regression with Jeffreys prior:*

   (a) By Corollary 6.4, $\pi_{\text{Jeff}}(\theta) \propto \sigma^{-2}$.

   (b) Posterior distributions: Set $s_{xx} = \sum_{i=1}^{n}(x_i - \bar{x})^2 = n\overline{x^2} - n(\bar{x})^2$, $s_{xy} = \sum_{i=1}^{n}(x_i - \bar{x})(y_i - \bar{y}) = n\overline{xy} - n\bar{x}\bar{y}$ with $n\bar{x} = \sum_{i=1}^{n} x_i$, $n\bar{y} = \sum_{i=1}^{n} y_i$ and $n\overline{x^2} = \sum_{i=1}^{n} x_i^2$, $n\overline{xy} = \sum_{i=1}^{n} x_i y_i$. It holds that

$$
\mathbf{X}^T = \begin{pmatrix} 1 \dots 1 \\ x_1 \dots x_n \end{pmatrix}^T,
$$

$$
\mathbf{X}^T\mathbf{X} = \begin{pmatrix} n & n\bar{x} \\ n\bar{x} & n\overline{x^2} \end{pmatrix}, \quad
(\mathbf{X}^T\mathbf{X})^{-1} = \frac{1}{s_{xx}}\begin{pmatrix} \overline{x^2} & -\bar{x} \\ -\bar{x} & 1 \end{pmatrix}.
$$

   The least-squares estimates are $\widehat{\alpha} = \bar{y} - \bar{x}\widehat{\beta}$, $\widehat{\beta} = \frac{s_{xy}}{s_{xx}}$.

i. Applying Theorem 6.7 and (6.40) we obtain

$$\begin{pmatrix} \alpha \\ \beta \end{pmatrix} \Big| \mathbf{y}, \sigma^2 \sim \mathsf{N}_2 \left(\begin{pmatrix} \widehat{\alpha} \\ \widehat{\beta} \end{pmatrix}, \sigma^2 (\mathbf{X}^T \mathbf{X})^{-1} \right). \tag{10.4}$$

ii. The marginal distribution of β in (10.4) is $\beta | \mathbf{y}, \sigma^2 \sim \mathsf{N}(\widehat{\beta}, \frac{\sigma^2}{s_{xx}})$.

iii. Applying Lemma 6.1 on page 132 to (10.4) with

$$\Sigma_{21} \Sigma_{11}^{-1} = -\frac{\bar{x}}{\overline{x^2}}, \quad \Sigma_{22} - \Sigma_{21}^2 \Sigma_{11}^{-1} = \frac{1}{n\overline{x^2}}, \quad \mu_{2|1} = \frac{1}{\overline{x^2}} (\overline{xy} - \alpha \bar{x})$$

gives

$$\beta | (\alpha, \sigma^2, \mathbf{y}) \sim \mathsf{N} \left(\frac{\overline{xy}}{\overline{x^2}} - \alpha \frac{\bar{x}}{\overline{x^2}}, \frac{\sigma^2}{s_{xx}} \right).$$

Note that, the mean in the posterior corresponds to the least-squares estimator of β when the intercept α is known.

iv. Set RSS $= \sum_{i=1}^{n} (y_i - \widehat{y}_i)^2$, with $\widehat{y}_i = \widehat{\alpha} + x_i \widehat{\beta}$. Applying Theorem 6.7 and (6.40),

$$\sigma^2 | \mathbf{y} \sim \mathsf{InvGamma} \left(\frac{n-2}{2}, \frac{\mathsf{RSS}}{2} \right). \tag{10.5}$$

v. From (ii) and (iv) it follows that $(\beta, \sigma^2) | \mathbf{y} \sim \mathsf{NIG}(n-2, \mathsf{RSS}, \widehat{\beta}, s_{xx}^{-1})$. Applying Lemma 6.5 we get

$$\beta \sim \mathsf{t}_1(n-2, \widehat{\beta}, \widehat{\sigma}^2 s_{xx}^{-1}), \quad \text{with } \widehat{\sigma}^2 = \frac{\mathsf{RSS}}{n-4}.$$

3. *Simple linear regression with variance as parameter of interest:*

(a) In case of known α and β it is an i.i.d. sample problem with $\epsilon_i = y_i - \alpha - \beta x_i$. Hence,

$$\pi(\sigma^2 | \mathbf{y}, \beta, \alpha) \propto \frac{1}{\sigma^2} \left(\frac{1}{\sigma^2} \right)^{\frac{n}{2}} \exp \left(-\frac{1}{2\sigma^2} \sum_{i=1}^{n} \epsilon_i^2 \right).$$

This is the kernel of $\mathsf{InvGamma}(\frac{n}{2}, \frac{b}{2})$ with $b = \sum_{i=1}^{n} (y_i - \alpha - \beta x_i)^2$.

(b) The posteriors are different. In Problem 2 (b) the regression parameters are estimated and the degree of freedom is reduced by the number of estimated parameters.

4. *Transformation of simple linear regression model to a centered model:* We eliminate the intercept α, by

$$\bar{y} = \alpha + \beta \bar{x} + \bar{\varepsilon}, \quad y_i - \bar{y} = \alpha - \alpha + \beta(x_i - \bar{x}) + \varepsilon_i - \bar{\varepsilon}.$$

(a) The transformation is given by $\xi_i = \varepsilon_i - \frac{1}{n} \sum_{j=1}^{n} \varepsilon_j$. Set $\mathbf{1}_n = (1, \ldots, 1)^T$ the n-dimensional column vector of ones. Then

$$\xi = \varepsilon - \frac{1}{n} \mathbf{1}_n \mathbf{1}_n^T \varepsilon = \mathbf{P} \varepsilon, \quad \mathbf{P} = \mathbf{I}_n - \frac{1}{n} \mathbf{1}_n \mathbf{1}_n^T$$

where $\mathbf{P}$ is a projection matrix.

(b) Applying the projection properties we obtain a singular covariance matrix
$$\text{Cov}(\xi) = \mathbf{P}\text{Cov}(\varepsilon)\mathbf{P}^T = \sigma^2\mathbf{P}, \quad \det(\mathbf{P}) = 0.$$

(c) Deleting the last observation:
$$\xi_{(-n)} = \varepsilon_{(-n)} - \frac{1}{n}\mathbf{1}_{n-1}\mathbf{1}_n^T\varepsilon = \mathbf{A}\varepsilon_{(-n)} - \frac{1}{n}\varepsilon_n\mathbf{1}_{n-1}$$

with $\mathbf{A} = \mathbf{I}_{n-1} - \frac{1}{n}\mathbf{1}_{n-1}\mathbf{1}_{n-1}^T$, where $\mathbf{A}$ is not a projection matrix. Further
$$\text{Cov}(\xi_{(-n)}) = \sigma^2(\mathbf{A}\mathbf{A} + \frac{1}{n^2}\mathbf{1}_{n-1}\mathbf{1}_{n-1}^T) = \sigma^2\Sigma$$

with $\Sigma = \mathbf{I}_{n-1} - \frac{1}{n}\mathbf{1}_{n-1}\mathbf{1}_{n-1}^T$. Let e be an eigenvector orthogonal to $\mathbf{1}_{n-1}$. Then $\Sigma e = e$, with corresponding eigenvalue 1. Consider the eigenvector $v = a\mathbf{1}_{n-1}$. Then $\Sigma v = n^{-1}v$ and the eigenvalue is $\frac{1}{n} \neq 0$. Thus $\det(\Sigma) = \frac{1}{n} > 0$.

(d) Consider model (6.159) with deleted last observation and $\xi_{(-n)} \sim N(0, \sigma^2\Sigma)$, where
$$\Sigma = \mathbf{I}_{n-1} - \frac{1}{n}\mathbf{1}_{n-1}\mathbf{1}_{n-1}^T, \quad \Sigma^{-1} = \mathbf{I}_{n-1} + \mathbf{1}_{n-1}\mathbf{1}_{n-1}^T$$
$$(\mathbf{x}_{(-n)} - \bar{x}\mathbf{1}_{n-1})^T\Sigma^{-1}(\mathbf{x}_{(-n)} - \bar{x}\mathbf{1}_{n-1}) = s_{xx}$$
$$(\mathbf{x}_{(-n)} - \bar{x}\mathbf{1}_{n-1})^T\Sigma^{-1}(\mathbf{y}_{(-n)} - \bar{y}\mathbf{1}_{n-1}) = s_{xy}$$

i. Model (6.159) with deleted last observation and Jeffreys prior: Applying Theorem 6.7 and using the expressions above we have $\beta|\mathbf{y}, \sigma^2 \sim N_1\left(\frac{s_{xy}}{s_{xx}}, \frac{\sigma^2}{s_{xx}}\right)$ and $\sigma^2|\mathbf{y} \sim \text{InvGamma}\left(\frac{n-2}{2}, \frac{RSS}{2}\right)$, which is the same result as in model (10.5).

ii. Model (6.159) with deleted last observation and conjugate prior, $\sigma^2 = 1$: Applying Corollary 6.1 and using the expressions above, we have $\beta|\mathbf{y} \sim N_1(\mu_1, \sigma_0^2)$ with $\mu_1 = \frac{\lambda_2 s_{xy} + \gamma_b}{\lambda_2 s_{xx} + 1}$ and $\sigma_0^2 = \frac{\lambda_2}{\lambda_2 s_{xx} + 1}$. This posterior is different from the posterior (6.71), given in Example 6.9 on page 151.

5. *Precision parameter*: The conjugate family includes the normal-gamma distributions, defined as follows. A vector valued random variable X with sample space $\mathcal{X} \subseteq \mathbb{R}^p$ and a positive random scalar λ have a normal-gamma distribution
$$(X, \lambda) \sim \text{NGam}(a, b, \mu, \mathbf{P})$$

iff
$$X|\lambda \sim N_p(\mu, \lambda^{-1}\mathbf{P}^{-1}), \quad \lambda \sim \text{Gamma}\left(\frac{a}{2}, \frac{b}{2}\right).$$

The kernel of the density of $\text{NGam}(a, b, \mu, \mathbf{P})$ is given as
$$\lambda^{\frac{a+p-2}{2}} \exp\left(-\frac{\lambda}{2}(b + (X - \mu)^T\mathbf{P}(X - \mu))\right).$$

Set $\lambda = \sigma^{-2}$. Suppose that $\theta = (\beta, \lambda) \sim \mathsf{NGam}(a_0, b_0, \gamma_0, \mathbf{P}_0)$. Then

$$\pi(\theta|\mathbf{y}) \propto \lambda^{\frac{a_0+p-2}{2}} \exp\left(-\frac{\lambda}{2}(b_0 + (\beta - \gamma_0)^T \mathbf{P}_0(\beta - \gamma_0))\right)$$

$$\times \lambda^{\frac{n}{2}} \exp\left(-\frac{\lambda}{2}(\mathbf{y} - \mathbf{X}\beta)^T(\mathbf{y} - \mathbf{X}\beta)\right).$$

Applying (6.25) given in Lemma 6.4 on page 138 we obtain $\theta|\mathbf{y} \sim \mathsf{NGam}(a_1, b_1, \gamma_1, \mathbf{P}_1)$ where $a_1 = a_0 + n$,

$$b_1 = b_0 + (\mathbf{y} - \mathbf{X}\gamma_1)^T(\mathbf{y} - \mathbf{X}\gamma_1) + (\gamma_0 - \gamma_1)^T \mathbf{P}_0(\gamma_0 - \gamma_1),$$

$$\mathbf{P}_1 = \mathbf{P}_0 + \mathbf{X}^T\mathbf{X}, \quad \gamma_1 = \mathbf{P}_1^{-1}(\mathbf{P}_0\gamma + \mathbf{X}^T\mathbf{y}).$$

6. *Jeffreys prior for the scale parameter*: Set $\eta^T = (\eta_1^T, \eta_2) = (\beta^T, \sigma)$ and $\theta^T = (\theta_1^T, \theta_2) = (\beta^T, \sigma^2)$. From Theorem 6.6 we know that

$$I(\theta) = \begin{pmatrix} \frac{1}{\theta_2}\mathbf{X}^T\mathbf{X} & 0 \\ 0 & \frac{n}{2\theta_2^2} \end{pmatrix}.$$

Using transformation $\theta_1 = h_1(\eta) = \eta_1$, $\theta_2 = h_2(\eta) = \eta_2^2$, we get the Jacobian matrix

$$\mathsf{J} = \begin{pmatrix} \mathbf{I}_p & 0 \\ 0 & 2\eta_2 \end{pmatrix}.$$

Applying (3.31), with $I(\eta) = \mathsf{J}^T I(\theta)\mathsf{J}$ at $\theta = h(\eta)$, we obtain

$$I(\eta) = \begin{pmatrix} \mathbf{I}_p & 0 \\ 0 & 2\eta_2 \end{pmatrix} \begin{pmatrix} \frac{1}{\eta_2^2}\mathbf{X}^T\mathbf{X} & 0 \\ 0 & \frac{n}{2\eta_2^4} \end{pmatrix} \begin{pmatrix} \mathbf{I}_p & 0 \\ 0 & 2\eta_2 \end{pmatrix} = \frac{1}{\eta_2^2}\begin{pmatrix} \mathbf{X}^T\mathbf{X} & 0 \\ 0 & 2n \end{pmatrix}.$$

Using Definition 3.4 we obtain $\pi(\eta) \propto \eta_2^{-(p+1)}$. Under the independence assumption we calculate separately $\pi(\eta_1) \propto \mathrm{const}$, and $\pi(\eta_2) \propto \eta_2^{-1}$, so that $\pi(\eta) \propto \eta_2^{-1}$.

7. *Three sample problem*:

(a) Reformulation as linear model is given by

$$\mathbf{y} = \begin{pmatrix} X^T \\ Y^T \\ Z^T \end{pmatrix} = \begin{pmatrix} \mathbf{1}_m & 0 & 0 \\ 0 & \mathbf{1}_m & 0 \\ 0 & 0 & \mathbf{1}_m \end{pmatrix} \begin{pmatrix} 1 & 0 \\ 0 & 1 \\ -1 & -1 \end{pmatrix} \begin{pmatrix} \mu_1 \\ \mu_2 \end{pmatrix} + \epsilon,$$

with $\epsilon \sim \mathsf{N}_{3m}(0, \sigma^2\Sigma)$

$$\Sigma = \begin{pmatrix} \lambda_1\mathbf{I}_m & 0 & 0 \\ 0 & \lambda_2\mathbf{I}_m & 0 \\ 0 & 0 & \lambda_3\mathbf{I}_m \end{pmatrix} \quad \text{and} \quad \mathbf{X} = \begin{pmatrix} \mathbf{1}_m & 0 \\ 0 & \mathbf{1}_m \\ -\mathbf{1}_m & -\mathbf{1}_m \end{pmatrix}.$$

(b) Applying Theorem 6.4 we obtain the posterior

$$\begin{pmatrix} \mu_1 \\ \mu_2 \end{pmatrix} \mid \sigma^2, X, Y, Z \sim N_2(\gamma_1, \sigma^2\Gamma_1), \quad d = \frac{1}{n((2+\frac{1}{n})^2 - 1)},$$

$$\Gamma_1 = d \begin{pmatrix} 2+\frac{1}{n} & -1 \\ -1 & 2+\frac{1}{n} \end{pmatrix}, \quad \gamma_1 = d \begin{pmatrix} (2+\frac{1}{n})(\bar{x} - \bar{z}) - (\bar{y} - \bar{z}) \\ (2+\frac{1}{n})(\bar{y} - \bar{z}) - (\bar{x} - \bar{z}) \end{pmatrix}.$$

(c) Posterior based on two samples: Applying Theorem 6.3 with $\tau^2 = \sigma^2$ we obtain

$$\begin{pmatrix} \mu_1 \\ \mu_2 \end{pmatrix} \mid \sigma^2, X, Y \sim N_2\left(\begin{pmatrix} \frac{n\bar{Y}}{n+1} \\ \frac{n\bar{X}}{n+1} \end{pmatrix}, \sigma^2 \begin{pmatrix} \frac{1}{n+1} & 0 \\ 0 & \frac{1}{n+1} \end{pmatrix} \right).$$

(d) Comparison of posteriors covariance matrices: The posterior based on three samples has a smaller covariance matrix. It holds that $\mathbf{I}_2 - (n+1)\Gamma_1 = \frac{n+1}{d}D$ with

$$D = \begin{pmatrix} -(1+\frac{1}{n}) & 1 \\ 1 & -(1+\frac{1}{n}) \end{pmatrix}, \quad \det(D) = (1+\frac{1}{n})^2 - 1 > 0.$$

8. *Parallel regression lines:*

   (a) Set $x = (x_1, \ldots, x_n)^T$, $y = (y_1, \ldots, y_n)^T$, $z = (z_1, \ldots, z_n)^T$, $\varepsilon = (\varepsilon_1, \ldots, \varepsilon_n)^T$, and $\xi = (\xi_1, \ldots, \xi_n)^T$. Reformulation of (6.161) as linear model is given by

$$\mathbf{y} = \begin{pmatrix} y \\ z \end{pmatrix} = \begin{pmatrix} \mathbf{1}_n & 0 & x \\ 0 & \mathbf{1}_n & x \end{pmatrix} \begin{pmatrix} \alpha \\ \gamma \\ \beta \end{pmatrix} + \begin{pmatrix} \varepsilon \\ \xi \end{pmatrix},$$

   with error distribution $N_{2n}(0, \sigma^2\mathbf{I}_{2n})$.

   (b) Set $n\bar{y} = \sum_{i=1}^n y_i$, $n\bar{z} = \sum_{i=1}^n z_i$, and $n\overline{x^2} = \sum_{i=1}^n x_i^2$. We have

$$\mathbf{X}^T\Sigma^{-1}\mathbf{X} = \begin{pmatrix} n & 0 & 0 \\ 0 & n & 0 \\ 0 & 0 & 2n\overline{x^2} \end{pmatrix} \text{ and } \mathbf{X}^T\Sigma^{-1}\mathbf{y} = \begin{pmatrix} n\bar{y} \\ n\bar{z} \\ x^T(y+z) \end{pmatrix}.$$

   Applying Corollary 6.1

$$\begin{pmatrix} \alpha \\ \gamma \\ \beta \end{pmatrix} \mid \sigma^2, \mathbf{y} \sim N_3\left(\begin{pmatrix} \frac{n}{n+\lambda}\bar{y} \\ \frac{n}{n+\lambda}\bar{z} \\ d\,x^T(y+z) \end{pmatrix}, \sigma^2 \begin{pmatrix} \frac{1}{n+\lambda} & 0 & 0 \\ 0 & \frac{1}{n+\lambda} & 0 \\ 0 & 0 & d \end{pmatrix} \right),$$

so that

$$\beta|\sigma^2, \mathbf{y} \sim \mathsf{N}(dx^T(y+z), \sigma^2 d) \text{ with } d = (2n\overline{x^2} + \lambda)^{-1}.$$

(c) Applying Theorem 6.5, especially (6.59), (6.66), (6.69) gives $\sigma^2|\mathbf{y} \sim$ InvGamma$(\frac{a+2n}{2}, \frac{b_1}{2})$ with

$$b_1 = b + y^T y + z^T z - \frac{1}{n+\lambda}(n\bar{y})^2 - \frac{1}{n+\lambda}(n\bar{z})^2 - d(x^T(y+z))^2.$$

(d) The results above give $(\beta, \sigma^2)|\mathbf{y} \sim \mathsf{NIG}(a+2n, b_1, dx^T(y+z), d)$. Lemma 6.5 delivers

$$\beta|\mathbf{y} \sim \mathsf{t}(a+2n, d\,x^T(y+z), \frac{b_1}{a+2n}d).$$

9. *Linear mixed model*: Set $n\overline{xy} = \sum_{i=1}^n x_i y_i$, $n\overline{xz} = \sum_{i=1}^n x_i z_i$. A conjugate prior is $\mathsf{N}(\beta_0, \sigma_0^2)$. Applying Theorem 6.10 with $\sigma^2 = 1$ gives

$$\begin{pmatrix} \beta \\ \gamma \end{pmatrix} | \mathbf{y} \sim \mathsf{N}_2(\alpha_1, \Psi_1), \quad \Psi_1 = d\begin{pmatrix} n+1 & -n\overline{xz} \\ -n\overline{xz} & n+\sigma_0^{-2} \end{pmatrix}, \quad \alpha_1 = d\begin{pmatrix} m_1 \\ m_2 \end{pmatrix}$$

with

$$d = \left((n+1)(n+\sigma_0^{-2}) - n^2\overline{xz}^2\right)^{-1}, \quad m_1 = (n+1)(\overline{xz} + \sigma_0^{-2}\beta_0) - n^2\overline{xz}\,\overline{xy}.$$

Hence $\beta|\mathbf{y} \sim \mathsf{N}(d\,m_1, d\,(n+1))$.

10. *Two correlated lines*: Set $x = (x_1, \ldots, x_n)^T$, $y = (y_1, \ldots, y_n)^T$, $z = (z_1, \ldots, z_n)^T$, $\varepsilon = (\varepsilon_1, \ldots, \varepsilon_n)^T$, and $\xi = (\xi_1, \ldots, \xi_n)^T$.

(a) Reformulation of (6.162) as multivariate model is given by

$$\mathbf{Y} = \begin{pmatrix} y & z \end{pmatrix} = \mathbf{XB} + \begin{pmatrix} \varepsilon & \xi \end{pmatrix}, \text{ with } \mathbf{X} = \begin{pmatrix} 1_n & x \end{pmatrix}, \mathbf{B} = \begin{pmatrix} \alpha_y & \alpha_z \\ \beta_y & \beta_z \end{pmatrix}, \tag{10.6}$$

where

$$\mathbf{Y}|\theta \sim \mathsf{MN}_{n,2}(\mathbf{XB}, \mathbf{I}_n, \Sigma) \text{ with } \Sigma = \begin{pmatrix} \sigma_1^2 & \sigma_{12} \\ \sigma_{12} & \sigma_2^2 \end{pmatrix}.$$

(b) Set $n\overline{xy} = \sum_{i=1}^n x_i y_i$, $n\overline{xz} = \sum_{i=1}^n x_i z_i$. Applying Theorem 6.11 gives the posterior $\mathsf{NIW}(\nu_1, \mathbf{B}_1, \mathbf{C}_1, \Sigma_1)$ with $\nu_1 = \nu + n$,

$$\mathbf{B}_1 = \begin{pmatrix} \tilde{\alpha}_y & \tilde{\alpha}_z \\ \tilde{\beta}_y & \tilde{\beta}_z \end{pmatrix} = \begin{pmatrix} \frac{c_1}{1+nc_1}\bar{y} & \frac{c_1}{1+nc_1}\bar{z} \\ \frac{c_2}{1+nc_2}\overline{xy} & \frac{c_2}{1+nc_2}\overline{xz} \end{pmatrix}, \quad \mathbf{C}_1 = \begin{pmatrix} \frac{c_1}{1+c_1 n} & 0 \\ 0 & \frac{c_2}{1+c_2 n} \end{pmatrix}. \tag{10.7}$$

Further $\mathsf{E}(\Sigma|\mathbf{Y}) = \frac{1}{\nu_1-3}\Sigma_1$. Set $\tilde{y}_i = \tilde{\alpha}_y + \tilde{\beta}_y x_i$ and $\tilde{z}_i = \tilde{\alpha}_z + \tilde{\beta}_z x_i$. Then

$$\mathsf{E}(\sigma_{12}|\mathbf{Y}) = \frac{1}{\nu+n-3}\left(\sum_{i=1}^n (y_i - \tilde{y}_i)(z_i - \tilde{z}_i) + \frac{1}{c_1}\tilde{\alpha}_y\tilde{\alpha}_y + \frac{1}{c_2}\tilde{\beta}_y\tilde{\beta}_z\right).$$

10.6 Solutions for Chapter 7

1. *Sample from* $\mathsf{Gamma}(\nu, \theta)$: Likelihood and prior, $\theta \sim \mathsf{Gamma}(\alpha, \beta)$, are given by

$$\ell(\theta|\mathbf{x}) \propto \theta^{n\nu} \exp(-\sum_{i=1}^{n} x_i \theta), \quad \pi(\theta) \propto \theta^{\alpha-1} \exp(-\beta\theta).$$

The posterior, $\pi(\theta|\mathbf{x}) \propto \pi(\theta)\ell(\theta|\mathbf{x})$, is

$$\pi(\theta|\mathbf{x}) \propto \theta^{n\nu+\alpha-1} \exp(-(\sum_{i=1}^{n} x_i + \beta)\theta),$$

i.e., $\theta|\mathbf{x} \sim \mathsf{Gamma}(n\nu + \alpha, \sum_{i=1}^{n} x_i + \beta)$.

(a) Applying (7.20)

$$\widehat{\theta}_{L_2}(\mathbf{x}) = \mathsf{E}(\theta|\mathbf{x}) = \frac{n\nu + \alpha}{\sum_{i=1}^{n} x_i + \beta}.$$

(b) Applying mode's formula

$$\widehat{\theta}_{\mathrm{MAP}}(\mathbf{x}) = \mathsf{mode}(\theta|\mathbf{x}) = \frac{n\nu + \alpha - 1}{\sum_{i=1}^{n} x_i + \beta}.$$

(c) Set F^{-1} for the quantile function of $\mathsf{Gamma}(n\nu + \alpha, \sum_{i=1}^{n} x_i + \beta)$.

  i. For each $a \in [0, \alpha]$ the quantiles $q_{\mathrm{low}}(a) = F^{-1}(a)$ and $q_{\mathrm{upp}}(a) = F^{-1}(a + 1 - \alpha)$ give α–credible interval $[q_{\mathrm{low}}(a), q_{\mathrm{upp}}(a)]$.
  ii. Find a^* such that $q_{\mathrm{upp}}(a) - q_{\mathrm{low}}(a)$ is minimal.
  iii. The HPD interval is given by $[q_{\mathrm{low}}(a^*), q_{\mathrm{upp}}(a^*)]$.

2. *Sample from* $\mathsf{N}(\mu, \sigma^2)$, $\theta = (\mu, \sigma^2)$: Let $\theta \sim \mathsf{NIG}(a, b, \gamma, \lambda)$ be conjugate prior. Then $\theta|\mathbf{x} \sim \mathsf{NIG}(a_1, b_1, \gamma_1, \lambda_1)$, as given in Example 7.3.

(a) Set $s_{xx} = \sum_{i=1}^{n}(x_i - \overline{x})^2$ and $n\overline{x} = \sum_{i=1}^{n} x_i$. The likelihood function is

$$\ell(\theta|\mathbf{x}) \propto (\sigma^2)^{-\frac{n}{2}} \exp\left(-\frac{1}{2\sigma^2}(s_{xx} + n(\mu - \overline{x})^2)\right)$$

and $\widehat{\mu}_{\mathrm{MLE}} = \arg\max_\mu \ell(\mu, \sigma^2|\mathbf{x}) = \overline{x}$. Further

$$\widehat{\sigma}^2_{\mathrm{MLE}} = \arg\max_{\sigma^2} \ell(\overline{x}, \sigma^2|\mathbf{x}) = \arg\max_{\sigma^2}\left(\frac{1}{\sigma^{2n}} \exp\left(-\frac{s_{xx}}{2\sigma^2}\right)\right) = \frac{1}{n} s_{xx}.$$

(b) Set $z_i = x_i - \mu$. Then $\overline{x} - \mu = \overline{z}$, $s_{zz} = \sum_{i=1}^{n}(z_i - \overline{z})^2$, and $\mathsf{E}z_i = 0$, where

$$\mathsf{Cov}(\widehat{\mu}_{\mathrm{MLE}}, \widehat{\sigma}^2_{\mathrm{MLE}}|\theta) = \frac{1}{n}\mathsf{E}\left(\overline{z}\sum_{i=1}^{n}(z_i - \overline{z})^2\right) = 0,$$

since $\mathsf{E}(z_i z_j z_k) = 0$ for all $i, j, k = 1, \ldots, n$, $\mathsf{E}\overline{z}^3 = 0$, and $\mathsf{E}(\overline{z} z_i^2) = 0$.

(c) $\theta|\mathbf{x} \sim \text{NIG}(a_1, b_1, \gamma_1, \lambda_1)$ implies $\mu|\mathbf{x}, \sigma^2 \sim \text{N}(\gamma_1, \sigma^2\lambda_1)$. The mode of a normal distribution is the expectation, thus

$$\widehat{\mu}_{\text{MAP}} = \gamma_1 = \frac{1}{n+\lambda}(\lambda\gamma + n\overline{x}).$$

Further $\sigma^2|\mathbf{x} \sim \text{InvGamma}(\frac{a_1}{2}, \frac{b_1}{2})$ and the mode formula gives

$$\widehat{\sigma}^2_{\text{MAP}} = \frac{b_1}{a_1+2}, \quad b_1 = b + s_{xx} + \frac{n}{n+1}\overline{x}^2.$$

Set $\sum_{i=1}^{n} x_i^2 = n\overline{x^2}$ and use $s_{xx} + n\overline{x}^2 = n\overline{x^2}$. We have for $c_0 = b + \frac{\lambda_1^2}{\lambda^2}\gamma^2$, $c_1 = -2\frac{\lambda_1^2}{\lambda^2}\gamma$, $c_2 = -\lambda_1^2 n^2$, and $c_3 = n$, that

$$\widehat{\sigma}^2_{\text{MAP}} = \frac{1}{a+n+2}(c_0 + c_1\overline{x} + c_2\overline{x}^2 + c_3\overline{x^2}).$$

(d) Set $\mu = 0$ and $\sigma^2 = 1$. Then $X_1, \ldots, X_n$ i.i.d. $\text{N}(0,1)$, with $\text{E}\overline{x} = 0$, $\text{E}\overline{x}^3 = 0$, $\text{E}\overline{x^2}\,\overline{x} = 0$, and $\text{E}\overline{x}^2 = \frac{1}{n}$. For $\gamma \neq 0$

$$\text{Cov}(\widehat{\mu}_{\text{MAP}}, \widehat{\sigma}^2_{\text{MAP}}) = \text{E}\left(\frac{n\overline{x}}{a+\lambda}\widehat{\sigma}^2_{\text{MAP}}\right) = -2\gamma\frac{\lambda+n}{\lambda^2(a+n+2)} \neq 0.$$

3. *Simple linear regression*: Applying (6.64) and (6.62) we get $\theta|\mathbf{y} \sim \text{N}_2(\gamma_1, \sigma^2\Gamma_1)$, where $\gamma_1 = (\tilde{\alpha}, \tilde{\beta})^T$ and

$$\Gamma_1 = \begin{pmatrix} \frac{\lambda_1}{1+\lambda_1 n} & 0 \\ 0 & \frac{\lambda_2}{1+\lambda_2 s_{xx}} \end{pmatrix}, \quad \tilde{\alpha} = \frac{\lambda_1 n\overline{y} + \gamma_a}{1 + n\lambda_1}, \quad \tilde{\beta} = \frac{\lambda_2 n\overline{xy} + \gamma_b}{1 + \lambda_2 s_{xx}}.$$

(a) We have $\mu(z)|\mathbf{y} \sim \text{N}(\tilde{\mu}(z), \sigma^2 s^2(z))$ with $\tilde{\mu}(z) = \tilde{\alpha} + \tilde{\beta}z$ and

$$s^2(z) = \frac{\lambda_1}{1 + \lambda_1 n} + \frac{z^2\lambda_2}{1 + \lambda_2 s_{xx}}.$$

The HPD α_0-credible interval is given by

$$[\tilde{\mu}(z) - \mathsf{z}_{(1-\frac{\alpha_0}{2})}\,\sigma s(z), \tilde{\mu}(z) + \mathsf{z}_{(1-\frac{\alpha_0}{2})}\,\sigma s(z)],$$

where $\mathsf{z}_{(1-\frac{\alpha_0}{2})}$ is the $(1 - \frac{\alpha_0}{2})$–quantile of $\text{N}(0,1)$, i.e., $\Phi(\mathsf{z}_\gamma) = \gamma$.

(b) Applying Lemma 6.2 on page 132 to

$$\begin{pmatrix} y_f \\ \theta \end{pmatrix}\bigg|\mathbf{y} \sim \text{N}_3\left(\begin{pmatrix} 1 & z \\ 1 & 0 \\ 0 & 1 \end{pmatrix}\gamma_1, \sigma^2\begin{pmatrix} s^2(z)+1 & (1,z)\Gamma_1 \\ \Gamma_1^T\begin{pmatrix} 1 \\ z \end{pmatrix} & \Gamma_1 \end{pmatrix}\right)$$

gives

$y_f|\mathbf{y} \sim \text{N}\left(\tilde{\mu}(z), \sigma^2(s^2(z)+1)\right)$. The prediction interval is given by

$$[\tilde{\mu}(z) - \mathsf{z}_{(1-\frac{\alpha_0}{2})}\,\sigma\sqrt{s^2(z)+1}, \tilde{\mu}(z) + \mathsf{z}_{(1-\frac{\alpha_0}{2})}\,\sigma\sqrt{s^2(z)+1}].$$

4. *Predictive distribution in binomial model*: In Example 2.11 the posterior is calculated as $\theta|X \sim \text{Beta}(\alpha, \beta)$ with $\alpha = \alpha_0 + x$ and $\beta = \beta_0 + n - x$.

(a) Applying (7.33), the predictive distribution,

$$
\begin{aligned}
\pi(x_f|x) &= \int_0^1 \binom{n_f}{x_f} \theta^{x_f}(1-\theta)^{(n_f-x_f)} \frac{1}{B(\alpha,\beta)} \theta^{\alpha-1}(1-\theta)^{\beta-1} \, d\theta \\
&= \binom{n_f}{x_f} \frac{1}{B(\alpha,\beta)} \int_0^1 \theta^{x_f+\alpha-1}(1-\theta)^{(n_f-x_f+\beta-1)} \, d\theta \\
&= \binom{n_f}{x_f} \frac{B(\alpha+x_f, \beta+n-x_f)}{B(\alpha,\beta)},
\end{aligned}
$$

is a beta-binomial distribution,

$$
x_f|x \sim \text{BetaBin}(n_f, \alpha_0 + x, \beta_0 + n - x).
$$

(b) A possible R code:

```
library(VGAM); k<-0:5; dbetabinom.ab(k,1,4,3)
```

5. *Two independent lines, interested in averaged slope*:

(a) Set $x = (x_1, \ldots, x_n)^T$, $y = (y_1, \ldots, y_n)^T$, $z = (z_1, \ldots, z_n)^T$, $\varepsilon = (\varepsilon_1, \ldots, \varepsilon_n)^T$, and $\xi = (\xi_1, \ldots, \xi_n)^T$. Reformulation of (7.70) as linear model is given by

$$
\mathbf{y} = \begin{pmatrix} y \\ z \end{pmatrix} = \begin{pmatrix} \mathbf{1}_n & x & 0 & 0 \\ 0 & 0 & \mathbf{1}_n & x \end{pmatrix} \begin{pmatrix} \alpha_y \\ \beta_y \\ \alpha_z \\ \beta_z \end{pmatrix} + \begin{pmatrix} \varepsilon \\ \xi \end{pmatrix}, \tag{10.8}
$$

with error distribution $N_{2n}(0, \sigma^2 \mathbf{I}_{2n})$.

(b) We have

$$
\mathbf{X}^T \Sigma^{-1} \mathbf{X} = n\mathbf{I}_4 \text{ and } \mathbf{X}^T \Sigma^{-1} \mathbf{y} = n(\bar{y}, \overline{xy}, \bar{z}, \overline{xz})^T.
$$

Applying Corollary 6.1, we get

$$
\begin{pmatrix} \alpha_y \\ \beta_y \\ \alpha_z \\ \beta_z \end{pmatrix} \bigg| \sigma^2, \mathbf{y} \sim N_4 \left(\begin{pmatrix} \tilde{\alpha}_y \\ \tilde{\beta}_y \\ \tilde{\alpha}_z \\ \tilde{\beta}_z \end{pmatrix}, \sigma^2 \begin{pmatrix} \frac{\lambda_1}{1+n\lambda_1} & 0 & 0 & 0 \\ 0 & \frac{\lambda_2}{1+n\lambda_2} & 0 & 0 \\ 0 & 0 & \frac{\lambda_3}{1+n\lambda_3} & 0 \\ 0 & 0 & 0 & \frac{\lambda_4}{1+n\lambda_4} \end{pmatrix} \right),
$$

with

$$
\tilde{\alpha}_y = \frac{n\lambda_1 \bar{y} + m_1}{1 + n\lambda_1}, \quad \tilde{\beta}_y = \frac{n\lambda_2 \overline{xy} + m_2}{1 + n\lambda_2},
$$

$$\tilde{\alpha}_z = \frac{n\lambda_3 \bar{z} + m_3}{1 + n\lambda_3}, \quad \tilde{\beta}_z = \frac{n\lambda_4 \overline{xz} + m_4}{1 + n\lambda_4}.$$

Then $\eta|\sigma^2, \mathbf{y} \sim N(\tilde{\eta}, \sigma^2 s_y^2)$ with

$$\tilde{\eta} = \frac{1}{2}(\tilde{\beta}_y + \tilde{\beta}_z), \quad s_y^2 = \frac{1}{4}\left(\frac{\lambda_2}{1 + n\lambda_2} + \frac{\lambda_4}{1 + n\lambda_4}\right).$$

Further, $\sigma^2|\mathbf{y} \sim \mathsf{InvGamma}((a+2n)/2, b_1(\mathbf{y})/2)$ with $b_1(\mathbf{y})$ given in (6.67) as

$$b_1(\mathbf{y}) = b + \sum_{i=1}^{n}(y_i - \tilde{y}_i)^2 + \sum_{i=1}^{n}(z_i - \tilde{z}_i)^2$$
$$+ \frac{1}{\lambda_1}(m_1 - \tilde{\alpha}_y)^2 + \frac{1}{\lambda_2}(m_2 - \tilde{\beta}_y)^2$$
$$+ \frac{1}{\lambda_3}(m_3 - \tilde{\alpha}_z)^2 + \frac{1}{\lambda_4}(m_1 - \tilde{\beta}_z)^2$$

with $\tilde{y}_i = \tilde{\alpha}_y + \tilde{\beta}_y x_i$ and $\tilde{z}_i = \tilde{\alpha}_z + \tilde{\beta}_z x_i$. From Lemma 6.5 it follows that

$$\eta|\mathbf{y} \sim t(a + 2n, \tilde{\eta}, \frac{b_1(\mathbf{y})}{a + 2n}s_y^2).$$

(c) For known σ^2 the posterior is $\eta|\sigma^2, \mathbf{y} \sim N(\tilde{\eta}, \sigma^2 s_y^2)$. Thus

$$C_1(\mathbf{y}, \sigma^2) = \{\eta : \tilde{\eta} - z_{(1-\frac{\alpha}{2})}\sigma s_y \leq \eta \leq \tilde{\eta} + z_{(1-\frac{\alpha}{2})}\sigma s_y\}. \qquad (10.9)$$

(d) The Bayes L_2 estimate of σ^2 is the expected value of $\mathsf{InvGamma}((a + 2n)/2, b_1(\mathbf{y})/2)$. Thus

$$\tilde{\sigma}_1^2 = \frac{b_1(\mathbf{y})}{a + 2n - 2}.$$

(e) The HPD α-credible interval is given as

$$C_1(\mathbf{y}) = \{\eta : \tilde{\eta} - t_{(a+2n, 1-\frac{\alpha}{2})}s_1(\mathbf{y}) \leq \eta \leq \tilde{\eta} + t_{(a+2n, 1-\frac{\alpha}{2})}s_1(\mathbf{y})\}, \qquad (10.10)$$

with

$$s_1(\mathbf{y})^2 = \frac{b_1(\mathbf{y})}{a + 2n}s_y^2 = \frac{b_1(\mathbf{y})}{a + 2n}\frac{1}{4}\left(\frac{\lambda_2}{1 + n\lambda_2} + \frac{\lambda_4}{1 + n\lambda_4}\right)$$

where $t_{(df, \gamma)}$ is the γ-quantile of $t_1(df, 0, 1)$.

6. *Averaged observations*: Model (7.71) is a simple linear regression model with $\Sigma = \frac{1}{2}\mathbf{I}_n$.

(a) Applying Theorem 6.5 and Lemma 6.5,

$$\beta_u|\mathbf{u}, \sigma^2 \sim N\left(\tilde{\beta}_u, \sigma^2 \frac{c_2}{1 + 2nc_2}\right), \quad \sigma^2|\mathbf{u} \sim \mathsf{InvGamma}((a+n)/2, b_1(\mathbf{u})/2),$$

$$\beta_u | \mathbf{u} \sim \mathsf{t}\left(a + n, \tilde{\beta}_u, s_2(\mathbf{u})^2\right), \text{ with } s_2(\mathbf{u})^2 = \frac{b_1(\mathbf{u})c_2}{(a+n)(1+2nc_2)}.$$

Further $\tilde{u}_i = \tilde{\alpha}_u + \tilde{\beta}_u x_i$, where

$$\tilde{\alpha}_u = \frac{2nc_1\overline{u}}{1 + 2nc_1}, \quad \tilde{\beta}_u = \frac{2nc_2\overline{xu}}{1 + 2nc_2},$$

$$b_1(\mathbf{u}) = b + 2\sum_{i=1}^{n}(u_i - \tilde{u}_i)^2 + \frac{1}{c_1}\tilde{\alpha}_u^2 + \frac{1}{c_2}\tilde{\beta}_u^2.$$

(b) For known σ^2 the posterior is $\eta | (\sigma^2, \mathbf{y}) \sim \mathsf{N}(\tilde{\beta}_u, \sigma^2 s_u^2)$ with $s_u^2 = \frac{c_2}{1+2nc_2}$. Thus

$$C_2(\mathbf{u}, \sigma^2) = \{\eta : \tilde{\eta} - \mathsf{z}_{(1-\frac{\alpha}{2})}\,\sigma s_u \leq \eta \leq \tilde{\eta} + \mathsf{z}_{(1-\frac{\alpha}{2})}\,\sigma s_u\}.$$

(c) The Bayes L_2 estimate of σ^2 is the expected value of $\mathsf{InvGamma}((a + n)/2, b_1(\mathbf{u})/2)$. Thus

$$\tilde{\sigma}_2^2 = \frac{b_1(\mathbf{u})}{a+n-2}.$$

(d) The HPD α-credible interval for $\eta = \beta_u$ is

$$C_2(\mathbf{u}) = \{\eta : \tilde{\beta}_u - \mathsf{t}_{(a+n,1-\frac{\alpha}{2})}\, s_2(\mathbf{u}) \leq \eta \leq \tilde{\beta}_u + \mathsf{t}_{(a+n,1-\frac{\alpha}{2})}\, s_2(\mathbf{u})\},$$
$$(10.11)$$

where $\mathsf{t}_{(df,\gamma)}$ is the γ-quantile of students $\mathsf{t}_1(df, 0, 1)$ distribution with df degrees of freedom.

7. *Comparing the models in the two previous problems:*

(a) The prior of (η, σ^2) in model (7.70) is $\mathsf{NIG}(a, b, \frac{1}{2}(m_2+m_4), \frac{1}{4}(\lambda_2+\lambda_4))$. The prior of (η, σ^2) in model (7.71) is $\mathsf{NIG}(a, b, 0, c_2)$. Set $m_2 = m_4 = 0$ and $\lambda_2 = \lambda_4 = 2c_2$. Then (η, σ^2) has the prior $\mathsf{NIG}(a, b, 0, c_2)$ in both cases.

(b) For known σ^2 the HPD α-credible intervals for η coincide, since

$$\tilde{\beta}_u = \frac{c_2}{1 + 2nc_2} 2 \sum_{i=1}^{n} \frac{1}{2}(y_i + z_i)x_i = \frac{nc_2}{1 + 2nc_2}(\overline{xy} + \overline{xz}) = \frac{1}{2}(\tilde{\beta}_y + \tilde{\beta}_z) = \tilde{\eta}.$$

$$s_y^2 = \frac{1}{4}\left(\frac{\lambda_2}{1 + n\lambda_2} + \frac{\lambda_4}{1 + n\lambda_4}\right) = \frac{1}{2}\frac{2c_2}{1 + 2nc_2} = s_u^2.$$

This is expected, because $\mathbf{u}$ is a sufficient statistic for η. The orthogonal design and the diagonal prior covariance matrices imply that the intercepts have no influence on η in both models.

(c) Note that, $\mathbf{u}$ is not sufficient for σ^2. Assume additionally that $m_1 = m_3 = 0$ and $\lambda_1 = \lambda_3 = 2c_1$, $\lambda_2 = \lambda_4 = 2c_2$. Then

$$\tilde{\alpha}_u = \frac{2c_1}{1 + 2nc_1} \sum_{i=1}^{n} \frac{1}{2}(y_i + z_i) = \frac{nc_1}{1 + 2nc_1}(\overline{y} + \overline{z}) = \frac{1}{2}(\tilde{\alpha}_y + \tilde{\alpha}_z)$$

and $\tilde{u}_i = \tilde{\alpha}_u + \tilde{\beta}_u x_i = \frac{1}{2}(\tilde{y}_i + \tilde{z}_i)$. Set $R_y = \sum_{i=1}^{n}(y_i - \tilde{y}_i)^2$, $R_z = \sum_{i=1}^{n}(z_i - \tilde{z}_i)^2$, $R_u = \sum_{i=1}^{n}(u_i - \tilde{u}_i)^2$ and $R_{zy} = \sum_{i=1}^{n}(z_i - \tilde{z}_i)(y_i - \tilde{y}_i)$. Thus $2R_u = \frac{1}{2}(R_y + R_z) + R_{zy}$. The estimates $\tilde{\sigma}_1^2$ and $\tilde{\sigma}_2^2$ are different, mainly because $b_1(\mathbf{y})$ and $b_1(\mathbf{u})$ are different, i.e.,

$$b_1(\mathbf{y}) = b + R_y + R_z + \frac{1}{2c_2}(\tilde{\alpha}_y^2 + \tilde{\alpha}_z^2) + \frac{1}{2c_1}(\tilde{\beta}_y^2 + \tilde{\beta}_z^2)$$

$$b_1(\mathbf{u}) = b + \frac{1}{2}(R_y + R_z) + R_{zu} + \frac{1}{4c_1}(\tilde{\alpha}_y + \tilde{\alpha}_z)^2 + \frac{1}{4c_2}(\tilde{\beta}_y + \tilde{\beta}_z)^2.$$

(d) Comparing the intervals (10.10) and (10.11): The center of the intervals coincides. But $s_1(\mathbf{y})$ and $s_2(\mathbf{u})$ are different. Especially the degrees of freedom in student's distributions are different. In model (7.70) $2n$ observations are used, where in model (7.71) only n observations are used.

8. *Two dependent lines, interested in averaged slope*: The corresponding multivariate regression model is defined in (10.6). The posterior distribution is $\theta \sim \text{NIW}(\nu_1, \mathbf{B}_1, \mathbf{C}_1, \Sigma_1)$ with $\nu_1 = \nu + n$, $\mathbf{B}_1$, and $\mathbf{C}_1$ given in (10.7) and, using the notations R_y, R_z, R_{yz} above,

$$\Sigma_1 = \begin{pmatrix} \tilde{\sigma}_1^2 & \tilde{\sigma}_{12} \\ \tilde{\sigma}_{21} & \tilde{\sigma}_2^2 \end{pmatrix}$$

with

$$\tilde{\sigma}_1^2 = 1 + R_y + \frac{1}{\lambda_1}\tilde{\alpha}_y^2 + \frac{1}{\lambda_2}\tilde{\beta}_y^2, \quad \tilde{\sigma}_2^2 = 1 + R_z + \frac{1}{\lambda_1}\tilde{\alpha}_z^2 + \frac{1}{\lambda_2}\tilde{\beta}_z^2$$

$$\tilde{\sigma}_{12} = \tilde{\sigma}_{21} = R_{yz} + \frac{1}{\lambda_1}\tilde{\alpha}_y\tilde{\alpha}_z + \frac{1}{\lambda_2}\tilde{\beta}_y\tilde{\beta}_z.$$

(a) The Bayes estimates of β_y and β_z are $\tilde{\beta}_y$ and $\tilde{\beta}_z$, respectively. The parameter of interest is linear in β_y and β_z, i.e., $\tilde{\eta} = \frac{1}{2}(\tilde{\beta}_y + \tilde{\beta}_z)$. Note that estimates $\tilde{\eta}$ in model (7.70) with prior $\text{NIG}(a, b, 0, \text{diag}(\lambda_1, \lambda_2, \lambda_1, \lambda_2))$ and in model (7.71) with prior $\text{NIG}(a, b, 0, \frac{1}{2}\text{diag}(\lambda_1, \lambda_2))$ coincide.

(b) Set Σ as known. Applying Theorem 6.11 on page 176 gives $B|\mathbf{Y}, \Sigma \sim \text{MN}_{2,2}(\mathbf{B}_1, \mathbf{C}_1, \Sigma)$. The parameter of interest is $\eta = \mathbf{ABC}$, where $\mathbf{A} = \frac{1}{2}(0, 1)$, $\mathbf{C} = (1, 1)^T$. Using (7.73), we obtain

$$\eta \sim \text{MN}_{1,1}\left(\tilde{\eta}, \frac{\lambda_2}{4(1 + n\lambda_2)}, \sigma_1^2 + \sigma_2^2 + 2\sigma_{12}\right) \equiv \text{N}(\tilde{\eta}, s^2)$$

with

$$s^2 = \frac{\lambda_2(\sigma_1^2 + \sigma_2^2 + 2\sigma_{12})}{4(1 + n\lambda_2)}.$$

Thus

$$C_3(\mathbf{Y}, \Sigma) = \{\eta : \tilde{\eta} - z_{(1-\frac{\alpha}{2})} s \le \eta \le \tilde{\eta} + z_{(1-\frac{\alpha}{2})} s\}.$$

Under $\sigma_1 = \sigma_2 = \sigma$, $\sigma_{12} = 0$, and $\lambda_2 = \lambda_4$, we have $s^2 = s_y^2 \sigma^2$ and $C_3(\mathbf{Y}, \Sigma) = C_1(\mathbf{y}, \sigma^2)$ given in (10.9).

(c) Applying Theorem 6.11 on page 176 gives

$$B|\mathbf{Y} \sim t_{2,2}(\nu + n - 1, \mathbf{B}_1, \mathbf{C}_1, \Sigma_1).$$

Using (7.74) and (6.148) we get

$$\eta|\mathbf{Y} \sim t_{1,1}(\nu + n - 1, \tilde{\eta}, \frac{\lambda_2}{4(1 + n\lambda_2)}, 1_2^T\Sigma_1 1_2) \equiv t(\nu + n - 1, \tilde{\eta}, s(\mathbf{Y})^2),$$

with $s(\mathbf{Y})^2 = \frac{1}{\nu+n-1}\frac{\lambda_2}{4(1+n\lambda_2)}1_2^T\Sigma_1 1_2$ where

$$1_2^T\Sigma_1 1_2 = 2 + R_y + 2R_{yz} + R_z + \frac{1}{\lambda_1}(\tilde{\alpha}_y + \tilde{\alpha}_z)^2 + \frac{1}{\lambda_2}(\tilde{\beta}_y + \tilde{\beta}_z)^2.$$

(d) The HPD α-credible interval for η is

$$\{\eta : \tilde{\eta} - t_{(\nu+n-1,1-\frac{\alpha}{2})} s(\mathbf{Y}) \leq \eta \leq \tilde{\eta} + t_{(\nu+n-1,1-\frac{\alpha}{2})} s(\mathbf{Y})\}, \quad (10.12)$$

where $t_{(df,\gamma)}$ is the γ-quantile of $t_1(df, 0, 1)$ distribution with df degrees of freedom.

10.7 Solutions for Chapter 8

1. *Poisson distribution*:
   (a) Set $\theta \sim \text{Gamma}(\alpha, \beta)$. Then $\theta|\mathbf{x} \sim \text{Gamma}(\alpha + \sum_{i=1}^n x_i, \beta + n)$, since

$$\pi(\theta)\ell(\theta|\mathbf{x}) \propto \theta^{\alpha-1}\exp(-\theta\beta)\prod_{i=1}^n\left(\frac{\theta^{x_i}}{x_i!}\exp(-\theta)\right)$$
$$\propto \theta^{\alpha-1+\sum_{i=1}^n x_i}\exp(-\theta(\beta+n)).$$

   (b) Applying (8.4) we obtain

$$\varphi(\mathbf{x}) = \begin{cases} 1 & \text{if } P^\pi(\theta \geq 1|\mathbf{x}) < 0.5 \\ 0 & \text{if } P^\pi(\theta \geq 1|\mathbf{x}) \geq 0.5 \end{cases}.$$

   Since the median of a gamma distribution has no closed form, we cannot simplify this Bayes rule.
   (c) The posterior is $\text{Gamma}(17, 21)$ and $P^\pi(\theta \geq 1|\mathbf{x}) = 0.163$. We reject H_0.
2. *Gamma distribution*:
   (a) $\theta|\mathbf{x} \sim \text{Gamma}(\alpha_1, \beta_1)$ with $\alpha_1 = \alpha_0 + n\alpha$ and $\beta_1 = \beta_0 + \sum_{i=1}^n x_i$; see Problem 1 in Chapter 7.
   (b) The Bayes factor is defined in (8.18). We have

$$p(\mathbf{x}|\theta) = \frac{\theta^{\alpha n}}{\Gamma(\alpha)^n}\prod_{i=1}^n x_i^{\alpha-1}\exp(-\theta\sum_{i=1}^n x_i).$$

Using the integral $\int_0^\infty x^{a-1} \exp(-bx)\,dx = b^{-a}\Gamma(a)$, we obtain

$$m(\mathbf{x}) = \Gamma(\alpha)^{-n}\Gamma(\alpha_0)^{-1}\beta_0^{\alpha_0} \prod_{i=1}^{n} x_i^{\alpha-1}\Gamma(\alpha_1)\beta_1^{-\alpha_1}.$$

Hence

$$B_{01} = \frac{\Gamma(\alpha_0)}{\Gamma(\alpha_0 + n\alpha)} \frac{\theta_0^{\alpha n}}{\beta_0^{\alpha_0}} (\beta_0 + \sum_{i=1}^{n} x_i)^{\alpha_0 + n\alpha} \exp(-\theta_0 \sum_{i=1}^{n} x_i).$$

(c) Applying the table from Kass and Vos (1997), we get very strong evidence against the null hypothesis.

3. *Two sample problem, Delphin data:*

(a) In both models the parameter θ is related to location parameter. The non-informative prior is Jeffreys, $\pi(\theta) \propto$ const; see Example 3.21.

(b) First we consider the model $\{\mathcal{P}_0, \pi\}$. Only the distribution of the first m observations depends of θ. The posterior is $N(\gamma_{(0)}, \Gamma_{(0)})$ with

$$\gamma_{(0)} = \bar{x}^{(1)}, \quad \Gamma_{(0)} = \frac{1}{m}\sigma_1^2, \quad \bar{x}^{(1)} = \frac{1}{m}\sum_{i=1}^{m} x_i.$$

The second model $\{\mathcal{P}_1, \pi\}$ is the special case of a linear model. We apply Theorem 6.7 on page 157 , where $p = 1, \sigma^2 = 1, \mathbf{X} = (1, \ldots, 1, a, \ldots, a)^T$ and $\Sigma = \mathrm{diag}(\sigma_1^2, \ldots, \sigma_1^2, \sigma_2^2, \ldots, \sigma_2^2)$. Set $\bar{x}^{(2)} = \frac{1}{n-m}\sum_{i=m+1}^{n} x_i$. We obtain the posterior $N(\gamma_{(1)}, \Gamma_{(1)})$ with

$$\gamma_{(1)} = \frac{\sigma_1^{-2}m\,\bar{x}^{(1)} + \sigma_2^{-2}(n-m)a\bar{x}^{(2)}}{\sigma_1^{-2}m + \sigma_2^{-2}(n-m)a^2},$$

$$\Gamma_{(1)} = (\sigma_1^{-2}m + \sigma_2^{-2}(n-m)a^2)^{-1}.$$

(c) In the first model the fitted values are

$$\widehat{x}_i^{(0)} = \frac{1}{m}\sum_{i=1}^{m} x_i = \bar{x}^{(1)}, \quad i = 1, \ldots, m, \quad \widehat{x}_i^{(0)} = \mu, \quad i = m+1, \ldots, n,$$

with

$$\mathrm{RSS}^{(0)} = \frac{1}{\sigma_1^2}\sum_{i=1}^{m}(x_i - \bar{x}^{(1)})^2 + \frac{1}{\sigma_2^2}\sum_{i=m+1}^{n}(x_i - \mu)^2.$$

In the second model we have

$$\widehat{x}_i^{(1)} = \gamma^{(1)}, \quad i = 1, \ldots, m, \quad \widehat{x}_i^{(1)} = a\gamma^{(1)}, \quad i = m+1, \ldots, n,$$

with

$$\mathrm{RSS}^{(1)} = \frac{1}{\sigma_1^2}\sum_{i=1}^{m}(x_i - \gamma_{(1)})^2 + \frac{1}{\sigma_2^2}\sum_{i=m+1}^{n}(x_i - a\gamma_{(1)})^2.$$

(d) The Bayes factor is

$$B_{01} = \frac{m_0(\mathbf{x})}{m_1(\mathbf{x})}, \quad m_j(\mathbf{x}) = \int_{-\infty}^{\infty} p_j(\mathbf{x}|\theta)\, d\theta, \quad j = 0, 1.$$

We have

$$\frac{m_0(\mathbf{x})}{m_1(\mathbf{x})} = \frac{\exp(-\frac{1}{2\sigma_2^2}\sum_{i=m+1}^{n}(x_i - \mu)^2)A_1}{A_2}$$

with

$$A_1 = \int \exp\left(-\frac{1}{2\sigma_1^2}\sum_{i=1}^{m}(x_i - \theta)^2\right) d\theta$$

and

$$A_2 = \int \exp\left(-\frac{1}{2\sigma_1^2}\sum_{i=1}^{m}(x_i - \theta)^2 - \frac{1}{2\sigma_2^2}\sum_{i=m+1}^{n}(x_i - a\theta)^2\right) d\theta.$$

First consider A_1. Applying $\sum(x_i - \theta)^2 = \sum(x_i - \bar{x})^2 + m(\theta - \bar{x})^2$ and $\int \exp(-\frac{1}{2a}(x - b)^2)dx = \sqrt{2\pi a}$, we get

$$A_1 = \sqrt{2\pi}(\frac{\sigma_1^2}{m})^{\frac{1}{2}} \exp\left(-\frac{1}{2\sigma_1^2}\sum_{i=1}^{m}(x_i - \bar{x}^{(1)})^2\right).$$

For A_2, applying (6.86), we have

$$\frac{1}{\sigma_1^2}\sum_{i=1}^{m}(x_i - \theta)^2 + \frac{1}{\sigma_2^2}\sum_{i=m+1}^{n}(x_i - a\theta)^2 = \mathsf{RSS}^{(1)} + \Gamma_{(1)}^{-1}(\theta - \gamma_{(1)})^2.$$

Thus

$$A_2 = \sqrt{2\pi}(\Gamma_{(1)})^{\frac{1}{2}} \exp\left(-\mathsf{RSS}^{(1)}\right).$$

Summarizing,

$$B_{01} = (1 + \frac{n - m}{n}\frac{\sigma_1^2}{\sigma_2^2}a^2)^{\frac{1}{2}} \exp\left(\frac{1}{2}(\mathsf{RSS}^{(1)} - \mathsf{RSS}^{(0)})\right).$$

(e) We obtain $\mathsf{RSS}^{(0)} = 45.6062$, $\mathsf{RSS}^{(1)} = 32.7241$, $2\ln(B_{10}) = 22.98$. The evidence against H_0 is very strong. We prefer the second model.

4. *Corona example:* Set $\mathbf{y}^T = (y_{11}, \ldots, y_{1n}, y_{21}, \ldots y_{2n})$ and $\mathbf{x}_k = (x_{k,1}, \ldots, x_{k,n}, x_{k,n+1}, \ldots, x_{k,2n})^T$, $k = 1, \ldots, 4$. Then we have a linear model with design matrix

$$\mathbf{X} = \left(\begin{pmatrix} \mathbf{1}_n \\ \mathbf{0}_n \end{pmatrix} \begin{pmatrix} \mathbf{0}_n \\ \mathbf{1}_n \end{pmatrix} \quad \mathbf{x}_1 \quad \mathbf{x}_2 \quad \mathbf{x}_3 \quad \mathbf{x}_4 \right).$$

(a) We apply Corollary 6.1 on page 135 and obtain $\theta|y \sim N(\widehat{\theta}, \Gamma)$ with $\widehat{\theta} = (\widehat{\eta}_1, \widehat{\eta}_2, \widehat{\beta}_1, \ldots, \widehat{\beta}_4) = \frac{1}{\sigma^2}\Gamma X^T y$ and $\Gamma = (I_6 + \frac{1}{\sigma^2}X^T X)^{-1}$.

(b) Set $h = (1, -1, 0, 0, 0, 0)^T$ and $\sigma_\eta^2 = h^T\Gamma h$. Then

$$(\eta_1 - \eta_2)|y \sim N(\widehat{\eta}_1 - \widehat{\eta}_2, \sigma_\eta^2).$$

(c) Reject H_0 iff

$$\Phi\left(\frac{0.2 - (\widehat{\eta}_1 - \widehat{\eta}_2)}{\sigma_\eta}\right) - \Phi\left(\frac{-0.2 - (\widehat{\eta}_1 - \widehat{\eta}_2)}{\sigma_\eta}\right) \le 0.5,$$

where Φ is the distribution function of $N(0, 1)$.

5. *Test of parallelism:*

(a) See the notation and solution of Problem 8 of Chapter 6. We apply Theorem 6.5 on page 150 and obtain the posterior $NIG(a_1, b_1^{(0)}, \gamma^{(0)}, \Gamma^{(0)})$ with $a_1 = a + n$,

$$\Gamma^{(0)} = \begin{pmatrix} \frac{1}{n+1} & 0 & 0 \\ 0 & \frac{1}{n+1} & 0 \\ 0 & 0 & \frac{1}{2n+1} \end{pmatrix}, \quad \gamma^{(0)} = \begin{pmatrix} \frac{n}{n+1}\overline{y} \\ \frac{n}{n+1}\overline{z} \\ \frac{1}{2n+1}x^T(y+z) \end{pmatrix} = \begin{pmatrix} \widehat{\alpha}_y \\ \widehat{\alpha}_z \\ \widehat{\beta} \end{pmatrix}$$

and

$$b_1^{(0)} = b + \sum_{i=1}^{n}(y_i - \widehat{y}_i^{(0)})^2 + \sum_{i=1}^{n}(z_i - \widehat{z}_i^{(0)})^2 + (\gamma^{(0)})^T\gamma^{(0)} \qquad (10.13)$$

with $\widehat{y}_i^{(0)} = \widehat{\alpha}_y + \widehat{\beta}x_i$, $\widehat{z}_i^{(0)} = \widehat{\alpha}_z + \widehat{\beta}x_i$. Setting

$$R_1 = \sum_{i=1}^{n}(y_i - \widehat{\alpha}_y)^2, \quad R_2 = \sum_{i=1}^{n}(z_i - \widehat{\alpha}_z)^2,$$

we obtain

$$b_1^{(0)} = b + R_1 + R_2 + (\widehat{\alpha}_y)^2 + (\widehat{\alpha}_z)^2 - (2n+1)(\widehat{\beta})^2. \qquad (10.14)$$

(b) See the notation and solution of Problem 5(a), (b) of Chapter 6. We apply Theorem 6.5 and obtain the posterior $NIG(a_1, b_1^{(1)}, \gamma^{(1)}, \Gamma^{(1)})$ with $a_1 = a + n$,

$$\Gamma^{(1)} = \frac{1}{n+1}I_4, \quad \gamma^{(1)} = \begin{pmatrix} \frac{n}{n+1}\overline{y} \\ \frac{n}{n+1}\overline{z} \\ \frac{1}{n+1}x^T y \\ \frac{1}{n+1}x^T z \end{pmatrix} = \begin{pmatrix} \widehat{\alpha}_y \\ \widehat{\alpha}_z \\ \widehat{\beta}_y \\ \widehat{\beta}_z \end{pmatrix}.$$

Note that, the estimates of α_y and of α_z coincide in both models, because of the condition $\bar{x} = 0$. Further

$$b_1^{(1)} = b + \sum_{i=1}^{n}(y_i - \widehat{y}_i^{(1)})^2 + \sum_{i=1}^{n}(z_i - \widehat{z}_i^{(1)})^2 + (\gamma^{(1)})^T\gamma^{(1)} \qquad (10.15)$$

with $\widehat{y}_i^{(1)} = \widehat{\alpha}_y + \widehat{\beta}_y x_i$, $\widehat{z}_i^{(1)} = \widehat{\alpha}_z + \widehat{\beta}_z x_i$. We obtain

$$b_1^{(1)} = b + R_1 + R_2 + (\widehat{\alpha}_y)^2 + (\widehat{\alpha}_z)^2 - (n+1)(\widehat{\beta}_y)^2 - (n+1)(\widehat{\beta}_z)^2. \quad (10.16)$$

(c) Note that, $\widehat{\beta} = \frac{n+1}{2n+1}(\widehat{\beta}_y + \widehat{\beta}_z)$. Applying (10.14) and (10.16) we get

$$\frac{1}{n}(b_1^{(0)} - b_1^{(1)}) = \frac{n+1}{2n+1}(\widehat{\beta}_y - \widehat{\beta}_z)^2 - \frac{1}{n}\frac{2n+2}{2n+1}\widehat{\beta}_y\widehat{\beta}_z$$
$$= \frac{n+1}{2n+1}(\widehat{\beta}_y - \widehat{\beta}_z)^2 + o_p(1).$$

(d) We apply (8.20) and get

$$B_{01} = \frac{n+1}{\sqrt{2n+1}} \left(\frac{b_1^{(1)}}{b_1^{(0)}}\right)^{\frac{a+n}{2}}.$$

6. *Variance test in linear regression*:

(a) Under H_0 the variance is known. Thus the parameter is $\theta^{(0)} = \beta$. Jeffreys prior is $\pi_0(\beta) \propto$ const. Applying Theorem 6.7 on page 157 with $\Sigma = I_n$, we get the posterior $N(\widehat{\beta}, \sigma_0^2(X^TX)^{-1})$ with $\widehat{\beta} = (X^TX)^{-1}X^Ty$.

(b) Under H_1 the variance is unknown. Thus the parameter is $\theta^{(1)} = (\beta, \sigma^2)$. Applying Theorem 6.7 with $\Sigma = I_n$, we get the posterior $NIG(n - p, RSS, \widehat{\beta}, (X^TX)^{-1})$.

(c) In Definition 8.2 the Bayes factor is given as $B_{01} = \frac{m_0(y)}{m_1(y)}$ with

$$m_j(y) = \int_{\Theta_j} \ell_j(\theta^{(j)}|y)\pi_j(\theta^{(j)}) \, d\theta^{(j)} = \frac{\ell_j(\theta^{(j)}|y)\pi_j(\theta^{(j)})}{\pi_j(\theta^{(j)}|y)}, \quad j = 0, 1.$$

First we calculate $m_0(y)$. Set $\pi_0(\theta) = 1$. Applying Lemma 6.4 on page 6.4, we obtain

$$\ell_0(\theta^{(0)}|y)\pi_0(\theta^{(0)}) = (2\pi\sigma_0^2)^{-\frac{n}{2}} \exp(-\frac{1}{2\sigma_0^2}RSS) k_0(\beta|y),$$

with

$$k_0(\beta|y) = \exp\left(-\frac{1}{2\sigma_0^2}(\beta - \widehat{\beta})^TX^TX(\beta - \widehat{\beta})\right).$$

Since

$$\pi(\beta|y) = (2\pi\sigma_0^2)^{-\frac{p}{2}}|X^TX|^{\frac{1}{2}} k_0(\beta|y),$$

we get

$$m_0(\mathbf{y}) = (2\pi\sigma_0^2)^{-\frac{n-p}{2}}|\mathbf{X}^T\mathbf{X}|^{-\frac{1}{2}}\exp(-\frac{1}{2\sigma_0^2}\text{RSS}).$$

Now we calculate $m_1(\mathbf{y})$. Applying Lemma 6.4 we obtain

$$\pi_1(\theta^{(1)})\ell_1(\theta^{(1)}|\mathbf{y}) = (2\pi)^{-\frac{n}{2}}k_1(\theta^{(1)}|\mathbf{y}),$$

with

$$k_1(\theta^{(1)}|\mathbf{y}) = (\sigma^2)^{\frac{n+2}{2}}\exp(-\frac{1}{2\sigma^2}(\text{RSS} + (\beta - \widehat{\beta})^T\mathbf{X}^T\mathbf{X}(\beta - \widehat{\beta}))).$$

Since

$$\pi(\beta|\mathbf{y}) = (2\pi)^{-\frac{p}{2}}|\mathbf{X}^T\mathbf{X}|^{\frac{1}{2}}\frac{1}{\Gamma(\frac{n-p}{2})}\left(\frac{\text{RSS}}{2}\right)^{\frac{n-p}{2}}k_1(\theta^{(1)}|\mathbf{y}),$$

we get

$$m_1(\mathbf{y}) = (2\pi)^{-\frac{n-p}{2}}|\mathbf{X}^T\mathbf{X}|^{-\frac{1}{2}}\Gamma(\frac{n-p}{2})\left(\frac{\text{RSS}}{2}\right)^{-\frac{n-p}{2}}.$$

Summarizing,

$$\text{B}_{01} = \left(\frac{\text{RSS}}{\sigma_0^2}\right)^{-\frac{n-p}{2}}\Gamma(\frac{n-p}{2})^{-1}\exp(-\frac{1}{2\sigma_0^2}\text{RSS}).$$

(d) As $\text{B}_{10} = \text{B}_{01}^{-1}$ and $\widehat{\sigma}^2 = \frac{1}{n-p}\text{RSS}$, we obtain

$$2\ln(\text{B}_{10}) = (n-p)(x - \ln(x)) + \text{rest}(n), \quad x = \frac{\widehat{\sigma}^2}{\sigma_0^2},$$

with

$$\text{rest}(n) = \frac{1}{2}\ln(\Gamma(\frac{n-p}{2})) - (n-p)\ln(n-p).$$

The function $f(x) = x - \ln(x)$ has its minimum at $x = 1$. Using Stirling approximation we obtain $\lim_{n\to\infty}\text{rest}(n) = 0$.

7. *Test on correlation between two regression lines:*

(a) Under H_0 the lines are independent. We have the univariate model (10.8) given in Problem 5(a), Chapter 7. Using the notation and solution of Problem 5(b), Chapter 7, the posterior is $\text{NIG}(2 + 2n, b_1, \gamma, \frac{1}{n+1}\text{I}_4)$ with

$$\gamma^T = (\widehat{\alpha}_y, \widehat{\beta}_y, \widehat{\alpha}_z, \widehat{\beta}_z)^T = \frac{n}{n+1}(\overline{y}, \overline{xy}, \overline{z}, \overline{xz})^T.$$

Set $\widehat{y}_i = \widehat{\alpha}_y + x_i\widehat{\beta}_y$ and $\widehat{z}_i = \widehat{\alpha}_z + x_i\widehat{\beta}_z$. Then $b_1 = b_{1,y} + b_{1,z}$ with

$$b_{1,y} = 1 + \sum_{i=1}^{n}(y_i - \widehat{y}_i)^2 + \widehat{\alpha}_y^2 + \widehat{\beta}_y^2, \ b_{1,z} = 1 + \sum_{i=1}^{n}(z_i - \widehat{z}_i)^2 + \widehat{\alpha}_z^2 + \widehat{\beta}_z^2.$$

(b) Under H_1 the lines are dependent. We have model (10.6)

$$\mathbf{Y} = \mathbf{XB} + \epsilon, \epsilon \sim \mathrm{MN}_{n,2}(0, I_n, \Sigma).$$

Re-writing θ in matrix form $\theta = (\mathbf{B}, \Sigma)$ and using $\theta \sim \mathrm{NIW}(2, 0, \mathbf{I}_2, \mathbf{I}_2)$ we obtain the posterior $\mathrm{NIW}(2+n, \mathbf{B}_1, \frac{1}{n+1}\mathbf{I}_2, \Sigma_1)$ given in (10.7). Note that, the matrix $\mathbf{B}_1$ contains the same elements as the vector γ. Further, we have

$$\Sigma_1 = \begin{pmatrix} b_{1,y} & b_{1,yz} \\ b_{1,yz} & b_{1,z} \end{pmatrix}, \quad b_{1,yz} = \sum_{i=1}^{n}(y_i - \widehat{y}_i)(z_i - \widehat{z}_i) + \widehat{\alpha}_y\widehat{\alpha}_z + \widehat{\beta}_y\widehat{\beta}_z.$$

(c) Using the expectations of the corresponding posteriors, we obtain

$$\widetilde{\Sigma} = \begin{pmatrix} \widetilde{\sigma}^2 & 0 \\ 0 & \widetilde{\sigma}^2 \end{pmatrix}, \quad \text{with } \widetilde{\sigma}^2 = \frac{1}{2n}b_1, \quad \widehat{\Sigma} = \frac{1}{n-2}\Sigma_1.$$

Comparing both estimates we see that $\widetilde{\sigma}^2 = \frac{1}{2}(\widehat{\sigma}_1^2 + \widehat{\sigma}_2^2)$.

(d) In Definition 8.2 the Bayes factor is given as

$$B_{01} = \frac{m_0(\mathbf{y})}{m_1(\mathbf{y})}, \quad m_0(\mathbf{y}) = c_{0,\ell}\frac{c_{0,\mathrm{NIG}}}{c_{1,\mathrm{NIG}}}, \quad m_1(\mathbf{y}) = c_{1,\ell}\frac{c_{0,\mathrm{NIW}}}{c_{1,\mathrm{NIW}}},$$

where $c_{0,\ell}$ is the constant related to the likelihood under H_0, $c_{0,\mathrm{NIG}}$ is related to $\mathrm{NIG}(2, 2, 0, \mathbf{I}_4)$ and $c_{1,\mathrm{NIG}}$ to the posterior $\mathrm{NIG}(2+2n, b_1, \frac{1}{n+1}\mathbf{I}_4)$; and $c_{1,\ell}$ is the constant related to the likelihood under H_1, $c_{0,\mathrm{NIW}}$ is related $\mathrm{NIW}(2, \mathbf{0}, \mathbf{I}_2, \mathbf{I}_2)$ and $c_{1,\mathrm{NIG}}$ to $\mathrm{NIW}(2+n, \mathbf{B}_1, \frac{1}{n+1}\mathbf{I}_2, \Sigma_1)$. Especially

$$c_{0,\ell} = (2\pi)^{-n}, \quad c_{0,\mathrm{NIG}} = (2\pi)^{-1}, \quad c_{1,\mathrm{NIG}} = (2\pi)^{-1}\frac{(n+1)^2}{n!}\left(\frac{b_1}{2}\right)^{n+1},$$

$$c_{1,\ell} = c_{0,\ell}, \quad c_{0,\mathrm{NIW}} = (2^4\pi^3)^{-1},$$

$$c_{1,\mathrm{NIW}} = c_{0,\mathrm{NIW}}\frac{(n+1)^2}{\sqrt{\pi}\Gamma(\frac{n+2}{2})\Gamma(\frac{n+1}{2})}|\Sigma_1|^{\frac{n+2}{2}}.$$

Using the duplication formula, $\Gamma(2z) = (2\pi)^{-\frac{1}{2}}\Gamma(z)\Gamma(z+\frac{1}{2})$, with $2z = n+1$, we obtain

$$\frac{\sqrt{(2\pi)}n!}{\Gamma(\frac{n+2}{2})\Gamma(\frac{n+1}{2})} = \frac{n!}{\Gamma(n+1)} = 1.$$

Summarizing, we get

$$B_{01} = \sqrt{2}(n-2)(\frac{n-2}{n})^{n+1}\left(\frac{|\widehat{\Sigma}|}{|\widetilde{\Sigma}|}\right)^{\frac{n+1}{2}}|\widehat{\Sigma}|^{\frac{1}{2}}.$$

(e) It holds that

$$\frac{2}{n+1}\ln(\mathsf{B}_{10}) = \ln|\widetilde{\Sigma}| - \ln|\widehat{\Sigma}| + \text{rest}(n)$$

with

$$\text{rest}(n) = \frac{1}{2}\ln(\frac{n-2}{n}) + \frac{2}{n+1}\ln(n-2) + \frac{1}{n+1}(\ln(2) + \ln|\widehat{\Sigma}|) = o_{\mathsf{P}}(1),$$

since the Bayes estimate is bounded in probability.

10.8   Solutions for Chapter 9

1. *Stroke data*:

   (a) Treatment: $X_1 \sim \mathsf{Bin}(n_1, p_1)$; Control: $X_2 \sim \mathsf{Bin}(n_2, p_2)$ independent of each other.

   (b) Jeffrey priors: $p_1 \sim \mathsf{Beta}(0.5, 0.5)$ and $p_2 \sim \mathsf{Beta}(0.5, 0.5)$; see Example 3.20 on page 56.

   (c) Least favourable priors: $p_1 \sim \mathsf{Beta}(\sqrt{n_1}/2, \sqrt{n_1}/2)$ and $p_2 \sim \mathsf{Beta}(\sqrt{n_2}/2, \sqrt{n_2}/2)$; see Example 4.20 on page 109.

   (d) In Example 2.11 on page 16, it is shown that for prior $p \sim \mathsf{Beta}(\alpha, \beta)$, the posterior is $p|x \sim \mathsf{Beta}(\alpha + x, \beta + n - x)$.

   (e) Sampling of odds ratio:

   ```
   oddsratio<-function(N,a1,b1,a2,b2){
   theta<-rep(0,N); p1<-rbeta(N,a1,b1); p2<-rbeta(N,a2,b2);
   theta<-p1/p2; return(theta)}
   ```

   (f) Applying (8.4), we reject $\mathsf{H}_0 : \theta \leq 1$ under both priors.

2. *Independent MC*:

   (a) For $\alpha = 1$ and $\beta = 20$, the integral is

   $$\mu = \int \pi(\theta)\ell(\theta|x)d\theta = \int_0^1 \binom{n}{x} B(\alpha, \beta)^{-1}\theta^{\alpha+x-1}(1-\theta)^{\beta+n-x-1}\,d\theta$$

   (b) Method 1: Independent MC sampled from prior; Method 2: Independent MC sampled from $\mathsf{U}[0, 1]$; Method 3: Deterministic method implemented in R.

   (c) Method 2 applies factorization (9.5) with $p(\theta|x) = 1$.

      i. Draw $\theta^{(1)}, \ldots, \theta^{(N)}$ from $\mathsf{U}[0, 1]$.

      ii. Approximate μ by

   $$\widehat{\mu}(\mathbf{x}) = \binom{n}{x} B(\alpha, \beta)^{-1}\frac{1}{N}\sum_{j=1}^{N}(\theta^{(j)})^{\alpha+x-1}(1-\theta^{(j)})^{\beta+n-x-1}.$$

(d) Results are different since the deterministic integral is approximated by a random number in Methods 1 and 2.

(e) In Method 1 the generating distribution has variance 0.0021 and in Method 2 the variance is $\frac{1}{12} = 0.083$. Observe that, a prior with variance 0.0021 is not a good choice for a Bayes model.

3. *R code*:

(a) We have an i.i.d sample from $\mathsf{C}(\theta, 1)$ with prior $\theta \sim \mathsf{N}(2, 1)$.

(b) Rejection algorithm with trial $\mathsf{C}(3, 1)$ and constant $M = 5$.

(c) 10000 sampled from the posterior.

(d) Given current state $\theta^{(j)}$:

  i. Draw θ from $\mathsf{C}(3, 1)$ and compute

  $$r(\theta, \mathbf{x}) = \frac{\ell(\theta|\mathbf{x})}{\pi(\theta)}, \quad \ell(\theta|\mathbf{x}) = \prod_{i=1}^{n} f_{\mathsf{C}}(x_i|\theta, 1), \quad \pi(\theta) = f_{\mathsf{C}}(\theta|3, 1)$$

  where $f_{\mathsf{C}}(.|m, \lambda)$ denotes the density of Cauchy distribution with location parameter m and scale parameter λ.

  ii. Draw u independently from $\mathsf{U}[0, 1]$. Then

  $$\theta^{(j+1)} = \begin{cases} \theta & \text{if } u \leq r(\theta, \mathbf{x}) \\ \text{new trial} & \text{otherwise} \end{cases}.$$

4. *MCMC for logistic regression*:

(a) The prior is $\theta \sim \mathsf{C}(1, 1)$, such that $\pi(\theta) \propto \frac{1}{1+(\theta-1)^2}$. The Likelihood is $\ell(\theta|\mathbf{x}, \mathbf{y}) \propto \prod_{i=1}^{n} \left(\frac{\exp(-\theta x_i)^{y_i}(1-\exp(-\theta x_i))^{n-y_i}}{1+\exp(-\theta x_i)} \right)$ and trial is $T(\theta^{(j)}, \theta) = \frac{1}{2a}$ for $\theta^{(j)} - a \leq \theta \leq \theta^{(j)} + a$.

(b) R code:

```
MCMC.walk<-function(a,seed,N)
{
rand<-rep(NA,N);rand[1]<-seed;
for(i in 2:N)
 {rand.new<-rand[i-1]+a*runif(1,-1,1);
 lik.new<-exp(rand.new*s)/(prod(1+exp(rand.new*x)));
 p.new<-lik.new/(1+(rand.new-1)^2));
 lik.old<-exp(rand[i-1]*s)/(prod(1+exp(rand[i-1]*x)));
 p.old<-lik.old/(1+(rand[i-1]-1)^2));
 r<-min(1,p.new/p.old);
 if(runif(1)<r){rand[i]<-rand.new}else{rand[i]<-rand[i-1]}
 };
return(rand)
}
```

(c) Calculate M<-MCMC.walk(a,seed,N); acf(M) for $a = 0.5, 1, 2$. Best choice $a_0 = 1$. Calculate M1<- MCMC.walk(1,0,N); plot.ts(M1) and M2<- MCMC.walk(1,10,N); lines(1:N,M2,col=2). Burn-in time $k = 40$

(d) Carry out: MCMC.walk(a_0,0,N) obtain $\theta^{(1)}, \ldots, \theta^{(N)}$. Calculate $\widehat{\mu} = \frac{1}{N-k}\#\{\theta^{(j)} > 0.2, j > N - k\}$. We obtain $\widehat{\mu} = 0.0064$

5. *Gibbs sampling from* $N_2(\mu, \Sigma)$:

(a) Main steps: For a current state $(x^{(t)}, y^{(t)})$,

    i. Generate
$$x_{t+1}/y_t \sim N(\rho y_t, 1 - \rho^2).$$

    ii. Generate
$$Y_{t+1}/x_{t+1} \sim N(\rho x_{t+1}, 1 - \rho^2).$$

(b)
```
gibbs-norm<-function(rho,N)
{
  x=rep(0,N);y=rep(0,N)
    for(i in 1:N){x[i]<-rnorm(1,rho*y[i],(1-rho^2));
      y[i]<-rnorm(1,rho*x[i],(1-rho^2))
      }
  return(data.frame(x,y))
}
```

(c) After t iterations we get

$$
\begin{pmatrix} X^{(t)} \\ Y^{(t)} \end{pmatrix} \sim N_2\left(\begin{pmatrix} \rho^{2t-1}X^{(0)} \\ \rho^{2t}Y^{(0)} \end{pmatrix}, \begin{pmatrix} 1 - \rho^{4t-2} & \rho - \rho^{4t-1} \\ \rho - \rho^{4t-1} & 1 - \rho^{4t} \end{pmatrix} \right)
$$

The chain is not stationary, but the limit distribution is the right one, since for $|\rho| < 1$, $\lim_{t\to\infty} \rho^t = 0$.

6. *ABC for Iris data*:

(a) The data is an i.i.d. sample from a truncated bivariate normal distribution, TMVN(μ, Σ, a, b). This is $N_2(\mu, \Sigma)$ with

$$
\mu = \begin{pmatrix} \mu_1 \\ \mu_2 \end{pmatrix}, \quad \Sigma = \begin{pmatrix} \sigma_1^2 & \rho\sigma_1, \sigma_2 \\ \rho\sigma_1\sigma_2 & \sigma_2^2 \end{pmatrix}
$$

truncated to the support $[a_1, b_1] \times [a_2, b_3]$. Here we have $[0, \infty) \times [0, \infty)$. The parameter is $\theta = (\mu_1, \mu_2, \sigma_1, \sigma_2, \rho)$, m_1 is the average of the first variable and m_2 is the average of the second variable.

(b) The priors are $\sigma_1^2 \sim U[0.2, 2]$, $\sigma_2^2 \sim U[0.2, 2]$, $\rho \sim U[0, 0.9]$, $\mu_1 \sim$ TN$(0, \infty, 3, 1)$, and $\mu_1 \sim$ TN$(0, \infty, 2, 1)$, where TN$(0, \infty, m, v^2)$ is the normal distribution $N_1(m, v^2)$ truncated on $[0, \infty)$.

(c) The algorithm samples $\theta^{(1)}, \ldots, \theta^{(N)}$ independently from a distribution which approximates the posterior. The posterior has no closed form expression.

(d) It is an ABC Algorithm: Given current state $\theta^{(j)}$:

    i. Draw independently θ_1 from $\mathsf{TN}(0, \infty, 3, 1)$, θ_2 from $\mathsf{TN}(0, \infty, 2, 1)$, θ_3 from $\mathsf{U}[0.2, 2]$, θ_4 from $\mathsf{U}[0.2, 2]$, and θ_5 from $\mathsf{U}[0, 0.9]$.

    ii. For $i = 1, \ldots, 50$, generate new observations
$Z_{\mathrm{new},i} \sim \mathsf{TMVN}(\mu, \Sigma, a, b)$, where

$$\mu = \begin{pmatrix} \theta_1 \\ \theta_2 \end{pmatrix}, \quad \Sigma = \begin{pmatrix} \theta_3^2 & \theta_5\theta_3, \theta_4 \\ \theta_5\theta_3\theta_4 & \theta_4^2 \end{pmatrix}.$$

    iii. Calculate $\frac{1}{n}\sum_{i=1}^{n} Z_{\mathrm{new},i} = (m_{\mathrm{new},1}, m_{\mathrm{new},2})$, and $D = (m_1 - m_{\mathrm{new},1})^2 + (m_2 - m_{\mathrm{new},2})^2$.

    iv. For $D < \mathrm{tol}$ we update $\theta^{(j+1)} = \theta$; otherwise we go back to Step 1.

## Chapter 11

Appendix

Here we briefly list some of the most commonly used distributions in Bayesian inference. For details, we recommend other sources given in bibliography.

11.1 Discrete Distributions

Consider a discrete random variable X with sample space $\mathcal{X} \subseteq \mathbb{Z}$. We denote Poisson distribution $\mathsf{Poi}(\lambda)$, binomial distribution $\mathsf{Bin}(n, p)$, beta-binomial distribution $\mathsf{BetaBin}(n, \alpha, \beta)$, negative binomial distribution $\mathsf{NB}(r, p)$, and geometric distribution $\mathsf{Geo}(p)$.

Notation	$P(X = k)$	$\mathsf{E}X$	$\mathsf{Var}(X)$
$\mathsf{Poi}(\lambda)$	$\frac{\lambda^k}{k!} \exp(-\lambda)$	λ	λ
$\mathsf{Bin}(n, p)$	$\binom{n}{k} p^k (1-p)^{n-k}$	np	$np(1-p)$
$\mathsf{BetaBin}(n, \alpha, \beta)$	$\binom{n}{k} \frac{B(\alpha+k, \beta+n-k)}{B(\alpha+\beta)}$	$\frac{n\alpha}{\alpha+\beta}$	$\frac{n\alpha\beta}{(\alpha+\beta)^2} \frac{\alpha+\beta+n}{\alpha+\beta+1}$
$\mathsf{NB}(r, p)$	$\binom{k+r-1}{k} p^r (1-p)^k$	$\frac{r(1-p)}{p}$	$\frac{r(1-p)}{p^2}$
$\mathsf{Geo}(p)$	$p(1-p)^k$	$\frac{(1-p)}{p}$	$\frac{(1-p)}{p^2}$

DOI: 10.1201/9781003221623-11

Notation	R package	R function
Poi(λ)		dpois(k,lambda)
Bin(n,p)		dbinom(k,n,p)
BetaBin(n,α,β)	library(VGAM)	dbetabinom.ab
		(k,n,alpha,beta)
NB(r,p)		dnbinom(k,r,p)
Geo(p)		dgeom(k,p)

11.2 Continuous Distributions

Let X be a real-valued random variable with sample space $\mathcal{X} \subseteq \mathbb{R}$. We denote normal distribution $\mathsf{N}(\mu,\sigma^2)$, t-distribution $\mathsf{t}_1(f,\mu,\sigma^2)$, F-distribution F_{f_1,f_2}, exponential distribution $\mathsf{Exp}(\lambda)$, Cauchy distribution $\mathsf{C}(m,\sigma)$, Laplace distribution $\mathsf{La}(\mu,\sigma)$, beta distribution $\mathsf{Beta}(\alpha,\beta)$, gamma distribution $\mathsf{Gamma}(\alpha,\beta)$, and inverse-gamma distribution $\mathsf{InvGamma}(\alpha,\beta)$.

Notation	$f(x\|\theta) \propto$	EX	$Var(X)$
$\mathsf{N}(\mu,\sigma^2)$	$\exp\left(-\frac{1}{2}(\frac{x-\mu}{\sigma})^2\right)$	μ	σ^2
$\mathsf{t}_1(f,\mu,\sigma^2)$	$\left(1+\frac{1}{f}(\frac{x-\mu}{\sigma})^2\right)^{-\frac{f+1}{2}}$	μ	$\frac{f}{f-2}\sigma^2$ $f>2$
F_{f_1,f_2}	$x^{\frac{f_1}{2}-1}\left(1+\frac{f_1}{f_2}x\right)^{-\frac{f_1+f_2}{2}}$	$\frac{f_2}{f_2-2}$ $f_2>2$	$\frac{2f_2^2(f_1+f_2-2)}{f_1(f_2-2)^2(f_2-4)}$ $f_2>4$
$\mathsf{Exp}(\lambda)$	$\lambda\exp(-\lambda x)$	$\frac{1}{\lambda}$	$\frac{1}{\lambda^2}$
$\mathsf{C}(m,\sigma)$	$\left(1+(\frac{x-m}{\sigma})^2\right)^{-1}$	$-$	$-$
$\mathsf{La}(\mu,\sigma)$	$\exp\left(-\left|\frac{x-\mu}{\sigma}\right|\right)$	μ	$2\sigma^2$
$\mathsf{Beta}(\alpha,\beta)$	$x^{\alpha-1}(1-x)^{\beta-1}$	$\frac{\alpha}{\alpha+\beta}$	$\frac{\alpha\beta}{(\alpha+\beta)^2(\alpha+\beta+1)}$
$\mathsf{Gamma}(\alpha,\beta)$	$x^{\alpha-1}\exp(-x\beta)$	$\frac{\alpha}{\beta}$	$\frac{\alpha}{\beta^2}$
$\mathsf{InvGamma}(\alpha,\beta)$	$x^{-\alpha-1}\exp(-\frac{\beta}{x})$	$\frac{\beta}{\alpha-1}$ $\alpha>1$	$\frac{\beta^2}{(\alpha-1)^2(\alpha-2)}$ $\alpha>2$

Notation	R package	R function
$N(\mu, \sigma^2)$		dnorm(x,mu,sigma)
$t_1(f, \mu, \sigma^2)$	library(metRology)	dt.scaled(x,f,mu,sigma)
F_{f_1,f_2}		df(x,f1,f2)
$Exp(\lambda)$		dexp(x,lambda)
$C(m, \sigma)$		dcauchy(x,m,sigma)
$La(\mu, \sigma)$	library(ExtDist)	dLaplace(x,mu,sigma)
$Beta(\alpha, \beta)$		dbeta(x,alpha,beta)
$Gamma(\alpha, \beta)$		dgamma(x,alpha,beta)
$InvGamma(\alpha, \beta)$	library(invgamma)	dinvgamma(x,alpha,beta)

11.3 Multivariate Distributions

Let $\mathbf{X}$ be a random vector with sample space $\mathcal{X} \subseteq \mathbb{R}^p$. We denote multivariate normal distribution $N_p(\gamma, \Sigma)$ and multivariate t-distribution $t_p(f, \mu, \Sigma)$.

Further we list distributions of (X, λ) where X is a vector valued random variable with sample space $\mathcal{X} \subseteq \mathbb{R}^p$ and λ is a positive random scalar. We denote normal-gamma distribution $NGam(\alpha, \beta, \mu, \Sigma^{-1})$ and normal-inverse-gamma distribution $NIG(\alpha, \beta, \mu, \Sigma)$.

We also refer to the toolbox given in Chapter 6: Lemma 6.1, Lemma 6.2, Theorem 6.1, and Lemma 6.5.

Notation	$f(\mathbf{X}\|\theta) \propto$	$\mathbf{EX}$	$\mathrm{Cov}(\mathbf{X})$
$N_p(\mu, \Sigma)$	$\exp\left(-\frac{1}{2}\|\mathbf{X} - \mu\|_\Sigma^2\right)$	μ	Σ
$t_p(f, \mu, \Sigma)$	$\left(1 + \frac{1}{f}\|\mathbf{X} - \mu\|_\Sigma^2\right)^{-\frac{f+p}{2}}$	μ	$\frac{f}{f-2}\Sigma, \ f > 2$
$\mathrm{NGam}(\alpha, \beta, \mu, \mathbf{P})$	$\dfrac{\exp\left(-\frac{\lambda}{2}\left(\beta + \|\mathbf{X}-\mu\|_\Sigma^2\right)\right)}{\lambda^{-\frac{p+\alpha-2}{2}}}$	$\mathbf{EX} = \mu$	$\mathrm{Cov}(\mathbf{X}) = \frac{\beta}{\alpha-2}\Sigma$
$\mathbf{P} = \Sigma^{-1}$		$E\lambda = \frac{\alpha}{\beta}$	$\mathrm{Var}(\lambda) =$
			$\frac{2\alpha}{\beta^2(\alpha-1)}, \ \alpha > 2$
			$\mathrm{Cov}(\mathbf{X}, \lambda) = 0$
$\mathrm{NIG}(\alpha, \beta, \mu, \Sigma)$	$\dfrac{\exp\left(-\frac{1}{2\lambda}\left(\beta + \|\mathbf{X}-\mu\|_\Sigma^2\right)\right)}{\lambda^{\frac{p+\alpha+2}{2}}}$	$\mathbf{EX} = \mu$	$\mathrm{Cov}(\mathbf{X}) = \frac{\beta}{\alpha-2}\Sigma$
		$E\lambda = \frac{\beta}{\alpha-2}$	$\mathrm{Var}(\lambda)$
		$\alpha > 2$	$= \frac{2\beta^2}{(\alpha-2)^2(\alpha-1)}$
			$\mathrm{Cov}(\mathbf{X}, \lambda) = 0$

where $\|\mathbf{X} - \mu\|_\Sigma^2 = (\mathbf{X} - \mu)^T \Sigma^{-1}(\mathbf{X} - \mu)$.

Notation	R package	R function
$N_p(\mu, \Sigma)$	library(mvtnorm)	dmvnorm(X,mu,Sigma)
$t_p(f, \mu, \Sigma)$	library(mvtnorm)	dmvt(X,df=f,mu,Sigma)
$\mathrm{NGam}(\alpha, \beta, \mu, \mathbf{P})$	library(lestat)	mnormalgamma
		(mu,P,2*alpha,2*beta)
$\mathrm{NIG}(\alpha, \beta, \mu, \Sigma)$	library(PIGShift)	dmvnorminvgamma
		(xx,2*alpha,2*beta,mu,Sigma)

11.4   Matrix-Variate Distributions

Let $\mathbf{Z}$ be a random matrix, with sample space $\mathcal{Z} \subseteq \mathbb{R}^{m \times k}$. We denote matrix-variate normal distribution $\mathrm{MN}_{m,k}(\mathbf{M}, \mathbf{U}, \mathbf{V})$ and matrix-variate t-distribution $t_{m,k}(\nu, \mathbf{M}, \mathbf{U}, \mathbf{V})$.

Notation	$f(\mathbf{Z}\mid\theta) \propto$	$E\mathbf{Z}$	$\mathrm{Cov}(\mathrm{vec}(\mathbf{Z}))$
$\mathsf{MN}_{m,k}(\mathbf{M},\mathbf{U},\mathbf{V})$	$\exp\left(-\frac{1}{2}\mathrm{tr}\left(\mathbf{Q}\right)\right)$	$\mathbf{M}$	$\mathbf{V}\otimes\mathbf{U}$
$\mathsf{t}_{m,k}(\nu,\mathbf{M},\mathbf{U},\mathbf{V})$	$\lvert \mathbf{I}_k + \mathbf{Q}\rvert^{-\frac{\nu+m+k-1}{2}}$	$\mathbf{M}$	$\frac{1}{\nu-2}(\mathbf{V}\otimes\mathbf{U})$ $\nu > 2$

where $\mathbf{Q}(\mathbf{Z}) = \mathbf{V}^{-1}(\mathbf{Z}-\mathbf{M})^T\mathbf{U}^{-1}(\mathbf{Z}-\mathbf{M})$.

Let $\mathbf{W}$ be a positive definite random matrix with sample space $\mathcal{W} \subset \mathbb{R}^{k\times k}$. We denote Wishart distribution $\mathsf{W}_k(\nu,\mathbf{V})$ and inverse-Wishart distribution $\mathsf{IW}_k(\nu,\mathbf{V})$.

Notation	$f(\mathbf{W}\mid\theta) \propto$	$E\mathbf{W}$
$\mathsf{W}_k(\nu,\mathbf{V})$	$\lvert\mathbf{W}\rvert^{\frac{\nu-k-1}{2}}\exp\left(-\frac{1}{2}\mathrm{tr}\left(\mathbf{W}\mathbf{V}^{-1}\right)\right)$	$\nu\mathbf{V}$
$\mathsf{IW}_k(\nu,\mathbf{V})$	$\lvert\mathbf{W}\rvert^{-\frac{\nu+k+1}{2}}\exp\left(-\frac{1}{2}\mathrm{tr}\left(\mathbf{W}^{-1}\mathbf{V}\right)\right)$	$\frac{1}{\nu-k-1}\mathbf{V}$

Further

Notation	$\mathrm{Cov}(\mathrm{vec}(\mathbf{W}))$
$\mathsf{W}_k(\nu,\mathbf{V})$	$\nu(\mathbf{I}_{k^2}+\mathbf{K}_{k,k})(\mathbf{V}\otimes\mathbf{V})$
$\mathsf{IW}_k(\nu,\mathbf{V})$	$\frac{2}{(\nu+1)v^2(v-2)}\mathrm{vec}(\mathbf{V})\mathrm{vec}(\mathbf{V})^T$ $+\frac{2}{(\nu+1)v(v-2)}(\mathbf{I}_{k^2}+\mathbf{K}_{k,k})(\mathbf{V}\otimes\mathbf{V}),\ \nu > 2$

where $\mathbf{K}_{k,k}$ is the commutation matrix of $\mathbf{A} \in \mathbb{R}^{k\times k}$ defined by $\mathbf{K}_{k,k}\mathrm{vec}(\mathbf{A}) = \mathrm{vec}(\mathbf{A})^T$ and given as $\mathbf{K}_{k,k} = \sum_{j=1}^{k}\sum_{i=1}^{k}\mathbf{H}_{ij}\otimes\mathbf{H}_{ij}^T$, $\mathbf{H}_{ij}$ is a matrix with 1 at place (i,j) and zeros elsewhere. We refer to Kollo and von Rosen (2005) for the moments of $\mathsf{W}_k(\nu,\mathbf{V})$ and $\mathsf{IW}_k(\nu,\mathbf{V})$.

We also give the joint distribution of $(\mathbf{Z},\mathbf{W})$, i.e., normal-inverse-Wishart distribution, $\mathsf{NIW}(\nu,\mathbf{M},\mathbf{U},\mathbf{V})$.

Notation	$f(\mathbf{Z}, \mathbf{W}\mid\theta) \propto$	$E(\mathbf{Z}, \mathbf{W})$
$\mathrm{NIW}(\nu, \mathbf{M}, \mathbf{U}, \mathbf{V})$	$\|\mathbf{W}\|^{-\frac{\nu+k+m+1}{2}} \exp\left(-\frac{1}{2}\mathrm{tr}\left(\mathbf{W}^{-1}\mathbf{V}\left(\mathbf{I}_k + \mathbf{Q}\right)\right)\right)$	$E\mathbf{Z} = \mathbf{M}$
	$\mathbf{Q} = \mathbf{V}^{-1}(\mathbf{Z}-\mathbf{M})^T\mathbf{U}^{-1}(\mathbf{Z}-\mathbf{M})$	$E\mathbf{W} = \frac{1}{\nu-k-1}\mathbf{V}$

Finally we list the corresponding R functions.

Notation	R package	R function
$\mathrm{MN}_{m,k}(\mathbf{M}, \mathbf{U}, \mathbf{V})$	library(matrixNormal)	dmatnorm(Z,M,U,V)
$\mathrm{t}_{m,k}(\nu, \mathbf{M}, \mathbf{U}, \mathbf{V})$	library(mniw)	dMT(Z,nu,M,U,V)
$\mathrm{W}_k(\nu, \mathbf{V})$	library(mniw)	dwish(W,nu,V)
$\mathrm{IW}_k(\nu, \mathbf{V})$	library(mniw)	diwish(W,nu,V)
$\mathrm{NIW}(\nu, \mathbf{M}, \mathbf{U}, \mathbf{V})$	library(mniw)	rmniw(n,M,U,V,nu)
		random numbers

Bibliography

J. Albert. *Bayesian Computation with R.* Springer, 2005.

H. Augier, L. Benkoël, J. Brisse, A. Chamlian, and W. K. Park. Necroscopic localization of mercury-selenium interaction products in liver, kidney, lung and brain of Mediterranean striped dolphins (*Stenella coeruleoalba*) by silver enhancement kit. *Cell. and Molec. Biology*, 39:765–772, 1993.

M.A. Beaumont. Approximative Bayesian Computation. *Annual Review of Statistics and Its Application*, 6:379–403, 2019.

D.R. Bellhouse. The reverend Thomas Bayes, FRS: A biography to celebrate the tercentenary of his birth. *Statistical Science*, 19(1):3–32, 2004.

J. O. Berger and J. M. Bernardo. On the developement of reference priors. *Bayesian Statistics*, 4:35–60, 1992a.

J. O. Berger and J. M. Bernardo. Ordered group reference priors with application to the multinomial problem. *Biometrika*, 79(1):25–37, 1992b.

J.O. Berger. *Statistical Decision Theory and Bayesian Analysis.* Springer, New York, 1980.

J.O. Berger, J.M. Bernado, and D. Sun. The formal definition of reference priors. *The Annals of Statistics*, 37(2):905–938, 2009.

J. M. Bernardo. Reference posterior distributions for Bayesian inference. *Journal of the Royal Statistical Society, Ser. B*, 41(2):113–147, 1979.

J.M. Bernardo. Reference analysis. In D.K. Dey and C.R. Rao, editors, *Handbook of Statistics 25*, pages 17–90. Elsevier, 2005.

C.M. Bishop. *Pattern Regognition and Machine Learning.* Springer, 2006.

G.E.P. Box and G.C. Tiao. *Bayesian Inference in Statistical Analysis.* Addison-Wesley, 1973.

L.D. Broemeling. *Bayesian Methods for Repeated Measures.* CRC Press, 2016.

Ming-Hui Chen, Qi-Man Shao, and J.G. Ibrahim. *Monte Carlo Methods in Bayesian Computation.* Springer, 2002.

B.S. Clarke and A.R. Barron. Jeffreys' prior is asymptotically least favourable under entropy risk. *Journal of Statistical Planning and Inference*, 41:37–60, 1994.

M.J. Crowder and D.J. Hand. *Analysis of Repeated Measures.* Chapman & Hall, 1990.

W.W. Daniel and C.L. Cross. *Biostatistics: A Foundation for Analysis in the Health Sciences, 10th ed.* Wiley, 2013.

E. Demidenko. *Mixed models: Theory and Applications with R.* Wiley, 2013.

P. Diaconis and D. Freedman. On the consistency of Bayes estimates. *The Annals of Statistics*, 14(1):1–26, 1986.

N. R. Draper and H. Smith. *Applied Regression Analysis.* Wiley, 1966.

J.A. Dupuis. Bayesian estimation of movement and survival probabilities from capture-recapture data. *Biometrika*, 82(4):761–772, 1995.

G.H. Givens and J.L. Hoeting. *Computational Statistics.* Wiley, 2005.

A.K. Gupta and D.K. Nagar. *Matrix Variate Distributions.* CRC Press, 2000.

T.J. Hastie, R.J. Tibshirani, and M. Wainright. *Statistical Learning with Sparsity: The LASSO and Generalizations.* CRC Press, 2015.

W.K. Hastings. Monte Carlo sampling methods using Markov chains and their applications. *Biometrika*, 57:97–109, 1970.

A.E. Hoerl and R.W. Kennard. Ridge regression: Applications to nonorthogonal problems. *Technometrics*, 12:55–67, 1970.

W. James and C. Stein. Estimation with quadratic loss. In *Proc. 4th Berkeley Symp. Math. Statist. Probab.*, volume 1, pages 361–380, Berkeley, 1960. Univ. of California Press.

H. Jeffreys. An invariant form for the prior probability in estimation problems. *Proceedings of the Royal Statistical Society of London*, 186:453–461, 1946.

R. E. Kass and P.W. Vos. *Geometrical Foundations of Asymptotic Inference.* Wiley, 1997.

R. E. Kass and L. Wasserman. An invariant form for the prior probability in estimation problems. *Journal of the American Statistical Association*, 91 (435):1343–1370, 1996.

R.E. Kass and A.E. Raftery. Bayes factors. *Journal of the American Statistical Association*, 90(430):773–795, 1997.

K-R. Koch. *Introduction to Bayesian Statistics, 2nd ed.* Springer, 2007.

T. Kollo and D. von Rosen. *Advanced Multivariate Statistics with Matrices.* Springer, 2005.

Se Yoon Lee. Gibbs sampler and coordinate ascent variational inference: A set theoretical review. *Communications in Statistics-Theory and Methods*, 51(6):1549–1568, 2022.

H. Liero and S. Zwanzig. *Introduction to the Theory of Statistical Inference.* CRC Press, 2011.

F. Liese and K.-J. Miescke. *Statistical Decision Theory.* Springer, 2008.

Pi-Erh Lin. Some characterization of the multivariate t distribution. *Journal of Multivariate Analysis*, 2:339–344, 1972.

B. W. Lindgren. *Statistical Theory.* Chapman & Hall, 1962.

D.V. Lindley. On a measure of the information provided by an experiment. *The Annals of Mathematical Statistics*, 27(4):986–1005, 1956.

D.V. Lindley and A.F.M. Smith. Bayes estimates for the linear model. *Journal of the Royal Statistical Society, Ser. B*, 34:1–41, 1972.

Jun S. Lui. *Monte Carlo Strategies in Scientific Computing*. Springer Series in Statistics. Springer, 2001. ISBN 0-387-95230-6.

J.R. Magnus and H. Neudecker. The elimination matrix: Some lemmas and applications. *SIAM Journal on Algebraic Discrete Methods*, 1(4):422–449, 1980.

J.R. Magnus and H. Neudecker. *Matrix Differential Calculus with Applications in Statistics and Econometrics*. Wiley, 1999.

K. V. Mardia, J. T. Kent, and J. M. Bibby. *Multivariate Analysis*. Academic Press, 1979.

A.M. Mathai and H.J. Haubold. *Special Functions for Applied Scientists*. Springer, 2008.

N. Metropolis, A. Rosenbluth, M.N. Rosenbluth, A.H. Teller, and E. Teller. Equations of state calculations by fast computing machines. *Journal of Chemical Physics*, 21(6):1087–1091, 1953.

T.Tin. Nguyen, Hien D. Nguyen, F. Chamroukhi, and G.L. McLachlan. Bayesian estimation of movement and survival probabilities from capture-recapture data. *Cogent Mathematics & Statistics*, 7, 2020.

N. Polson and L. Wasserman. Prior distributions for the bivariate binomial. *Biometrika*, 77(4):901–904, 1990.

J. Press and J. M. Tamur. *The Subjectivity of Scientists and the Bayesian Approach*. Wiley, 2001.

H. Raiffa and R. Schlaifer. *Applied Statistical Decision Theory*. Harvard University, Boston, 1961.

C. P. Robert. *The Bayesian Choice*. Springer, 2001.

C.P. Robert and G. Casella. *Introducing Monte Carlo Methods with R*. Springer, 2010.

L. Schwartz. On Bayes procedures. *Z. Wahrscheinlichkeitstheorie*, 4:10–26, 1965.

G. Schwarz. Estimating the dimension of a model. *The Annals of Statistics*, 6(2):461–464, 1978.

S. Searle, G. Casella, and C.E. McCulloch. *Variance Components*. Wiley, 2006.

S.R. Searle. *Linear Models*. Wiley, 1971.

S.R. Searle and M.J. Gruber. *Linear Models, 2nd ed*. Wiley, 2017.

G.A.F. Seber. *A Matrix Handbook for Statisticians*. Wiley, 2008.

G.W. Snedecor and W.G. Cochran. *Statistical Methods, 8th ed.* Iowa State University Press, 1989.

D.J. Spiegelhalter, N.G. Best, P.C. Bradley, and A. Van der Linde. Bayesian measures of model complexity and fit. *Journal of the Royal Statistical Society Ser. B*, 64(4):583–639, 2002.

S. M. Stigler. Thomas Bayes's Bayesian inference. *Journal of the Royal Statistical Society. Ser. A*, 145(2):250–258, 1982.

T.J. Tibshirani. Regression shrinkage and selection via the lasso. *Journal of the Royal Statistical Society, Ser. B*, 58:267–288, 1996.

L. Tierney, Robert E.K., and J.B. Kadane. Fully exponential laplace approximation to expectations and variances of nonpositive functions. *Journal of the American Statistical Association*, 84(407):710–716, 1989.

A. W. van der Vaart. *Asymptotic Statistics.* Cambridge University Press, 1998.

A. Zellner. *An Introduction to Bayesian Inference in Econometrics.* Wiley, 1971.

Hui Zou and T. Hastie. Regularization and variable selection via the elastic net. *Journal of the Royal Statistical Society, Ser. B*, 67(2):301–320, 2005.

S. Zwanzig and B. Mahjani. *Computer Intensive Methods in Statistics.* CRC Press, 2020.

Index

Autoregressive model, 208

Bayes decision rule, 84
Bayes factor, 222
Bayes Theorem, 14
Bayesian linear models
    mixed, 166
        estimation, 162
    univariate, 131
        posterior, 134, 135
Brute-force, 241

Comparison of models, 228
Completing the squares, 137
Credible region, 199
Cumulant generating function, 43

Decision, 78
    admissible, 81
    inadmissible, 81
    admissible, 99, 100, 102, 103, 105
    Bayes, 101–103, 108
    minimax, 98–100, 105, 108
    randomized, 97, 101
Decision rule, 78
    test, 217
Decision space, 78
DIC, 233
Distribution
    F, 289, 324
    t, 186, 324
    Bernoulli, 8
    beta, 16, 184, 324
    beta-binomial, 35, 213, 268, 308, 323
    binomial, 86, 102, 109, 114, 323
    Cauchy, 21, 75, 88, 287, 324
    data generating, 11, 183, 269

Dirichlet, 27, 39, 41, 62
exponential, 125, 324
exponential family, 40
Fisher's z, 290
gamma, 18, 75, 125, 187, 212, 288, 324
geometric, 75, 288, 323
Haldane, 289
inverse-gamma, 19, 95, 145, 200, 324
inverse-Wishart, 173, 327
Laplace, 57, 190, 324
log-normal, 75, 287
matrix-variate t, 174, 178, 215, 326
matrix-variate normal, 170, 215, 326
multinomial, 6, 39, 41, 61, 73, 76, 290
multivariate t, 148, 325
multivariate normal, 134, 325
    conditional, marginal, 132
    joint, 132
    joint posterior, 146
negative binomial, 208, 323
noncentral chi squared, 83
normal, 6, 42, 76, 77, 82, 96, 291, 324
normal-gamma, 181, 302, 325
normal-inverse-gamma, 145, 147, 325
normal-inverse-Wishart, 173, 174, 327
parameter generating, 11
Pareto, 75, 287
Poisson, 206, 323
Poisson weighted mixture, 83

Printed in the United States
by Baker & Taylor Publisher Services